Springer Series in Computational Neuroscience

Volume 9

For further volumes:
http://www.springer.com/series/8164

James M. Bower
Editor

20 Years of Computational Neuroscience

Springer

Editor
James M. Bower
Health Sciences Center
University of Texas
San Antonio, TX, USA

ISBN 978-1-4614-1423-0 ISBN 978-1-4614-1424-7 (eBook)
DOI 10.1007/978-1-4614-1424-7
Springer New York Heidelberg Dordrecht London

Library of Congress Control Number: 2013942496

Printed on acid-free paper

Springer is part of Springer Science+Business Media (www.springer.com)

Foreword

A Venue for Establishing Training and Collaborations among Computational Neuroscientists

National Institutes of Health (NIH) Director Dr. Francis Collins was interviewed by Charlie Rose in 2010, where he remarked "Computational biologists will be the 'breakthrough' artists of the future." Taking this one step further, Dr. James Bower, the co-founder and long-term co-organizer of the annual Computational Neuroscience (CNS) meeting, has often suggested that "*every NIH-funded grant should include a computational analysis and/or modeling component.*"

The reasoning behind these statements stems from the nature of empirical research and the needs of today's biomedical science community. As neuroscientists, we all know that the most exciting, and also most daunting challenge facing us is to understand the intricate structures and dynamic functions of the nervous system. However, given the vast amount of extremely complex data that is routinely being collected, this understanding cannot be readily achieved without using a combination of experimental and computational approaches. Computational neuroscience requires either a close collaboration between experimentalists and theorists, or, perhaps even better, the participation of *bilingual* scientists—a cadre of well-trained individuals who are fluent in both experimental and theoretical "languages" and scientific perceptions.

In the past 2 decades, under the leadership of Drs. Bower, Miller, Linster, Jung, De Schutter, and others, the annual CNS meetings have served as a forum for close interactions between experimental and computational neuroscientists. Formal presentations during the main meetings, provocative discussions during the post-meeting workshops, and informal dialogues during poster sessions provided ample opportunities for the exchange of ideas that have led to much fruitful collaboration.

More importantly, the CNS meetings have served as a venue of the training of a new generation of bilingual scientists. From the very beginning, the meeting organizers have put an emphasis on providing a stage for young scientists, and this unique feature of the meeting has persisted to this day. Over the years, countless young investigators (many of them had early training in quantitative sciences and later became interested in neuroscience) gave their very first scientific presentations at these meetings—one of the first steps in preparation for an academic research career. Many of those early attendees are now well-established investigators with their own laboratories and successful careers. They are actively mentoring the next generation of CNS meeting attendees.

The pre-meeting tutorial is another platform for cross-education and enrichment, which provides training opportunities for both junior and established experimental scientists to learn computational approaches, and vice versa. Another such opportunity lies in the keynote speeches by invited outside speakers, who often provide wisdom and insight from different angles and an outsider's point of view.

Another unique feature of the CNS meeting is its emphasis on promoting the participation of female scientists in a field that was once male-dominated. The increasingly large number of women serving on the Board of Directors and Executive Program Committee, as well as presenters and meeting participants, demonstrates the success of this approach.

The NIH has long recognized the importance of the field of computational neuroscience and has strongly invested in supporting a wide array of computational neuroscience projects, research training and related meetings. As early as 1988, the National Institute of Mental Health (NIMH) and the then-named National Institute of Neurological and Communication Disorders and Stroke jointly released the *Mat hematical/Computational/Theoretical Neuroscience Research Awards Program Announcement*, and the NIMH has maintained an active Theoretical and Computational Neuroscience Program ever since.

During 1999 and 2000, three NIH institutes [the National Institute of Neurological Disorders and Stroke (NINDS), the National Institute on Drug Abuse (NIDA), and the National Institute on Alcohol Abuse and Alcoholism (NIAAA)] organized workshops to evaluate needs and identify opportunities on how to further nurture this emerging field. These workshops provided suggestions for future research directions, and identified challenging needs in the field, including encouraging equal level collaborations, early training and cross-training of bilingual computational scientists, creating career paths for quantitatively trained scientists in neuroscience, and enhancing peer reviewers' appreciation of multidisciplinary approaches.

Based on these recommendations, nine NIH institutes joined force with five National Science Foundation (NSF) directorates, and developed the *Collaborative Research in Computational Neuroscience* (*CRCNS*) program. Over the past decade, this program has funded almost 200 collaborative projects, involving well over 400 principal investigators, and also providing training opportunities for several hundred graduate students and postdoctoral fellows. NIH's devotion to this field has also been shown as one of the priorities of the NIH Blueprint for Neuroscience Research—a cooperative effort among 15 NIH institutes, centers, and offices. In 2006, and again in 2010, the NIH Blueprint issued initiatives on *Training in Computational Neuroscience* and supported several training grants. Many of the beneficiaries of these NIH programs are previous and current participants of the CNS meetings.

Computational neuroscience research has expanded from basic science, and the study of small circuits, to include translational research and clinical research; from cellular and network levels to include subcellular, molecular, and genetic levels and up through systems, and disease levels. The NIMH has supported the CNS meeting since its infancy, and over the past 2 decades many other NIH institutes, including NINDS, NIDA, NIAAA, and the National Institute of Biomedical Imaging and Bioengineering (NIBIB) have co-supported this meeting as well. The NIH has played an important role in nurturing the growth of the field of computational neuroscience.

Dennis L. Glanzman
Program Chief Theoretical and
Computational Neuroscience Program,
National Institute of Mental Health

Yuan Liu
Chief, Office of International Activities
National Institutes of Health/National
Institute for Neurological Disorders and Stroke

Disclaimer

This foreword was prepared by Dennis Glanzman and Yuan Liu in their personal capacity. The opinions expressed in this article are the authors' own and do not reflect the view of the National Institutes of Health (NIH), the Department of Health and Human Services, or the United States government.

Contents

1 **Introduction: Origins and History of the CNS Meetings** 1
John Miller and James M. Bower

2 **A Pictorial History of the Early Days of Computational Neuroscience: The CNS Meeting Posters** 13
James M. Bower

3 **History of Neural Simulation Software** 33
David Beeman

4 **Learning from the Past: Approaches for Reproducibility in Computational Neuroscience** 73
Sharon M. Crook, Andrew P. Davison, and Hans E. Plesser

5 **The Emergence of Community Models in Computational Neuroscience: The 40-Year History of the Cerebellar Purkinje Cell** 103
James M. Bower

6 **Calcium: The Answer to Life, the Universe, and Everything** 141
Kim Blackwell

7 **The More We Look, the More Biological Variation We See: How Has and Should This Influence Modeling of Small Networks?** 159
Ronald L. Calabrese

8 **20 Years of "Noise": Contributions of Computational Neuroscience to the Exploration of the Effect of Background Activity on Central Neurons** 167
Alain Destexhe

9 **Still Looking for the Memories: Molecules and Synaptic Plasticity** 187
Upinder S. Bhalla

10 **20 Years of the Dynamics of Memory: The Long and Winding Road Linking Cellular Mechanisms to Behavior** 207
Michael E. Hasselmo

11 **Spatiotemporal Coding in the Olfactory System** 229
Christiane Linster and Thomas A. Cleland

12 **20 Years of Learning About Vision: Questions Answered, Questions Unanswered, and Questions Not Yet Asked** 243
Bruno A. Olshausen

13 **Reinforcement Learning Models Then-and-Now: From Single Cells to Modern Neuroimaging** 271
P. Read Montague

Index 279

Contributors

David Beeman Department of Electrical, Computer, and Energy Engineering, University of Colorado at Boulder, Boulder, CO, USA

Upinder S. Bhalla National Centre for Biological Sciences, Tata Institute of Fundamental Research, Bangalore, India

Kim Blackwell Molecular Neuroscience Department, Krasnow Institute of Advanced Study, Fairfax, VA, USA

James M. Bower Health Sciences Center, University of Texas, San Antonio, TX, USA

Ronald L. Calabrese Department of Biology, Emory University, Atlanta, GA, USA

Thomas A. Cleland Department of Psychology, Computational Physiology Laboratory, Cornell University, Ithaca, NY, USA

Sharon M. Crook School of Mathematical and Statistical Sciences and School of Life Sciences, Center for Adaptive Neural Systems, Arizona State University, Tempe, AZ, USA

Andrew P. Davison Unité de Neuroscience Information et Complexité (UNIC), CNRS, Gif-sur-Yvette, France

Alain Destexhe Unité de Neuroscience Information et Complexité (UNIC), CNRS, Gif-sur-Yvette, France

Michael E. Hasselmo Department of Psychology and Graduate Program for Neuroscience, Center for Memory and Brain, Boston University, Boston, MA, USA

Christiane Linster Department of Neurobiology and Behavior, Computational Physiology Laboratory, Cornell University, NY, USA

John Miller Department of Cell Biology and Neuroscience, Montana State University, Bozeman, MT, USA

P. Read Montague Human Neuroimaging Lab, The Virginia Tech Carilion Research Institute, Roanoke, VA, USA

Department of Physics, Virginia Tech, Roanoke, VA, USA

Wellcome Trust Centre for Neuroimaging, University College London, London, UK

Bruno A. Olshausen Helen Wills Neuroscience Institute and School of Optometry, University of California—Berkeley, Berkeley, CA, USA

Redwood Center for Theoretical Neuroscience, UC Berkeley, Berkeley, CA, USA

Hans E. Plesser Department of Mathematical Sciences and Technology, Norwegian University of Life Sciences, Aas, Norway

Chapter 1
Introduction: Origins and History of the CNS Meetings

John Miller and James M. Bower

Abstract Since their official inception in 1992, the annual Computational Neuroscience (CNS) meetings have served as a format for the presentation and discussion of a broad range of research employing theoretical and experimental methods to study the functional organization and operation of an equally broad range of nervous systems. As the CNS meetings have now entered their third decade, this volume as a whole considers how the understanding of several of the subjects consistently highlighted in those meetings has advanced and changed over the last 20 years. Given the influence of the CNS meetings on many of this volume's authors, as well as the field of CNS as a whole, we thought it might be appropriate to provide a brief historical perspective and "back story" on the meeting's origins now more than 20 years ago. This chapter is therefore a narrative and combined personal recollection from the two scientists who worked together to conceive the CNS meetings.

Early Days

The first Computational Neuroscience (CNS) meeting was held from July 14th to 18th, 1992, at the University of California's Conference Center on Lister Hill in San Francisco and included 116 presented papers and 215 participants. Looking back at that meeting now, it is clear that not only the science but also much of the character of the subsequent CNS meetings was already established in that first meeting.

J. Miller
Department of Cell Biology and Neuroscience, Montana State University, Bozeman, MT, USA

J.M. Bower (✉)
Health Sciences Center, University of Texas, San Antonio, TX, USA
e-mail: bower@uthscsa.edu

J.M. Bower (ed.), *20 Years of Computational Neuroscience*, Springer Series in Computational Neuroscience 9, DOI 10.1007/978-1-4614-1424-7_1,

The original impetus for this new series of meetings was to provide an open forum to specifically consider the computational structure of nervous systems. While by 1992 a number of meetings had been established that sought to link computational studies of the nervous system to more engineering related fields like neural networks (see below), CNS*92 was the first 'open' meeting to specifically focus on computational research intended to understand brains for their own sake. Thus, as stated in the original call for papers for CNS*92:

> This is an interdisciplinary conference intended to address the broad range of research approaches and issues involved in the new field of Computational Neuroscience. The meeting is intended to bring together experimental and theoretical neurobiologists along with engineers, computer scientists, cognitive scientists, and physicists to consider the functioning of biological nervous systems. Peer reviewed papers will be presented on a range of subjects related to understanding how nervous systems compute.

In addition to the clear focus on understanding nervous systems as the primary objective of the meeting, the organizers were also very aware that this was a new field and accordingly that its growth would be especially dependent on nurturing the young scientists entering the field. While a few universities in the late 1980s had begun to organize graduate programs in CNS, in most research laboratories, computational techniques were being introduced by individual isolated students who we felt would benefit from a place to come, present their work, and commune with like-minded colleagues. Quoting from the original NIH grant application to support the CNS meetings:

> The meetings organizers are particularly committed to providing graduate students, postdoctoral fellows, and young research faculty the opportunity to attend this meeting. We believe that this is generally important for the future growth of computational approaches to neuroscience, (as) it is often the case that individual students and postdocs are the instigators of computational research in their laboratories, and thus are often isolated.

This focus on building a community of graduate students, postdocs, and young research faculty had several important structural consequences for the meeting. First, from its inception, a significant percentage of the CNS operating budget and fund raising efforts were devoted to providing travel support for students. In fact, in the early days of the meeting senior invited speakers were often asked to pay their own travel expenses so that more resources could be made available for student participants. As a result of this focus on young investigators, many of today's leading computational neuroscientists, while still students, gave their first major science presentations at CNS. That includes several of the contributors to this current volume.

Second, while CNS papers have always been peer reviewed, the large majority of papers have always been accepted on the assumption that the meeting itself provided an important opportunity for scientific feedback, especially for students. Most of the accepted papers were presented in poster sessions in order to maximize the opportunity for this feedback. In fact, in the early days, an important factor for inviting senior faculty was the likelihood they would be willing to spend time interacting with students in front of their posters. To facilitate the culture of the poster sessions, the organizing committee made sure they were well lubricated and, importantly, had no scheduled end. As a result, it was not unusual in early CNS meetings to find groups of participants sitting on the floor talking and debating until the wee small hours of the morning. To accommodate and support this obsessive behavior, the starting time for the morning oral sessions on each day was delayed more and more. To our delight, meeting evaluations consistently rated the poster sessions as the best feature of the meeting.

Another early objective of the meeting was to provide the opportunity for participants to publish their science in the meeting proceedings. Once peer reviewed had accepted the paper to the meeting, the authors were free to write up their results for publication in the conference proceedings as they saw fit. It seemed like almost every year the question was raised as to whether "real" peer review should be applied to the published papers themselves, for example, by requiring full papers to be submitted before the meeting. The organizers always argued that, given the youth of the field and its participants, the final published papers should benefit from the feedback at the conference. In addition the organizers always felt that once the science was accepted, it should be up to the individual researchers to present the work as they saw fit. After all, it was their reputation they were establishing. Looking back at the conference proceedings now, and as outlined in other chapters in this book, the publication process resulted in the publication of a number of important early papers in the field that may not have passed traditional peer review at the time. In fact, in those early days it was difficult to get computational papers published in more traditional forums. This is one reason why the CNS meeting organizers also worked to organize the Journal of Computational Neuroscience, which is, in fact, another spin off of the CNS meetings.

With respect to the proceedings themselves, during the first several years of the meeting, the publisher Kluwer Academic Press insisted that it would help sales if we came up with a new name for the volume each year. Accordingly, "Computation and Neural Systems 1992," became "Computation in Neurons and Neural Systems" in 1993, and then "The Neurobiology of Computation" in 1994. In 1995, a new publisher, Academic Press, finally allowed us to simply refer to the proceedings volume as "Computational Neuroscience: Trends in Research 1995" ... 1996 ... 1997. By 2002, the proceedings volume had reached almost 1,200 pages in length.

As already stated, a core function of the CNS meeting and its design was to promote interactions between its young and enthusiastic participants. Therefore, the 3 days of the formal meetings were followed by 2 days of workshops designed for yet more discussion and interaction. The early tradition of the meeting was to have these workshops at a separate site, and preferably a site remote enough to allow participants to focus on absorbing and debating what they had heard at the meeting with minimal outside distraction. For the first meeting, we chose the Marconi Center located on a remote site on the Pt. Reyes peninsula north of San Francisco. While the Marconi center was the site where Guglielmo Marconi first broadcast a radio signal across the Pacific in 1914, in 1992 it had no Internet connections, no computers, and almost no way for code hackers and computational neurobiologists to exercise their fingers. It is still one of the more amusing "CNS scenes" watching 150 tech savvy and already tech-dependent CNSers all playing Frisbee on the lawn in order to have something to do with their hands. This first workshop also established a tradition that lasted for the next 10 years of not allowing formal workshop presentations. The meeting organizers provided flip charts and not much more, and many of the workshops took place outside, a tradition that was revisited at CNS 2010, when at least one workshop took place while "tubbing" down the Guadalupe River in Texas. In 1992, after 2 days of workshops, the conference participants were very happy to be bussed back to civilization. The "summer camp" atmosphere of the CNS meeting at the Asilomar Conference Center outside Monterey California in 2002 is explicitly captured in the poster for that year's meeting (see Chap. 2).

The other important CNS tradition established in the first meeting, which has lived on now for 20+ years, is the effort spent to assure that the CNS banquet is a memorable event with a location and character appropriate for the culture of the host city. The organizers were very intent on assuring that long after the particular scientific results were forgotten, memories of the CNS banquet would live on.

This is perhaps true for no banquet more than the first held in 1992.

Given the meeting was in San Francisco, and given the strong "hands on" orientation of the organizers, the obvious choice for the site of the first CNS banquet was the world famous Exploratorium Science Museum on the Embarcadero. Built after World War II by Frank Oppenheimer (the brother of Robert Oppenheimer), the Exploratorium was already the most famous "hands on" science museum in the world. A visitor to the museum then and now could put their hands on the exhibits, including many in 1992 that had to do with human perception and brain science. Perhaps the most remarkable of these was the "tactile dome" in which patrons entered into a completely dark space and climbed up, down, and around through various rooms and spaces designed to provide different tactile sensations. While the Exploratorium was the obvious choice for the banquet site, the culture of San Francisco also required that the food served be well above the average banquet fare. Accordingly, Frank Eeckman, the conference organizer in charge of logistics, was asked to find a caterer worthy of San Francisco. Unbeknownst to either of us, Frank took this charge very seriously, and arranged for catering by one of the BEST caterers in San Francisco, oysters on the half shell, curried shrimp, wonderful sourdough breads, fresh asparagus, and all in unlimited amounts as one after another catering truck replenished the tables adorned with ice sculptures. Further, also taking seriously our instructions that the CNS banquet should be a memorable (if not exactly precisely a remembered) event, Frank had arranged for an open bar. His and our assumption was that computational neurobiologists were not likely to over indulge. We were wrong. Worse the "hands on" and innovative nature of the museum combined with the equally inventive nature of the meeting attendees meant that once the party was in full force, several of the party faithful took "hands on" on face value and started modifying, or as they put it at the time, "improving" the exhibits. One of the meeting organizers (whose identity we continue to protect) ended up in the tactile dome, where, sliding into a pit of Ping-Pong balls in the complete dark, he happened upon a biological experiment fully underway. The upshot of all this chaos was a staggering catering bill, and a clear indication from the Exploratorium staff that, perhaps, the CNS meeting should seek another venue for future banquets (a few years later we had sushi at the San Francisco Aquarium instead). The next several months were spent trying to figure out how to pay for the banquet before the University of California followed through on its threat to confiscate John Miller's house to pay the outstanding bill. Those of you who have attended CNS meetings since will note that free drinks are limited to two tickets each, in many cases you can thank your faculty mentors' youthful indiscretion for that.

The Back Story

What should be clear from the previous account is that the CNS meeting had a very particular and intentional design from the start. The primary reason is that the first CNS meeting was actually a confluence of the experience of the conference organizers with several previous meetings, some of that experience good and some not so good. That is the back story of the CNS meetings.

[Jim's story] As briefly recounted by Dave Beeman in his article in this volume on the origin and development of neuronal simulators, my first efforts to build realistic models of the nervous system started while I was a postdoctoral fellow in Lew Haberly's laboratory in Madison Wisconsin in 1983. During that same period of time, I had co-taught a course with Dr. Josh Chover, then chairman of the Department of Mathematics at UW on methods for the analysis of multi-neuronal data. During this course, one of Josh's long time friends, Ed Posner, from the Jet propulsion Laboratory and Caltech, visited Madison and gave a talk on the exciting new developments in "neural network engineering" happening at Caltech. The week before Ed's visit, Caltech and AT&T's Bell Labs had organized a "Hop-fest" at Caltech, centered on the "'neural-like' 'Hopfield network'" that Caltech Professor John Hopfield had just published. Because I had just accepted a faculty position in the Department of Biology at Caltech, I was invited to dinner with Ed, who subsequently invited me to the follow-up "Hop-fest" that took place at the Miramar Hotel in Santa Barbara, California a few months later.

For a young faculty member interested in neuronal-modeling, what turned out to be the first "Neural Network" meeting of the modern era was a remarkable event, full of an extraordinary level of excitement and anticipation among a broad range of computational scientists. The "Hopfield network" was regarded by the participants as a breakthrough return to research on "'nervous system like' 'neural network'" engineering models after MIT professors Minsky and Papert put a damper on the field with their famous book on perceptrons (Minsky and Papert 1969). Reflected the renewed wide ranging interest in what were regarded as nervous system-like engineering solutions, the 40 participants in the Santa Barbara meeting represented a remarkable mix of scientists and government officials, including representatives from powerful federal funding agencies like the DOD, CIA, NSA, etc. Of those participants, only two made any claim to being real biologists, myself and Terry Sejnowski. In fact, this was the first time I met Terry and, as I recall, the first time I heard his "net-talk" network "babbling" in what seemed to be a remarkable human-like fashion as it "learned" to produce speech from text. While Terry had been scheduled to give a talk at the meeting, I was only there as an observer until a scheduled speaker didn't show up, and I was asked to talk about modeling the actual brain. This turned out to be the first time I presented my work with Matt Wilson modeling the olfactory cortex (Wilson and Bower 1988) and I remember distinctly that it was news to many in the room that synaptic inputs could also be inhibitory. Having described olfactory cortex as having an extensive set of "associative" connections, I also remember being asked if olfactory cortex might be a Hopfield network. Although it was several years before I calculated (on a bar napkin) that if a human brain was fully interconnected like a Hopfield network, it would be 10 km in diameter, I remember saying that no, it wasn't; going on to suggest that because all real neural networks are much more complex than the basic Hopfield network, it was likely that hard problems would need to be solved by much more complex networks. This was not an opinion that the meeting participants believed or wanted to hear in 1985.

Regardless of the biological significance of neural networks, what was absolutely clear from the meeting in Santa Barbara was that the neural network movement was going to grow and that many more people would want to attend the next meeting. It was therefore decided to expand the meeting the following year, and hold it at the Snowbird Ski resort outside of Salt Lake City, Utah, the favorite skiing venue for one of the Bell Lab organizers. I attended that meeting and this time gave a prearranged invited talk on basic neurobiology, continuing to insist that a real neurobiological connection to neural networks required that engineers, physicists and mathematicians actually learn about the brain. Like the "Hopfest," the Snowbird meetings was also closed, with all speakers being invited by the organizing committee. However, the number of people who wanted to attend the Snowbird meetings continued to grow and quickly outstripped the capacity of the resort hotel. As a result, in the second year of the Snowbird meeting the decision was made to organize a more open meeting. In what I took at the time as an ecumenical gesture, it was suggested that I co-organize the meeting with my Caltech colleague Yasir Abu Mustafa, a well known learning theorist. The meeting we organized was the first NIPS (Neural Information Processing) meeting in Denver, CO, a meeting that will soon celebrate its 25th continuous year. I wrote the meeting announcement to emphasize the meetings interest in engineering as well as neuroscience.

To telescope events, by the end of the second NIPS meeting, I was growing increasingly uncertain as to whether the optimistic fusion of neurobiology with engineering really had legs. Not surprisingly, the principle focus of the engineers was on engineering, and the neurobiologists, including my friend John Miller, who I had invited to participate in the second NIPS meeting, found most of the talks either irrelevant to neurobiology or naive in their neurobiological claims. The meeting also became wrapped up in politics especially when the newly formed Neural Network Society decided to organize its own meeting. I decided that it was time to consider founding a computational meeting specifically focused on the nervous system by itself. As John recounts in his own history, a meeting at the Neuroethology meeting in Berlin, followed by an invitation to attend the workshop on CNS John was organizing in San Francisco, led to our mutual decision to organize the first CNS meeting.

[John's story] In some ways complementary to Jim, and certainly reflected in the interdisciplinary nature of the CNS meetings, my own training spanned a large range of disciplines and approaches, ultimately pointing me toward the application of engineering and modeling approaches to studies of neural function. My undergraduate training was in physics, at U.C. Berkeley. On a lark, I took a Sensory Neurophysiology class during my senior year from a young professor named Frank Werblin, who got me interested in cracking simple neural circuits using neurophysiology and engineering analysis. After doing graduate work on the neurophysiology of the stomatogastric ganglion with Al Selverston at U.C. San Diego, I did a postdoctoral project with Wil Rall and John Rinzel at the NIH, where I soaked up their perspectives and approaches toward compartmental neuron modeling. During that postdoc, I also benefitted from interactions with Bob Burke at NIH and Gordon Shepherd at Yale, picking up knowledge and inspiration from the cutting edge electrophysiological and quantitative neuroanatomical studies they were pursuing. We all subsequently collaborated on the development of several complex compartmental neural models, all of which used a program called "NET-2," which was an early equivalent to the electronic simulation program "SPICE" (way back in 1985, we used compartmental models to study the implications of active membrane on dendritic spine heads, and made predictions that have only recently been verified by Roberto Araya and his colleagues (Miller et al. 1985; Araya et al. 2007)). So I came out of my postdoctoral studies with my training (and attention) distributed at uneven depths over a pretty broad terrain, but inspired to focus on quantitative analysis of synaptic integration in neurons with complex dendritic architectures.

Starting as a young assistant professor at Berkeley in 1981 "inspired" me to focus even more, and also exposed me to other researchers with similar and complimentary interests. In retrospect, one of the guiding lights during my early career at Berkeley, and a very important (but not-so-familiar) figure in quantitative systems neurophysiology in general, was Ted Lewis in the Department of Electrical Engineering. Ted was a senior Professor, and was way ahead of his time in applying advanced engineering and control theory approaches to the study of operational aspects of the auditory system. It was my interactions with Ted that ultimately led to my involvement in the establishment of the CNS meetings, which will also answer the question why was CNS*2010 identified as the 20th anniversary meeting (since, as Jim noted above, the first official CNS meeting was held in 1992). The CNS meetings were a direct descendent of a series of two workshops that were held the preceding 2 years at U.C. Berkeley. In 1989, Ted Lewis and I, along with Frank Eeckman and Muriel Ross at Lawrence Livermore National Labs, decided that it would be interesting to organize an invited workshop built around our common interest in the nature of the processing tasks executed by nerve cells and systems, the codes by which information is represented during the execution of these tasks, and the structure of the neural machinery through which the computational algorithms are implemented. Although we always enjoyed communicating with one another during random encounters at committee meetings or other specialized scientific conferences, we lamented that there was no single meeting that took the general field of "Computational Neuroscience" as its core theme. At the time, there were several other excellent smaller conferences that were meeting on an annual basis, featuring excellent CNS research. However, these all tended to focus on specific subdisciplines or technological approaches: e.g., meetings on Vision, Audition, or the application of back-propagation to tune artificial neural networks, or computational brain models based on Adaptive Resonance Theory as mentioned above by Jim. A notable exception to that trend was the International Congress of Neuroethology, which hosted presentations of interdisciplinary research on a wide variety of vertebrate and invertebrate preparations, framed within the context of natural behaviors. While these meetings featured many excellent talks at the interfaces between neuroscience, engineering, applied mathematics, and computational modeling, they only took place every 3 years at the far ends of expensive plane tickets. It was actually at the second International Congress of Neuroethology in Berlin, in September 1989, where Jim Bower and I met, were inspired by some great presentations, and began to hatch schemes that eventually led to our mutual involvement in the CNS meetings (as well as John's ultimate move from Berkeley to Jim's old *alma mater*: Montana State University in Bozeman). There was also the Annual Society for Neuroscience meeting, but the size and complexity of that meeting, and the fact that many computational scientists didn't attend, were limiting. Accordingly, Frank and Muriel came up with the idea of running a workshop on CNS, and played major roles in organizing and raising the necessary funding. The Berkeley workshops in 1990 and 1991 were extremely popular and successful from a scientific standpoint, and seemed to fill a very important niche. At that point, Jim and I decided to "incorporate" the workshops as the "CNS Meetings," and continue them on a regular (and more financially stable!) basis.

The Ongoing CNS Culture

From the previous brief histories, the distinct cultural origins of the CNS meetings should be clear. Instead of a closed meeting with the meeting organizers determining invited presentations, we pushed hard in the direction of few invited speakers and a meeting consisting mostly of submitted papers. Instead of a meeting dominated by the current "dons" of the field, our strong sense was that the real growth of CNS should be fostered from the ground up, with strong support for student participation and presentations. If something new was really starting, then students were probably in a better position to recognize and pursue the new directions than more seasoned faculty members anyway. To preserve its scientific and political integrity, the meeting had a very strong policy that members of the organizing committee not be allowed to give oral presentations themselves and that the program committee and organizers change frequently and include young faculty. We put in place what we considered to be a strong and fair peer review system predisposed to accept rather than to reject papers. In addition to the promotion of young scientists, the CNS meeting has also always placed particular emphasis on diversity and, as a result, the CNS organizing committee has, for 20 years, included an approximately even mix of men and women. In the early days, the CNS meeting even included day care options for young parents. Early on we decided it would increase the ability of students to attend if the meeting changed locations each year, and that the meeting should have a designated local organizer to help with logistics, but also to identify an appropriate location of the meeting as well as the all important banquet. By tradition, the CNS meeting venue is often old and sometimes a bit funky, but strongly reflective of local culture. The point being again that the CNS meeting should be fun, interesting, and anything but generic. The CNS meeting has also, from the outset, been highly multinational, and now explicitly alternates between North America and Europe every other year.

Programmatically, the CNS meetings have always been crafted to attract grad students, postdocs, and early-career researchers from a variety of intersecting fields and give them the opportunity to interact and learn from each other. The main meeting sessions are held over a period of 3 days, with no concurrent sessions. The large majority of presentations are selected from those submitted in response to an open call for abstracts, with the best submitted papers, often authored by students, offered longer oral presentations. Approximately two talks per day are reserved for longer invited seminars, given by international leaders in the field. In the early meetings, these speakers were often chosen based on their likely receptivity to the use of computational techniques, the idea being that they should learn by attending the meeting as well. As the field has grown, distinguished invited speakers are often now full fledged computational neurobiologist in their own right, many of whom, again, gave their first major talks at a CNS meeting as students. Speakers are expected to stay for the entire meeting, providing student attendees in particular the opportunity to meet with leading figures in the field within an extremely interactive atmosphere. We steadfastly maintain time for questions at the end of all oral presentations, and dedicated a significant proportion of the meeting time toward smaller "break-out" workshops, organized by meeting participants themselves. In the early days the topics of the workshops were actually chosen during the meeting to directly reflect the content and important issues raised in the meeting. We also encouraged a "no-holds-barred" attitude toward incorporating extreme mathematical and theoretical rigor in all presentations. And to encourage (and facilitate) the interdisciplinary nature of the early meetings, we added 1-day pre-meeting tutorial sessions: in the early days offering one in "neuroscience for non-neuroscientists," and a concurrent one in "computational analysis for neuroscientists." One of us (JPM) remembers organizing (along with his grad student Frederic Theunissen) an exciting "hybrid" pre-meeting tutorial at CNS*94 intended for both groups, on "applications of information theory to CNS." Only a handful of people in the room had even heard about information theory at the time.

Reflecting the first meeting, CNS continues to represent an extraordinary diversity of specific problems, preparations, and methods used in computational research. Through the years, it has became obvious that no one approach to CNS is ideally suited to all problems, and that all researchers interested in the structure and operation of nervous systems can benefit from a deeper understanding of the values and limitations of a variety of theoretical and modeling strategies. Likewise, no one preparation is ideally suited for all analyses, and the meetings have seen the presentation of a huge variety of vertebrate and invertebrate studies. With all of these factors in mind, this meeting was always intended to facilitate cross-fertilization between experimentalists and theorists using a wide variety of preparations and approaches, and to help those researchers discover and articulate the general principles that emerge. We believe that each of these objectives, designs, and properties of the CNS meeting are responsible for its continuing success and extension into a third decade. In addition, the CNS meeting now benefits from the establishment of the Organization for Computational Neurosciences (OCNS) which has provided important financial and leadership stability, and whose organization itself reflects many of the design features of the meeting itself.

Of course, the ultimate success of any scientific meeting, or any human endeavor, depends not only on the strength of its program and scientific content but also on the level of engagement of its participants. From the outset, CNS meeting attendees have been willing to get down and party scientifically and otherwise. All of that said, however, the other essential ingredient in the success of the meeting has been the extraordinary people (and we don't mean ourselves) who have spent hours even years of their lives supporting the meeting. Of the large number of people in this category, several are worthy of special mention. First, it is not at all clear that CNS*92 would have happened had it not been good fortune that a seasoned Belgian meeting organizer, Chris Plougart, was not already indirectly (through family relations) associated with Jim's laboratory at Caltech. Her previous experience with meeting organization was invaluable in establishing the basic administrative structure for the meeting. All participants in the next 10 years of the meeting also know that the meeting would have stopped in its tracks had it not been for the extraordinary skills and efforts of Judy Macias, Jim's secretary at Caltech. For 10 years, Judy Macias was synonymous with the CNS meeting, managing every component of the meeting from the most minute to the most absurd. Finally, it is important to acknowledge one other important, even critical reason for the meetings success, and that is the unwavering assistance, guidance, and support of Dennis Glanzman initially and then Dennis and Yuan Liu at the National Institutes of Health together. Dennis actually attended the first CNS meeting and it was at his suggestion that the second meeting be held in Washington, DC. Designed to expose other government officials to this developing field, discussions with Dennis about the second meeting inspired the first of the CNS meeting posters which are now included together in chapter two of this volume with commentaries for the first time. While other governmental agencies, and notably the National Science Foundation, have provided support for the CNS meeting through the years, at 20 years, it is our understanding that the CNS meeting currently has the record as the scientific meeting with the longest continuous funding from NIH. Continuing in the tradition he himself helped to establish, the CNS meeting has always openly encouraged attendance by program officers and others interested in CNS, providing them a spot in the agenda to present the interests and new funding opportunities of their agencies. In this case, the invitation to Yuan Liu to the CNS meeting in Montana in 1997 proved a personal life changing experience for both Dennis and Yuan. We have always thought of the CNSers as being a family operation. In the case of Dennis and Yuan, it literally is.

Finally, in retrospect and looking back, it is rewarding to look at the list of early student attendees of the CNS meetings and find a veritable roll-call of the current "rich and famous" mid-career and senior computational neuroscientists. In addition, the meetings have always been a lot of fun. Through the hard work of a lot of different people, we still regard it as remarkable that the CNS meeting continues to live up to its original objectives as listed in the first grant submitted to NIH:

> to provide an annual open forum for the discussion of progress in CNS, broadly defined... to support the increase in the quantity and quality of research being carried out in the field of computational neuroscience... to stimulate and facilitate interdisciplinary collaborative research... to provide a forum for young researchers to present their research and get professional feedback... to provide for rapid publication of current work in computational neurobiology through a well-organized set of meeting proceedings.

References

Minsky M, Papert S (1969) Perceptrons. MIT Press, Cambridge

Wilson M, Bower JM (1988) A computer simulation of olfactory cortex with functional implications for storage and retrieval of olfactory information. In: Anderson D (ed) Neural information processing systems. AIP Press, New York, NY, pp 114–126

Miller JP, Rall W, Rinzel J (1985) Synaptic amplification by active membrane in dendritic spines. Brain Res 325:325–330

Araya R, Nikolenko V, Eisenthal KB, Yuste R (2007) Sodium channels amplify spine potentials. Proc Natl Acad Sci USA 104(30):12347–12352

Proceedings Volumes of the Conference on Analysis and Modeling of Neural Systems (AMNS)

Eeckman F (ed) (1992) Analysis and modeling of neural systems. Kluwer, Boston

Eeckman F (ed) (1993) Neural systems: analysis and modeling. Kluwer, Boston

CNS Proceedings Volumes

Eeckman F, Bower JM (eds) (1993) Computation and neural systems 1992. Kluwer, Boston
Eeckman F (1994) Computation in neurons and neural systems. Kluwer, Boston
Bower JM (ed) (1995) The neurobiology of computation: proceedings of the annual computational neuroscience meeting. Kluwer, Boston
Bower JM (ed) (1996) Computational neuroscience: trends in research 1995. Academic, New York
Bower JM (ed) (1997) Computational neuroscience: trends in research 1997. Plenum, New York
Bower JM (ed) (1998) Computational neuroscience: trends in research 1998. Plenum, New York
Bower JM (ed) (1999) Computational neuroscience: trends in research 1999. Elsevier, New York
Bower JM (ed) (2000) Computational neuroscience: trends in research 2000. Elsevier, New York
Bower JM (ed) (2001) Computational neuroscience: trends in research 2001. Elsevier, New York
Bower JM (ed) (2002) Computational neuroscience: trends in research 2002. Elsevier, New York

Chapter 2
A Pictorial History of the Early Days of Computational Neuroscience: The CNS Meeting Posters

James M. Bower

Abstract In addition to its oral history, the CNS meeting also has a graphical or pictorial history represented by the series of posters produced for CNS*93–CNS*03, and explicitly captured in the poster produced for CNS*10, which represented the meeting's transition into its 20th year. These posters, reproduced in annotated form here for the first time, hang in laboratories around the world and in the halls of the U.S. National Institutes of Health, and represent in their own way, the history of computational neuroscience.

Chapter 1 in this volume recounts an oral history of the origins of the CNS meeting. This chapter in some sense tells the same story, through the posters produced for the CNS meetings (Fig. 2.1). The posters reproduced here reflect those produced from 1993–2003, as well as the explicitly historical poster produced to celebrate the beginning of the 20th year of CNS meetings. The posters reproduced here were created in close collaboration with two artists, Erica Oller (CNS*93–CNS*2000) and Bonnie Callahan (CNS*2001–CNS*2003 and CNS*2010) and can be found on the walls of laboratories around the world as well as in the halls of the U.S. National Institutes of Health.

So, why reproduce the posters here? In keeping with the theme of this book, this art itself reflects the growth and development of the field. Even a cursory examination should make clear that each is allegorical, blending some characteristic of each meeting's venue with some perceived aspect of the field of computational neuroscience. It is important to note that because neither artist had any direct association with the field of computational neuroscience, all implied interpretations regarding the field should be entirely attributed to me. This, in fact, is another motivation for publishing the posters with explanations. Through the years I have been quite

J.M. Bower (✉)
Health Sciences Center, University of Texas, San Antonio, TX, USA
e-mail: bower@uthscsa.edu

J.M. Bower (ed.), *20 Years of Computational Neuroscience*, Springer Series in Computational Neuroscience 9, DOI 10.1007/978-1-4614-1424-7_2,

Fig. 2.1 So many years, so many posters

amazed with interpretations I have heard as to their significance, especially with respect to the presumed representation of particular individuals. While it does state in the text at the bottom of the poster produced for CNS*93 that "any similarity to computational neurobiologists either living or dead may be intended," (Fig. 2.3)

Fig. 2.2 Who dat?

in fact, in the majority of posters very few of the figures represented real people. It clearly was my intent, however, to capture some sense of the CNS meetings and the computational neuroscience enterprise. To this point, some years ago I was visiting MIT and noticed the original CNS*93 poster (Fig. 2.3) hanging on the wall behind a receptionist. She did not know that I had anything to do with the design, so when I asked her where the rather odd poster behind her desk came from she said, "I have no idea, but they sure got the craziness around here right." I couldn't have asked for anything more.

In the brief descriptions of the posters that follow, I will mostly provide the most general context, refraining from detailing every buried meaning and significance. Like the brain itself, I believe that one can generally discover more, by looking closely (see photo of students at CNS*09 in Berlin: Fig. 2.2). To encourage that exploration, or just to decorate more laboratory walls, high-resolution images of each poster can be obtained online: http://www.genesis-sim.org/CNSposters.

We were far too busy organizing the first CNS meeting to generate a poster. Accordingly, the series of CNS posters began with the second meeting held in Washington, DC, with a poster designed to represent the complex state of computational neuroscience "hung" on the beautiful and complex dendrite of the cerebellar Purkinje cell (Fig. 2.3). The explicit reason we took the meeting to Washington was

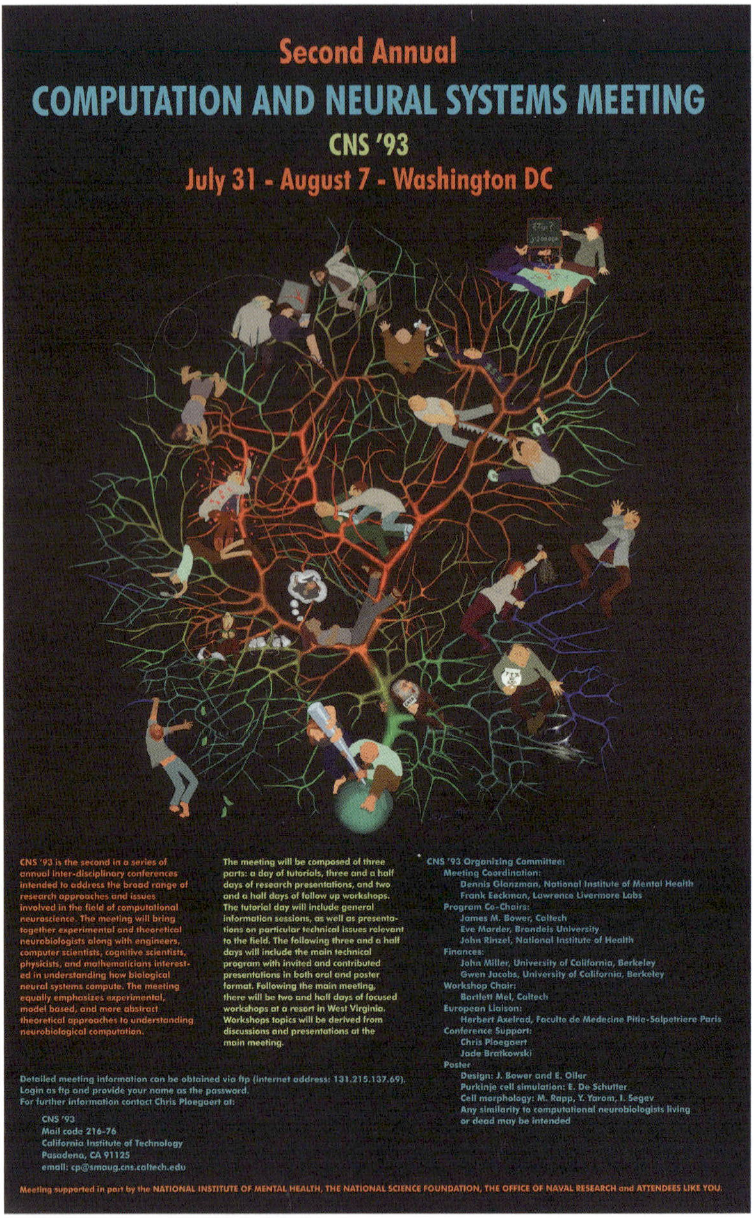

Fig. 2.3 Beware theorists

to expose federal funding agencies to computational neuroscience as a field deserving support. One can see a funding fight taking place in the geographical center of the dendritic tree while a poor young theorist (no despite the beard not a young, svelte Bill Bialek) is grasping at a few dollars, while a fully funded experimentalist and his rats sit contently just above.

The poster was also intended to contrast the still largely detached experimental and theoretical efforts. Therefore, we see two industrious scientists struggling to impale the Purkinje cell soma and a third clearly in love with his section of the dendrite. However, also note the fellow spreading TTX apparently at random, unguided by theoretical considerations. All theorists in the poster are located on the margins of the dendrite, with one in the lower right explicitly refusing to consider the complexities of real biology. Observing this conflict, a concerned experimentalist clutches both the core of the dendritic tree and his data. In the upper left three collaborating modelers seem confused as they examine their computer screen while the theorists in the most precarious position (other than the one about to fall out of the tree altogether) are sitting on the top right, engaged in translating the complex geometry of the Purkinje cell into the standard mathematical formulation for a Hopfield Neural Network. While clearly excited about the possibilities, they seem oblivious to the two experimentalists sawing through the branch (limb) on which they are sitting. In this case, the theorist enthusiastically directing this effort is Christof Koch (the only real person represented in this poster), my friend, and colleague at Caltech and then fellow director of the Methods in Computational Neuroscience Summer Course at the Marine Biological Laboratory in Woods Hole. As discussed in the introduction to this volume, the CNS meeting itself originated in part from a concern that the neural networks community might not be paying enough attention to the actual structure of the nervous system, although in fairness to Christof he always has.

Finally, I also want to mention explicitly one unfortunate misinterpretation that I know exists concerning this poster. It is my understanding that this is the only one of the CNS posters that is not hanging in the halls at NIH, as there was some concern that the somewhat darker skinned young female graduate student in the center of the tree wistfully dreaming of graduation could be interpreted as implying sloth or laziness. This is absolutely not the intention. Instead, I wanted to give the sense that even in the nasty political mix of science and funding, and abstract and realistic modeling, it is still possible, especially for young scientists, to dream and aspire.

CNS*94 and *95 were both held the Double Tree Hotel in beautiful Monterey, CA, at a time when we thought it would be logistically easier to have the meeting each year in the same place (and why not Monterey?). Accordingly, this poster was originally produced for CNS*94, but was so well received, and we felt so beautiful, that we decided to use it again for CNS*95. As a consequence, I can now confess, one reason for deciding to move the meeting in 1996 was to provide a new source of inspiration for the CNS poster.

Of all the CNS posters, this poster is at the same time the simplest and perhaps the most graphically complex (Fig. 2.4). The thematic simplicity is based once again on the perceived importance of hauling in federal funding to support a growing field. In fact, perhaps in part as a result of locating CNS*93 in Washington, DC, CNS*94 and *95 were supported by grants from no less than five federal agencies, four at the National Institutes of Health and an additional grant from the National Science Foundation. (Support for the original CNS meetings from the office of Naval Research had been withdrawn as commemorated in the CNS*98 poster as

Fig. 2.4 What a nice catch

discussed below.) Of course the complexity of this poster is manifest in the structure of the Kelp forest, which on closer inspection can actually be seen to represent the neuronal circuitry of the mammalian primary visual cortex, beautifully and carefully rendered by Erika Oller. Monterey Bay and especially the Monterey aquarium (where the meeting banquet was held) are famous for their kelp forests, and visual cortex was then and remains today one of the most studied subjects in computational neuroscience.

This poster is also probably the least political of all the posters, with one feature as an exception. When the poster was first designed, several members of the organizing committee felt that it contained too little explicit information about the meeting itself. In particular, they believed that including a list of speakers would attract more participants. As discussed in the introduction to this volume, our emphasis was always more on students than on invited speakers, and as should already be clear, my motivations for poster design did not only include advertising. However, in deference to the concern I added eight white "speaker fish" floating along the bottom. If you look very closely at their fins, you can actually see the names of the speakers (I think☺). One final historical note is that this poster includes a meeting email address. While younger scientists might find it hard to believe, in fact, the CNS meetings were one of the first to make use of the Internet for communication and advertising.

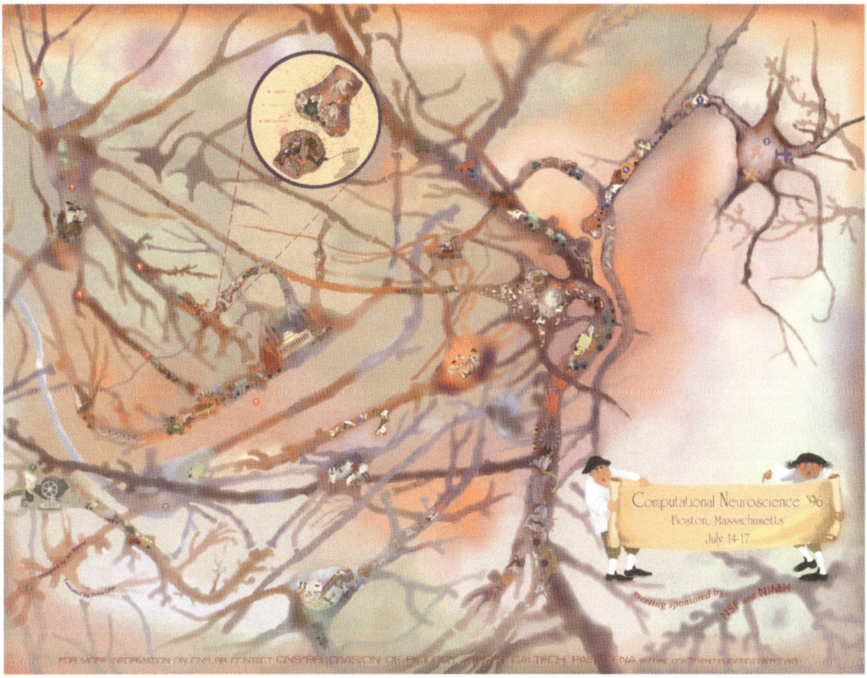

Fig. 2.5 Lost on the MTA

After 2 years on the west coast, the organizing committee decided to move the meeting back to the east coast, and what better place than Boston. Of all the CNS posters, this poster is the one whose symbolism has apparently been the hardest to decipher, even for people living in Boston (Fig. 2.5). While neurobiologically representing the complexities of intracellular trafficking in a small network of neurons, the poster itself is based on the also complex traffic patterns in Boston. Accordingly, in the upper left is Logan airport, with the T's representing the various stops on the MTA. Note also the large synaptic connection between Harvard and MIT, with characteristically (for each institution) attired pre- and postsynaptic faculty. With respect to computational neuroscience, in 1996 MIT was definitely presynaptic. There are many other bits of Boston both modern and historic represented in this poster (note for example the somewhat industrial effort at Boston University), but like molecular trafficking and the streets of Boston, the details are too complex and numerous to go into. Holding the CNS meeting in Boston fully established the importance of moving the meeting each year, as easier access for students resulted in a doubled meeting attendance and great vitality. The Boston meeting was also the first performance at a CNS meeting by Ramon and the K-halls (Fig. 2.6).

The "Computational Gang" definitely came to town in beautiful Big Sky Montana in the summer of 1997 (Fig. 2.7), however, in addition John Miller and I were also at the same time engaged in establishing the new Computational Biology Institute at nearby Montana State University, which shortly thereafter became John's new

Fig. 2.6 What would Ramon have thought?

academic home. In the poster, John is represented by the tall figure in the center with the orange pants and red shirt about to draw calculators from his holsters. As shown in the photograph, John's entrance to the meeting itself was on horseback, being "hauled in" as the Computational Neuroscience Unibomber (Fig. 2.8).

This poster also represents the field of experimental neuroscience as a kind of lawless western town in need of the structure and organization that, in principle, computational neuroscience could provide. What organization exists in this town is based on what part of the brain, or what type of organism is being studied, with the poster characterizing the different scientific cultures in each case. Thus, on the left, the "United Cerebral Evangelical Fellowship" is advertising a sermon titled "Oscillations to Higher Consciousness", while local maps can be obtained at the Hotel Hippocampus. The Crunchies and Squishies General Store is a fairly collegial happy place with lots of odd creatures hanging around, whereas next door, there is an all out fist fight raging at the Cerebellar Bar. At the end of the street are two banks: the Bank of NIMH, large and quite prosperous, and the Bank of NSF, quite a bit more modest. At this early stage of the computational neuroscience invasion, the few townspeople paying any attention seem either dubious or actively resistant.

Fig. 2.7 The Wild Wild West

Fig. 2.8 Not IACUC approved

The poster for CNS*98 held in Santa Barbara, CA, is the design that has apparently produced the most consternation and speculation regarding who is being represented doing what (Fig. 2.9). In 1998, "Santa Barbara" was one of the more famous US-produced daytime soap operas around the world. In fact, the Santa Barbara script

Fig. 2.9 Such a tangled web we weave

in the poster is taken directly from the trademark for that soap opera. Of course in Southern California, the city of Santa Barbara is also generally associated with a soap opera-like culture. Accordingly, this seemed a wonderful opportunity to illustrate the more "soap opera-like" features of computational neuroscience.

Over many years, I have heard numerous speculations regarding who is represented in the poster (as one would expect from a soap opera) but, in fact, only four figures are based on actual individuals. The figure on the left (with a nametag) commemorates the winner of the official tequila drink off (and the first runner up in the barrel races) at CNS*97 in Montana. Despite this fact, I understand this poster is hanging in the halls at NIH. The women playing catch-up behind him was the principle competitor in the actual tequila drink off, even though she was at a decided handicap having spent the entire previous evening in Montana playing drums for Ramon and the K-halls while standing up. Then there is a somewhat unseemly transaction going on at a table in the back between someone dressed in a naval uniform and someone being asked to sign a contract in exchange for cash. The naval officer actually represents a real program officer at the Office of Naval Research who several years earlier had cut grant funding for the CNS meetings because the meeting's participants had voted down his proposal to merge with a neural networks meeting. In this poster, for the first time, I included myself—I am riding the horse. But the computational neuroscientist drowning himself in the center of the poster was intended to more generally represent the plight of modelers trying to publish

Fig. 2.10 Visions of the future

their results. Specifically, he is clutching a rejection letter from Science Magazine, which had recently introduced a new process to "reject without review" papers judged by a small group of senior editors to be less interesting than other papers submitted during the same week. In my view this introduced a new and unfortunate level of scientific politics more appropriate to a soap opera.

The poster for CNS*99 commemorates both the city of Pittsburgh and the increasing growth in papers submitted to the meeting describing molecular and cellular level modeling (Fig. 2.10). As a welcoming gesture, the poster reflects not only

Fig. 2.11 It's only rock and roll …

Pittsburgh's famous history as a steel town but also what seemed to me to be the more industrial nature of many molecular and cellular research efforts. Produced by a cloning machine on the second floor, the basement is filled with graduate students laboring with pipet men in individual cubicles collectively producing a delicate thread of DNA. That DNA first winds through a huddle of postdoctoral fellows trying to make sense of the sequence by hand, before ascending to the head of the laboratory, who sits converting the DNA into cash. While perhaps a bit dark, there is hope in the form of a group of computational biologists breaking into the building from above. They have actually divided into two groups, one with decidedly greedy expressions headed straight for the head of the laboratory (and the cash), and the other parachuting from a loftier height on the way to the postdocs doing the analysis. I would note that nobody is heading to liberate the poor basement dwelling graduate students. To my amazement, I have seen this poster on the wall in many molecular biology laboratories. Of course, the CNS meeting has always warmly welcomed molecular and cellular experimentalists as well as theorists. "The photograph from CNS*99 shown in Fig. 2.11, commemorates another performance by Ramon and the K-halls, this time in a "Pittsburgh appropriate" bar discovered by a dedicated group of local CNS graduate students.

The new millennium meeting in Brugge, Belgium, represented an important step for the CNS meetings, as it was the first meeting held outside the United States. Today, the meeting alternates between the United States and a foreign site, but CNS 2000 was our first trip to Europe. It seemed fitting, therefore, to represent this important step as yet another invasion, this time backwards from the new world to the old (Fig. 2.12). Thus, the Santa Maria can be seen unloading computational neurobiologists (as well as computers, and *Rattus norvegicus*) to a decidedly old masters version of Belgium. Greeting the arriving hoard at the doc is one Erik De Schutter, dressed in a green hoody and standing in front of the conference flag. It is clear that the Europeans, in general, are not quite sure what to make of the scene. The Nina, the next ship to be unloaded, is carrying "Ramon and the K-halls" for

Fig. 2.12 Re-migration

their upcoming performance at a Celtic Bar in Brugge (Fig. 2.13). Unfortunately, an effort to quickly make a Stonehenge pillar for that performance was thwarted because the artist thought that I had mistakenly specified the pillar in inches rather than feet. This poster was the last drawn by Erika Oller, as her growing independent art career left her little time for such frivolities.

In addition to the artist transition to Bonnie Callahan, CNS 2001 also represented a transition for me as I had decided, after 10 years, to resign as meeting chairman. It was at this meeting that the initial steps to form the Organization for Computational Neuroscience (OCNS) were initiated as a way to continue the meeting as well as provide more general support for the field of computational neuroscience. Because CNS 2001 was also the meeting's 10th anniversary, I wanted to use the poster to commemorate the many individuals who had played an important role in the meeting's growth and also to reflect the meeting's fun/festival/summer camp-like culture they had helped to establish (Fig. 2.14). Asylomar on the California coast was the ideal venue.

In this poster, for the first time, everybody is somebody. This poster also includes more inside information than any poster in the series, most of which, to protect the guilty and the innocent, I won't go into. There are, however, several individuals and circumstances worth noting: First, this poster specifically includes a performance by Ramon and the K-halls now named simply "the K-halls," as Ramon has left the band (note the free standing microphone and the fellow slipping away in the distance carrying his cowboy boots). Second, the occupants of the deck on the right are all important figures in guiding and growing the CNS meeting. Especially worth

Fig. 2.13 Druids all

Fig. 2.14 Summer camp

Fig. 2.15 A tribute

noting are the meeting's cofounder John Miller (wearing his unibomber outfit), our Federal Government Liaisons Dennis Glanzman and Yuan Liu (who also wrote the introduction to this book), and Ranu Jung, who played a particularly important role in the birth of OCNS. A few other individuals scattered about are Bill Bialek (sage-like), Eve Marder (famous for mentoring her graduate students), Erik De Schutter (proud to be a member of the EU and also subsequently central to the growth and success of OCNS), and Valentino Braitenburg who only attended one CNS meeting (in Montana) but astonished everyone by not leaving the dance floor while the K-halls performed (for four hours). Dancing near Valentino is David Nicoladie Tam who is the only scientist to have attended every CNS meeting for the last 22 years, and Dave Beeman and his wife, once again reliving the 60s. Of particular importance is Judy Macias, dancing and holding the tambourine, who, as the conference secretary, was the heart and soul of the meeting for many years. This poster also includes, for the second time, Christof Koch, standing behind the fence along with several other prominent computational neuroscientists (behind the hedge) who for one reason or another had not yet attended a CNS meeting. Finally, behind this group is someone hanging from a gallows. I have heard many amusing speculations as to who that person might be, but, it turns out that Bonnie Callahan (the artist) snuck the image in at the last minute to represent what might happen to her if all the stories buried in this poster were brought to light. No risk there.

The new CNS meeting chair for 2002 was the Phil Ulinski aided by his lovely wife Mary (Fig. 2.15). For the first time in 10 years my only responsibility was for the poster and a not very inspiring K-Halls unplugged acoustic guitar session. Given that Chicago was at that time world famous for the Chicago Bulls, it seemed reasonable to use a basketball motif for the meeting (Fig. 2.16). Interestingly enough, the depicted basketball game has often been misinterpreted as a contest

Fig. 2.16 Michael with the ball

between real neurons and neuronal models. In fact, my intention was a competition between more abstract models and those based on actual neuronal morphology. In this case, no doubt reflecting my own scientific biases, the biologically realistic models are ahead 69–66 and have the ball. However, storm clouds are brewing in the distance. The game is being referred by NIH and NSF and is, in fact, still on going as there is no official game clock. Phil Ulinski, who sadly passed in 2010, played an important role in the CNS meetings, the growth of computational neuroscience, and the formation of the OCNS.

The poster for CNS 2003 was the first poster that was only available in digital form (Fig. 2.17). Capitulating, the poster also included the names of the invited speakers. Because the meeting was held in Spain, I thought that it was simply too wonderful an opportunity to make a final personal statement about the nature of computational neuroscience and neuroscience as a whole. In this case, instead of windmills, Don Quixote, dressed in the official medieval academic robes of a biologist, is tilting at Purkinje cells while poor Poncho Panza (his graduate student) is trying to make sense of the data using an ever more complex set of experimental and computational tools. However, in contrast to the poster for CNS*93 designed 10 years earlier, this time the electrode being inserted into the Purkinje cell is directly connected to computational tools, and the experimentalist is technically linked to the computational neurobiologist. While this is still clearly madness, there is now hope. Figure 2.18 provides some sense of the process involved in generating the CNS poster each year.

In the years from 2003 to 2010 meetings were held, and posters were made (Fig. 2.1), however, I did not return to CNS poster design until 2010, when Charlie

Fig. 2.17 Many knights-errant astride Ratinante

Wilson, Todd Troyer, and myself were appointed local co-organizers for the meeting in San Antonio, TX. While officially the 19th CNS meeting, the poster celebrates the full sequence of meetings, going back to its origins (Fig. 2.19). As shown, this poster was also produced in two slightly different forms, one to advertise the meeting (above) and one handed out during the meeting (below).

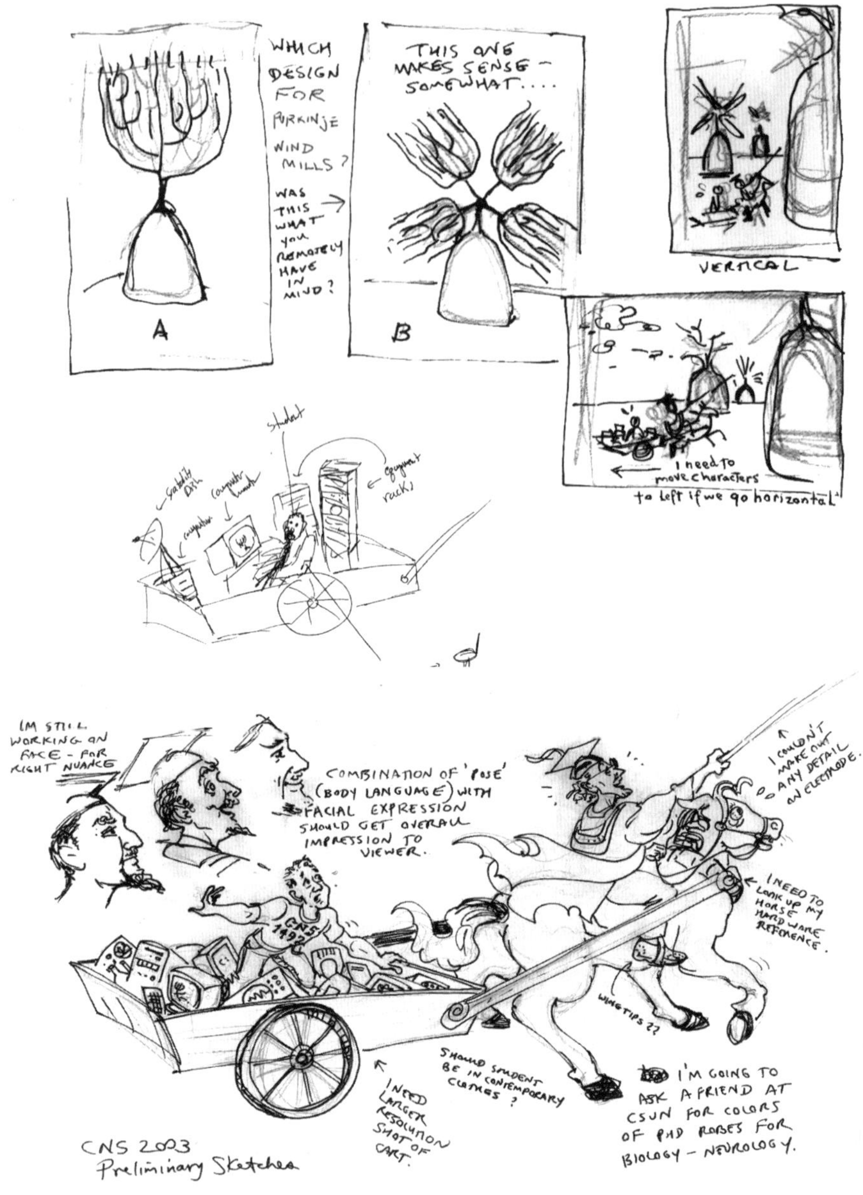

Fig. 2.18 The back and forth—creation of a CNS poster. At the top are the artist's choices for design of the central Purkinje cell in the poster for CNS 2003. In the center is the original sketch by JMB of the cart containing the long suffering graduate student. At the bottom is an early version of the cart by the artists with design notes. Iteration on poster design often continued for several months before the final version was ready for production. Artist renderings by Bonnie Calahan

Fig. 2.19 In sum and summary

The poster venue is the famous San Antonio River Walk running adjacent to the meeting site. In the pre-meeting version of the poster, the riverboat is being piloted by a somewhat shady figure, and the boat passengers (actually the OCNS organizing committee) seem clearly to be worried, this despite the fact that two additional lines are being held on the shore for stability. In the final version handed out at the meeting, the boat has moved a bit further down the river and the OCNS committee seems somewhat less concerned, perhaps because there isn't much that can be done about the situation now anyway. As with the poster for CNS*2001, all the figures represent real individuals who have played an important role in the development of the CNS meetings and computational neuroscience as a whole. Ranu Jung, at the far head of the boat before the meeting, is seen leaping out in the second poster, as she transitioned after a number of years as OCNS president. Erik De Schutter is seen climbing on, unaware that he would soon be arrested at the CNS banquet.

Fig. 2.20 The hall of fame and shame at CNS 2010

On the banks of the river, as on the San Antonio River Walk, are clubs, restaurants, and other establishments in this case representing the 18 previous CNS meetings. In the far distance, the river bifurcates at the eternal Purkinje cell into two streams, each presenting the meetings from which the original CNS meeting evolved. One stream represents the early Berkeley workshops organized by John Miller and colleagues (John standing on the bridge), and the other representing the original Neural Information Processing (NIPS) meetings. Old style signs made from the meetings' posters hang from each establishment commemorated 20+ years of meetings. From neurobiologists populating the dendrite of a Purkinje cell, to the complex serenity of a kelp forest, to the tangled complexity of Boston, to the wild west, to a return trip to the old country, and a summer camp in Asylomar, this final poster, for me, represents the process involved in the emergence of computational neuroscience as a stable, organized, and sophisticated science, as well as a certain wistfulness about its more playful and less certain past. Figure 2.20 shows the "walk down memory lane" at CNS 2010. Here is to another 20 years of CNS meetings.

Chapter 3
History of Neural Simulation Software

David Beeman

Abstract This chapter provides a brief history of the development of software for simulating biologically realistic neurons and their networks, beginning with the pioneering work of Hodgkin and Huxley and others who developed the computational models and tools that are used today. I also present a personal and subjective view of some of the issues that came up during the development of GENESIS, NEURON, and other general platforms for neural simulation. This is with the hope that developers and users of the next generation of simulators can learn from some of the good and bad design elements of the last generation. New simulator architectures such as GENESIS 3 allow the use of standard well-supported external modules or specialized tools for neural modeling that are implemented independently from the means of the running the model simulation. This allows not only sharing of models but also sharing of research tools. Other promising recent developments during the past few years include standard simulator-independent declarative representations for neural models, the use of modern scripting languages such as Python in place of simulator-specific ones and the increasing use of open-source software solutions.

Introduction

When Jim Bower first asked me if I would write a chapter on the history of realistic neural simulators, I refused. I reminded him that although I am at an age when scientists wrap up their long careers with a historical account full of advice for young researchers, I have only been involved with computational neuroscience for a little more than 20 years, and I am just getting started with serious cortical modeling.

D. Beeman (✉)
Department of Electrical, Computer, and Energy Engineering,
University of Colorado at Boulder, Boulder, CO 80309-0425, USA
e-mail: dbeeman@colorado.edu

J.M. Bower (ed.), *20 Years of Computational Neuroscience*, Springer Series in Computational Neuroscience 9, DOI 10.1007/978-1-4614-1424-7_3,

Furthermore, my experience has been almost entirely as a developer of tutorials and documentation for the GEneral NEural SImulation System (GENESIS). Not only that, but I missed out on the crucial first 2 years of GENESIS development.

However, Jim can be very persuasive, and I gave in after he told me that I could tell the story my way as a personal history of what I have learned during the past 22 years. Of course, the task was made easier by the fact that I share Jim's definition, considered very narrow by many, of what constitutes a "realistic neural simulator" (Bower 1992, 2005). What I offer is a somewhat GENESIS-centric and subjective view of some of the issues that came up during the last 20 years of development of GENESIS, NEURON, and other platforms for structurally realistic simulations of neurons and their networks.

More specifically, I finally agreed to write this chapter because it provides an opportunity to use the last 20 year history of the development of neural simulators to offer a few, hopefully useful, opinions on what the developers and users of the next generation of simulators can learn from some of the good and bad design elements of those from the last generation. My own current view is that it is time for a new generation of simulators that addresses a large range of issues beyond merely creating numerical solutions to mathematical models of neural systems. As will be discussed later, new simulation architectures such as that being developed for GENESIS 3 (G-3) offer options for extensibility, interoperability, and model sharing that will significantly extend the capacity and value of simulation technology, providing a foundation for the next 20 years and beyond (Cornelis et al. 2012a). My hope is that this chapter will help motivate and inform these efforts going forward.

My Seduction by Neuroscience

This chapter is in large part a personal recounting of my own experience in the design and development of GENESIS and it is therefore appropriate, I think, to provide the reader some background information on my own path to computational neuroscience. In the spring of 1989 I was 51 years old, and a happily tenured Professor of Physics at Harvey Mudd College, an undergraduate science and engineering college about 30 miles from Caltech. I enjoyed the atmosphere at Harvey Mudd because the students are very intelligent, self-motivated, and creative. In addition, without graduate students, I was able to concentrate on undergraduate education, and to take advantage of the great freedom I had to create and teach interdisciplinary special topics courses on nearly anything that interested me.

Although it might seem strange that, 20+ years later, I would have abandoned physics, given up my tenured faculty position at Harvey Mudd, and now be deeply involved in building models of the mammalian auditory cortex, in fact, looking back, the transition from teaching undergraduate physics to working on GENESIS makes some sense.

A core component of the GENESIS project from the outset, and one that I have been particularly involved with, is the development and application of simulation-based tutorials to engage students in learning. At Harvey Mudd, I had already developed several simulation-based tutorials for teaching concepts in upper division physics courses (e.g., Beeman and Boswell 1977). My transition to computational neuroscience as a subject of study was also linked to my teaching through my research interests, which at an undergraduate college, need to be tied closely together. That research involved the computer modeling of amorphous solids or disordered spin systems which served as a very good source of student projects, and a few publications (e.g., Maley et al. 1988; Thorpe and Beeman 1976; Alben et al. 1977).

Because of the spin system modeling I had been doing, I became aware of the publication by Hopfield (1982) applying spin glass models as a possible model for how networks of neurons might operate. As a result, I volunteered to give some lectures in an interdisciplinary course that was to be taught during my sabbatical in spring 1990. For that course, a biologist was to lecture on basic properties of neurons, a mathematician would present a mathematical approach to artificial neural networks, and I would give a more engineering-oriented approach to artificial neural networks, aided with tutorial simulations.

Of course I was also aware that John Hopfield was at that time a professor in the new Computation and Neural Systems program at Caltech just down the 210 freeway from Harvey Mudd. My first thought was to ask Dr. Hopfield about spending some time with his group at Caltech during my sabbatical year. However, our department received a weekly list of seminars at Caltech, which was posted outside the Physics Department office, and given my interest in simulation technology, I was intrigued by the description of a talk to be given by Matt Wilson on a simulator for biologically realistic neurons and neural networks called GENESIS. I went to the talk and became fascinated with the prospect of actually building models of real neurons. Although Matt's seminar emphasized the use of GENESIS as a research tool, I thought that I might also be able to use it in the course I was planning to teach at Harvey Mudd. I also thought that, through the simulator, I might be able to learn something about biological neurons myself. I had no idea at the time that this was the first step on the long slippery slope that has led so many physicists into neuroscience.

After the talk, as I spoke to Matt, Jim introduced himself and we talked about the possibility of using GENESIS as a basis for a simple tutorial on the properties of biological neurons. Jim told me of his strong commitment to developing tools for education, and we discussed possible collaborations. As a result, I spent my sabbatical year in Jim's laboratory, learning about GENESIS, and far more about neuroscience than I had ever intended.

History of Neural Modeling Prior to Fall 1989

Before I could write tutorial simulations about neuroscience, I needed to learn something about the subject. Of course, I was familiar with the history of artificial "neuron-like" networks, beginning with the Mcullough and Pitts (1943) model, but had no knowledge of modeling spiking neurons, nor very much about their physiology. I sat in on David van Essen's introductory neuroscience course, along with the first year graduate students in neuroscience, and also began learning how to use GENESIS. As always, I learn best by doing, so I started writing my first tutorial based on a simulation of a simple neuron model, as I learned about the Hodgkin–Huxley model, and about compartmental modeling of dendrites.

Of course, I read the classic paper by Hodgkin and Huxley (1952) to understand the modeling of action potentials. Their work, carried out in the early 1950s and described in a series of 1952 papers, won them the Nobel Prize in 1963 and exemplifies, in my view, the ideal combination of modeling and experimental measurements. I can't emphasize enough the importance of this connection between modeling and experimental work. Too many theoretically inclined people get lost in a world of their own when doing computer modeling and lose touch with the world of experiment. This is a common pitfall for theoretical physicists who hope to apply their expertise to the study of the brain. Likewise, experimentalists may end up mindlessly gathering data without having a clear idea of how it will advance theoretical understanding.

The importance of this connection is also clear in computational neuroscience and was essential to what I regard as the first great success of computational neuroscience, the Hodgkin–Huxley model (now 60 years old), which still stands as the basis for most neuronal cell models. Most neurobiologists recognize the importance of Hodgkin and Huxley's work and their development of the voltage clamp technique without realizing how important the modeling was to the work. Essentially, the model was what made them throw out their old way of looking at the changes in the membrane and introduce a new viewpoint. It is important to remember that at the time of their experiments, the modern concept of ion-selective channels controlling the flow of current through the membrane was only one of the several competing hypotheses. It was the model that ruled out these alternative ideas, and also predicted the results of experiments that were not used in formulating the model. The fascinating history of this pioneering synthesis of experiment and modeling has been told in reviews by Cole (1968), Rinzel (1990), Nelson and Rinzel (1998), and many others.

However, it is important to mention how Hodgkin and Huxley quantitatively explained the process by which action potentials are formed by the voltage-dependent activation and subsequent inactivation of sodium channels, terminated by a delayed activation of potassium channels. They did this not by fitting model parameters to those needed to produce action potentials, but by fitting them to an entirely different set of experimental data, obtained using the voltage clamp. Then, with no further changes in parameters, they were able to reproduce the action potential, correctly calculate the velocity of propagation, analyze the refractory period, and account for the phenomenon of post-inhibitory rebound or "anode break." All of this modeling was performed by integrating the coupled differential equations, step by step, on mechanical "hand-crank" calculators, following the method used 20 years before by the physicist Hartree (1932) to calculate atomic wave functions.

As a perhaps ironic aside, a few years after I had taken charge of the GENESIS Users Group (BABEL), I received an email from a postdoctoral student who pointed out what he claimed to be a serious bug in GENESIS. He found that using a hyperpolarizing current injection pulse in one of the GENESIS tutorial simulations produced the obviously impossible result of producing an action potential! I tactfully suggested that he read the original Hodgkin–Huxley papers. At that time, I had been extending Mark Nelson's "Squid" tutorial for use with the chapter that he and John Rinzel were writing for "The Book of GENESIS" (Bower and Beeman 1998), familiarly called "The BoG." It was only after I added the ability to plot the channel activation variables during a current pulse that I fully understood myself the action potential refractory period and the biological phenomenon of post-inhibitory rebound.

In my own efforts to understand the existing techniques for simulating real neurons, the next step was to understand why GENESIS broke a neuron into "compartments." I quickly learned of the other crucial development in the history of neural modeling, which was the introduction of compartmental modeling by Rall (1964). Rall had previously contributed a great deal to the understanding of postsynaptic potential (PSP) propagation in dendrites by applying the mathematical analysis of the attenuation of signals in the transatlantic telephone cable by William Thompson (Lord Kelvin). For example, neural modeling has benefited from the simplifications introduced by the "trees equivalent to a cylinder" transformation, in which Rall (1959, 1962a) demonstrated the conditions under which a branched dendritic tree can be collapsed into a linear cable. The "cable" theory of propagation in dendrites has been reviewed by Rall and Agmon-Smir (1998).

By using a lumped parameter model, dividing a branched dendritic tree into coupled chains of approximately equipotential compartments, Rall's method made it possible to explore realistic dendritic morphologies that could only be analyzed by using numerical methods and computer simulations. In one of the first applications of this method Rall (1967) modeled a linear chain model of a motor neuron with a soma and nine dendritic compartments, activated with an "alpha function" form of synaptic conductance having a linear rise and exponential decay with time. It had no voltage activated conductances.

Rall and Shepherd (1968) created the first model to combine compartmental modeling of dendrites into a cell model that generated action potentials. Their model with a soma and ten dendritic compartments used parameters taken from rabbit olfactory bulb mitral and granule cells. Because the Hodgkin–Huxley model parameters for the neurons being simulated were not yet known, a simpler and less computationally intensive model was used for the generation of action potentials with active conductances. These simulations were carried out on a Honeywell 800 computer during 1963 and 1964 at the NIH, at a time when realistic simulation of ionic currents was considered to be very time-consuming. Personally, I believe that Rall's pioneering modeling efforts have been on a par with those of Hodgkin and Huxley, and perhaps the only reason that he has not received a Nobel Prize is due to the sheer complexity of dendrites themselves, whose function we still don't understand. Certainly Rall's technical contribution to modeling is on the same level as that of Hodgkin and Huxley.

Around the time that Rall and Shepherd were building their first models of neurons, others were applying the Hodgkin–Huxley equations to single compartment neuron models. For example, Connor and Stevens (1971) performed one of the first computer simulations of the ionic currents and the resulting action potentials in giant molluscan neurons. This model added a transient potassium conductance ("A-current") to modified Hodgkin–Huxley fast sodium and delayed potassium currents, using parameters fitted to experiments.

Dodge and Cooley (1973) were the first to publish a description of a model that combined compartmental modeling with the Hodgkin–Huxley equations. They collapsed a large spinal motor neuron into a compartmentalized nonuniform equivalent cylinder by the method of Rall (1962b) and used fast sodium and delayed rectifier potassium channels with parameters modified to fit motor neuron voltage clamp data.

This model became the basis for later models by Traub (1977), who included calcium-dependent potassium channels in the dendrites. This led to a series of increasingly realistic hippocampal pyramidal cell models (Traub and Llinás 1979; Traub 1982; Traub et al. 1991, 1994) with active conductances in the dendrites. These were run on IBM mainframe computers, with programs initially written in PL/1 and later in FORTRAN.

The earliest network model with multi-compartmental spiking neurons of which I am aware was a simplified model of the cerebellar cortex of the frog by Pellionisz et al. (1977). These used 62-compartment Purkinje cell models having modified Hodgkin–Huxley conductances.

The Introduction of Neural Simulation Systems

In each of the cases mentioned to this point, the computational modeling was done with specific code written by the individual researchers. To my knowledge, there was no effort made to provide that code to anyone else or to generalize it beyond a particular model. There was also no explicit effort to use these simulations as a tool in neuroscience education. Their intended purpose was purely research and was based only in the individual research labs. Yet, the nervous system itself is made up of neurons that share many common components (e.g., ion-selective channels), raising the distinct possibility that a modeling system with common code and a common set of libraries could allow the sharing of components between different laboratories. In effect, this kind of sharing in physics has occurred for hundreds of years, in part because it is easier to share equations than complex biological models dependent on a whole system of equations. In principle, however, a common modeling platform for neural simulations could not only support the sharing of components, but also start to build a common set of models, which we have called "community models," supporting communication between different laboratories and research projects. Jim Bower's chapter in this volume talks about what may be the first such model, of the cerebellar Purkinje cell, originally developed in GENESIS and now implemented in numerous other simulation systems (including

NEURON) and also now being used as the basis for research in a growing number of laboratories. Of particular interest to me, such a general neural simulation system could also, in principle, be used to generate simulation-based tutorials for education in the tools of computational neuroscience, as well as neuroscience itself.

The use of an electrical network simulator was first suggested by Shepherd and Brayton (1979) for a simulation of a dendro-dendritic synapse circuit in the olfactory bulb. However, the first software that could be called a general simulator specifically for realistic neural models was a suite of FORTRAN programs called MANUEL developed by Don Perkel in 1981 (Perkel and Watt 1981). MANUEL generalized some of his earlier custom made programs into a package for constructing multi-compartmental neurons and small circuits. Jim Bower tells me that one of the first things he did after moving to Caltech was to visit Don Perkel in his research trailer parked behind a research building at the University of Irvine. Don had recently been fired as a professor by Stanford, but had used grant money to purchase a trailer, outfitted with computer equipment, and had made a temporary arrangement with UC Irvine renting space in their parking lot. Don was well ahead of his time. The MANUEL programs allowed a wide variety of physiological and anatomical properties to be specified and provided a versatile set of utilities for providing stimuli and recording the results. As a way to provide support for his pioneering efforts, MANUEL was available for a substantial fee, and was written in a Digital Equipment Corporation (DEC) variant of FORTRAN IV, and did not run on Unix systems. MANUEL was used by Peter Getting to make what was probably the first network model made out of realistic neurons to study the swim central pattern generator circuit of the mollusc *Tritonia diomedea* (Getting 1989). Ironically enough, Peter Getting had also not been given tenure at Stanford and ended up at the University of Iowa, where he built his model accurately reproducing the swim pattern and explaining the role of the various ionic conductances in determining the behavior of this small network. Sadly, both Don Perkel and Peter Getting shortly thereafter developed serious health problems that ended their research careers.

There was also an effort in the 1980s to use simulators built for modeling electric circuits, such as SPICE (Segev et al. 1985) and SABER (Carenvale et al. 1990) to construct models of neurons. In principle these simulation systems had the built-in tools needed for simulating the circuits used in compartmental models. They also had the advantage that they were widely used by electrical engineers and also had the advantage of being available for a wide range of computer operating systems. In the end, however, most of the development of these systems was focused on electrical circuit simulation and they were not further optimized for building neuronal models. Similarly, the growth in interest in neural networks for engineering purposes also resulted in the construction of several "neural network" simulation systems such as the Rochester Connectionist Simulator (Goddard et al. 1987) that were also promoted for their possible use in biological network simulations. These also turned out to be too specialized and restricted in their capacity to support full realistic models.

Early History of GENESIS and NEURON

It was during this same period of time, in the middle 1980s, that the development of both GENESIS and NEURON started as dedicated systems specifically for building realistic simulations of the nervous system. By 1989, when I first learned of GENESIS, both systems were well into the early phase of their development—each, however, starting from a different point of view and with somewhat different objectives.

The NEURON simulator had its beginnings in the laboratory of John W. Moore at Duke University, where Michael Hines developed an efficient implicit numerical integration algorithm for use in branched compartmental dendritic models (Hines 1984). The Hines method was initially implemented in CABLE, a simulator developed for modeling propagation of PSPs in dendrites (Hines 1989).

Although it was primarily being used for modeling dendritic structures at this time, it already had the capability of making multi-compartmental models of single cells. In addition to the standard voltage-activated Hodgkin–Huxley sodium and potassium channels, it could model basic mechanisms for calcium dynamics and calcium-dependent potassium channels. By 1990, the name had changed to NEURON, and its single cell modeling capabilities began to expand. Over the next 2 years, it gained a scriptable GUI, the ability to model small networks, and a method of loading and compiling user-specified channel kinetics.

A Personal History of GENESIS

Of course, my personal knowledge of the development of GENESIS is much more detailed than that of NEURON.

GENESIS had its origins during 1984, when Matt Wilson was studying for a Master's degree in electrical engineering at the University of Wisconsin. During this time, after having done postdoctoral studies in the laboratory of Rudolfo Llinás at New York University, Bower was finishing up his postdoctoral work in Lewis Haberly's lab at the University of Wisconsin, studying the olfactory cortex.

As Bower recalls, Matt had been hired to program data acquisition software for a new brain slice preparation he was setting up in the Haberly Laboratory. Jim had also brought to Wisconsin the data he had recorded from many cerebellar Purkinje cells at once while a postdoc at NYU. This data was unusual and complex as it was one of the first sets of multi-single neuron data ever obtained, consisting of recordings of 16–32 signals at once (Sasaki et al. 1989). Bower was interested in finding some means of analyzing the data other than cross-correlation analysis. After approaching Josh Chover, head of the Math Department at the University of Wisconsin, Chover and Bower decided to co-teach a course on statistical analysis of multiunit recording data, for which Matt became the teaching assistant.

During that course, Jim realized that understanding complex neurobiological data would eventually require a tight coupling between experimental and model-based studies (Bower 1991). As an experimentalist, he believed that the types of model that would be most useful were ones that as closely as possible approximated the actual morphological and physiological properties of the brain structures being studied. While it was several years before his laboratory began developing models of the cerebellum (e.g., Santamaria et al. 2007), he decided to see if a model of the olfactory cortex might help explain the pattern of oscillations he was studying in the Haberly laboratory. Working together, Jim and Matt generated the first network model of the olfactory cortex constructed on an IBM XT computer. The model consisted of a linear chain of 75 5-compartment neurons which almost as soon as it was constructed began oscillating with 40 Hz bursts at Theta frequency. Jim's recollection is that it took several weeks to figure out how the bursts were generated in the model.

When Jim came to Caltech in early 1985, as one of the cofounders of the interdisciplinary graduate degree program "Computation and Neural Systems," he encouraged Matt to apply as a doctoral student. After arriving from Wisconsin, Matt continued to elaborate the olfactory cortex model, and published a thesis in 1990 predicting the neural mechanisms underlying the 40 Hz and theta frequency oscillations in cerebral cortex (Gray et al. 1989; Wilson and Bower 1991, 1992).

According to Jim, while Matt was working on his own modeling studies, he asked Matt to generalize his simulation software so that it could be used as a general purpose simulator, rather than as a stand-alone single purpose model. Matt resisted at first, feeling that no computational modeler would ever want to use software that was written by someone other than themselves. Fortunately, Matt relented and began work on GENESIS (Wilson et al. 1989).

Later, as I attended the annual Computational Neuroscience meetings in the early to mid 1990s, and collected data to satisfy funding agencies that GENESIS indeed was being widely used outside the Bower laboratory, Matt's prediction that real programmers would write their own code seemed to be born out. During that decade, the number of poster presentations using GENESIS or NEURON was generally outnumbered by those that used custom software written for a particular simulation or category of simulation. However, today, the number of scientists using simulation systems continues to rise, and importantly, the simulators are increasingly providing an opportunity for nonprogrammers to engage in computational studies. They are also increasingly being used in graduate and even undergraduate education, replacing textbooks with dynamic simulation tutorials.

As a side note, as he continued to develop GENESIS, Matt became increasingly interested in the experimental side of neuroscience research, and the studies of the hippocampus being carried out by Bruce McNaughton. After he received his Ph.D. in 1990, he went to the McNaughton laboratory at the University of Arizona for his postdoctoral studies, bringing the multielectrode array design with him. Legend has it that he said that he would come only if he were not required to do further modeling. Matt is now Sherman Fairchild Professor in Neurobiology, Departments of Brain and Cognitive Sciences and Biology at MIT, researching the role of sleep in learning and memory.

With respect to the early history of GENESIS, it is also important to mention the contributions of Upinder S. Bhalla who developed the first GUI for GENESIS, XODUS (the X Oriented Display Utility for Simulations) (Bhalla 1998). Entering the Bower laboratory as a doctoral student in Neuroscience in 1986, "Upi's" principal focus was on multielectrode recording in the olfactory bulb of awake behaving animals, where, along with Matt Wilson, he designed a unique multielectrode array that has subsequently became the basis for many multi-single unit recording experiments in many laboratories. However, Upi was also using GENESIS to model olfactory bulb mitral and granule cells, and started adding graphical capabilities around 1987. The initial version of XODUS was based on the Unix X Window System, and added scripting commands to Matt's Script Language Interpreter (SLI).

Although it is getting somewhat ahead of the story, it is appropriate to mention that after completing his Ph.D. in April 1993, Upi began his postdoctoral studies with Ravi Iyengar at Mount Sinai School of Medicine in New York, contributing to GENESIS development over the Internet. By November 1995 he had added the kinetics library and Kinetikit GUI for modeling chemical kinetics and signaling pathways (Bhalla and Iyengar 1999). After assuming his faculty position at the National Centre for Biological Sciences in Bangalore in January 1996, he continued to study the systems biology of olfaction and memory, and to extend the capabilities of the GENESIS kinetics library (Bhalla 2000). In order to exchange biochemical signaling models, he established the Database of Quantitative Cellular Signaling (DOQCS), one of the first databases of models of signaling pathways in the brain (Bhalla 2003). DOQCS (http://doqcs.ncbs.res.in/) currently contains 76 models contributed by users world-wide. The model representation format used in DOQCS is based on GENESIS 2 SLI commands using the kinetics library; however, it is now being extended to include SBML (Hucka et al. 2003) and Matlab formats. Later, as the limitations of the 1980s and 1990s simulator architectures became apparent, Upi began a major reimplementation of GENESIS 2 as MOOSE (Ray and Bhalla 2008).

Because the technical basis for GENESIS was a model of the olfactory cortex, GENESIS differed from the design of NEURON in that, from the outset, it was designed to simulate neural structures at multiple levels of scale (Wilson et al. 1989). Due to the influence of the parallel computing group at Caltech headed by Geofry Fox, GENESIS was also from the outset, designed to be implemented on parallel computers (Nelson et al. 1989). Although NEURON had its origins as a simulator for single cell models, it acquired improved network modeling capabilities and a parallel implementation in the following years.

By the July 1990 public release of GENESIS version 1.1 with full source code, Upi had added the "Neurokit" graphical environment for editing and running single cell models to GENESIS. Neurokit was written entirely in the GENESIS scripting language, using XODUS. Figure 3.1 shows this first version of Neurokit running a mitral cell model, with the menus, cell view, and graph displayed. By 1991, Upi had added the Hines (1984) integration method to GENESIS and added a cell reader to read in cell model specifications from a GENESIS cell parameter (".p") file, greatly increasing its capabilities for large single cell models.

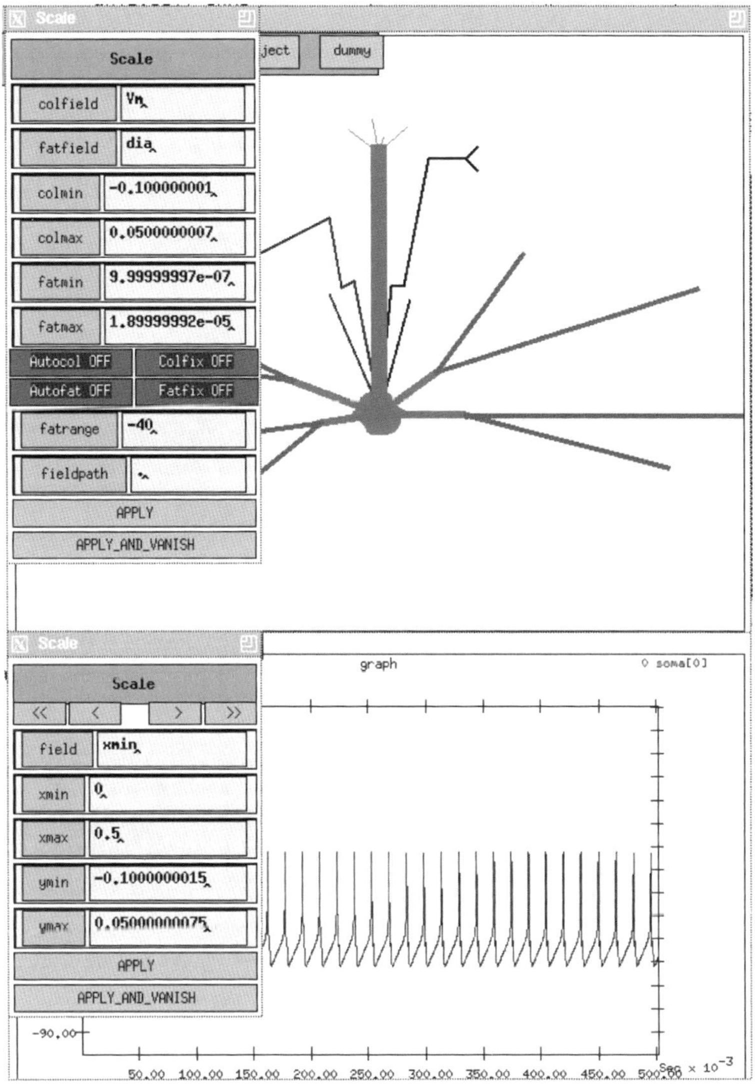

Fig. 3.1 An early GENESIS 1 version of Neurokit used to run and edit a mitral cell model. This used scripted XODUS objects to create menus, an animated cell view, and graph

GENESIS and NEURON Go Public

Returning to the chronology, the next major step for GENESIS and NEURON, and the initial availability for the use of both systems by those outside the founding laboratories, came as a result of the establishment of the Summer Course in Methods in Computational Neuroscience at the Marine Biological Laboratory (MBL) in

Woods Hole. As Jim Bower recounts the history, he and his Caltech colleague Christof Koch were sitting in Christof's backyard on a particularly hot and smoggy day in the summer of 1986 in Pasadena California, discussing what they could do the following year to be somewhere more pleasant with their families. Jim, who has spent time as a postdoctoral fellow doing summer research at the MBL, suggested that they propose offering a course in computational neuroscience, and spend the late summer in Woods Hole. That fall, the MBL accepted the course and the first in what has now become a series of courses around the world was offered. The first course of its kind, Jim's strong bias towards "hands on science learning" meant that the course was designed so that its central focus was on student projects based in a computer laboratory. For the first course, Jim and his feisty laboratory systems administrator John Uhley manually pulled the first Internet lines from the Woods Hole Oceanographic Institute in the tunnels under Water Street in Woods Hole to the MBL. Uhley installed 20 brand-new graphics workstations that were donated by the now defunct DEC in the laboratory, and literally the night before the opening day of the course, a public version of GENESIS was installed for the first time. Although Matt regarded this as version 0.001 of GENESIS, it already included a graphical user interface (thanks to Upi), powerful network creation commands, and an efficient method of summing spike events from multiple connections to a synaptically activated channel (Wilson and Bower 1989). The official release of GENESIS 1.0 coincided with the second Woods Hole course in July 1989 with usability greatly increased, due to continued work by Matt, Upi, Dave Bilitch, and John Uhley.

However, the first course in Woods Hole was not only the introduction of GENESIS but also of NEURON. As Jim recounts the story, on the very first day of the course, Michael Hines approached Jim and Christof about the possibility of installing CABLE on the laboratories computers as well, and giving students the option to use either GENESIS or CABLE (soon to be NEURON). Both Jim and Christof thought that this was a wonderful idea, and Michael Hines and NEURON became a regular part of the course from then on.

Federation and User Support

While I missed the first two Woods Hole courses, in the spring of 1990, as Matt was finishing his work at Caltech, my wife received a job offer that was too good to pass up in Boulder, Colorado. I then said goodbye to Harvey Mudd and came to the University of Colorado, supported as a consultant on GENESIS grants. This began the "federalization" of GENESIS development via the Internet, and a collaboration between GENESIS developers that still continues to this day. With the departure of Matt, Dave Bilitch gradually took over as the lead GENESIS software developer, coordinating our efforts with the crude Internet tools of the time: text-based email without attachments, ftp, and remote logins to the server "smaug" at Caltech via telnet. During the 1990s, as former members of the Bower laboratory formed research groups using GENESIS, and the expanding number of GENESIS users contributed to GENESIS capabilities, GENESIS development became increasingly distributed. To facilitate communication between users and developers, an email newsletter to the GENESIS users group was established in April 1991, and the GENESIS web site was established in June 1994.

When I entered the Bower laboratory in fall 1989, GENESIS development seemed to "just happen" by a small group interacting closely without a lot of formal organization. I would tell Upi "it would be nice if GENESIS could do ...." A couple of days later he would casually mention that GENESIS could now do it. I noticed this among the Bower lab group at Woods Hole. Someone noticed that something needed to be done, and just did it. This may not be a scalable plan for a major software development project, but it worked extremely well in those days. As GENESIS development has spread from a single laboratory to many, we have had to face the challenge of "federalizing" a large software development project among many user-developers.

From the beginning, GENESIS had the framework for interactive help, but not yet a lot of content. The "man page" was often a shout down the hall to Matt's office "Hey, man. I have a question." Gradually his answers evolved into additions that I made to the documentation as I saw the need for it. The interactive help was invoked in a terminal window at the "genesis >" prompt, and was plain text, formatted similarly to a Unix "man page." Upi added a printed manual with LaTeX source that covered basic syntax for the SLI, and the main GENESIS and XODUS objects and commands.

The GENESIS 1.0 release came with two tutorials that were created in April 1989. Mark Nelson contributed the "Squid" tutorial on the Hodgkin–Huxley model that is still in use today after years of enhancements by GENESIS users. The "MultiCell" tutorial was a simulation of two neurons having a soma and dendrite compartment with synaptic connections, with one being excited and the other being inhibited. The GUI had a control panel, graphs of membrane potential and channel conductances, and labeled text fields (called "dialogs" in XODUS) for changing the synaptic channel parameters. The extensive documentation with reference to a line-numbered version of the main scripts was the most useful early GENESIS documentation.

At an early GENESIS developers meeting, around the time that we launched the GENESIS web site in June 1994, we discussed ways to use a common source for the generation of plain text "help," a printed manual, and an Hypertext Markup Language (HTML) version for the web. I had been looking at an open-source package called "linuxdoc-sgml" that was then being used by The Linux Documentation Project (http://www.tldp.org) for doing this by writing the documentation, not in HTML, but in the much richer Standard Generalized Markup Language (SGML), that was the basis of HTML, and a few years later would become the basis for the now-popular eXtensible Markup Language (XML).

The near-unanimous decision was that I should use a well-supported commercial tool, FrameMaker to generate the documentation. As I was the one writing the documentation and I lived in Colorado, not Pasadena, I went home and wrote the documentation for the GENESIS 2.0 release in SGML. Now, FrameMaker no longer exists, but SGML lives on in the form of an open standard, XML. Actually, this move from commercial software to an open standard was only the first in a series of similar events in computational neuroscience.

Discussions of a major rewrite of GENESIS to create version 2.0 began in early 1993. After more than 2 years of development and several months of beta testing, GENESIS version 2.0 was finally released in August 1995. In addition to having a detailed reference manual in all three formats, it now ran under the Linux and FreeBSD operating systems, and with programming help from Maneesh Sahani, had a completely rewritten version of the XODUS graphical interface. The add-on library for Parallel GENESIS (PGENESIS) was released at the same time, allowing simulations to be spread over multiple processors or networks of workstations on a variety of hardware and software platforms.

In addition to their own research, one of the motivations for the further development of GENESIS was provided by its potential use as a tool in neuroscience education. "The BoG" mentioned previously, was written and edited by Jim Bower and myself as a step-by-step tutorial and interactive self-study for professionals, researchers, and students working in neuroscience. The free Internet edition (http://www.genesis-sim.org/GENESIS/bog/bog.html) and the printed version (Bower and Beeman 1998) use exercises and hands-on tutorials developed at the Woods Hole and later courses, with contributed chapters by researchers in computational neuroscience that are linked to the tutorials.

Expanding Simulator Capabilities

With the early history of simulator development now described, I will turn to the discussion of some of the issues in the construction of neural simulations that arose in the evolution of GENESIS and NEURON to their current state, their differences, and what I think this portends for the future.

It is easy to argue that the expanding base of the use of simulation systems is directly related to the expansion of their technical capabilities as well as their ease of use. During the 1990s, both GENESIS and NEURON expanded their graphical capabilities and repertoire of built-in tools specific to neural modeling, making the advantages of using a general simulator package obvious. These simulators then became the preferred method of constructing these types of models. At present, GENESIS and NEURON have very similar functionality for realistic neural modeling. However, there are some significant differences in the way that simulations are created and models are represented. An examination of the different approaches taken in their design may offer some insight into issues facing developers of the next generation of neural simulators.

Scripting and GUIs

I was attracted to the idea of using GENESIS to write tutorials because of its built-in graphical tools that I could use with the same scripting language that would be used to construct models. Matt never made much use of GUIs in his simulations, preferring to run his long network simulations in batch mode, sending the output to files for later analysis. Many modelers still follow this approach, using simulation scripts written with a text editor, and minimal graphics. However, a customizable GUI was necessary for developing tutorials. Later, I found out how important it could be for interpreting the results of parameter changes in a simulation during run time, allowing a quick exploration of a model.

At the time, most neural modelers, and even nonprogrammer users of personal computers, were at home in a command-line computer environment. Over the years, the expectations of modelers have changed, and a GUI is considered essential. However, using scripting commands to position graphical "widgets" (buttons, text fields, graphs, etc.) in a window is difficult and time-consuming. Newer graphical libraries provided for Java or Python, or the cross-platform wxWidgets library being used for G-3 provide more powerful tools than the comparatively low-level syntax used in XODUS. Nevertheless, the script code to set up a neural simulation GUI is often much longer than that needed to set up and run the simulation.

Generic GUIs such as the GENESIS Neurokit or the NEURON Cell Builder can be very useful for analyzing or tuning a single cell model, but rarely have the flexibility or unique features needed to perform and visualize the results of a research simulation, or to use as the basis of a tutorial simulation. In principle, Neurokit and the Purkinje Cell Tutorial can be modified by the user, as they are written in the XODUS extensions to the GENESIS SLI language. In practice, the scripts are far too complicated for most users to want to modify. G-3 allows for the future use of an Integrated Development Environment (IDE) such as Glade to let users create, size, and position graphical elements with a mouse, saving the layout in a standard format (Cornelis et al. 2012b).

When I began writing my first GENESIS tutorial simulation, the "Neuron" tutorial (Beeman 1994), I began to appreciate the object-oriented (OO) nature of the language that Matt had created for building simulations. I admit to being a somewhat lazy programmer who would rather hack at an existing example than to plan a program out from the beginning and start with a blank screen. I think that every computer program or simulation script that I have ever written has been a modified version of something else. Of course, I want to give a lot of thought to planning the structure of the program before I start, but I am likely to start with something that I have already written as a template. Then I fill in bits of code taken from examples or from other programs that I had written for something else. This wasn't too hard with my own FORTRAN, Pascal, or C code, but trying to merge pieces of someone else's code into my own and keep track of all the global variables and dependencies was often more work than starting over and writing it myself from the beginning.

The scripting language that Wilson developed for the GENESIS SLI had a syntax similar to C, and the ability to create simulation "elements" (or "objects" in modern terminology) from templates or "object types" (i.e., "classes"). This enabled neural models to be constructed using a "building block" approach. Simulations are constructed from modules that receive inputs, perform calculations on them, and then generate outputs. Model neurons are constructed from these basic components, such as dendritic compartments, and variable conductance ion channels. Compartments are linked to their channels and are then linked together to form multi-compartmental neurons of any desired level of complexity. Neurons may be linked together to form neural circuits. By keeping most of the variables and functions (methods or "actions") local to these elements, it was easy to pull in pieces of another model without having to understand very much about the large script in which it was embedded. This approach to building simulations has worked very well when building neural models.

Object-oriented programming concepts were well known among the artificial intelligence community at the time that GENESIS was written, but had not entered the mainstream of computer programming until the mid-1990s. The object-oriented programming language C++ was then in its infancy and would not become standardized until much later.

The parser for the scripting language devised by Matt Wilson was written by him in C. In order to make GENESIS as flexible as possible for creating different types of models, the object types were made as general as possible. For example, an "hh_channel" was not a particular Hodgkin–Huxley squid axon channel, but would model any channel that could be modeled with Hodgkin–Huxley type equations. The "tabchannel" and "tab2Dchannel" objects with tables for gate activation, introduced shortly afterwards, provided further generality. Later, De Schutter (De Schutter and Smolen 1998) added a library of GENESIS objects for modeling calcium diffusion.

By giving these objects different parameter values, creatively connecting them together in a script, and manipulating them with user-defined commands, a great amount of user-extensibility was achieved without having to do any programming outside of the scripting language. Inevitably, the time comes when a new object type or command needs to be defined and compiled into GENESIS. This requires some C programming ability of the user, although the process is simplified considerably by the detailed documentation and examples that are provided. Once the new version of GENESIS is compiled, the new functionality is available for any subsequent use of GENESIS.

NEURON is written in C, but made use of an existing scripting language and parser software by choosing HOC (Kernigan and Pike 1984). HOC has a C-like syntax and is also written in C. It was easily extended to include functions specific to modeling neurons, and over the years graphical commands and object-oriented programming concepts were added.

However, it proved difficult for users to add new channel mechanisms in HOC, so a high-level model description language, NMODL (Kohn et al. 1989) was incorporated into NEURON (Hines and Carnevale 2000). This made it much easier for users to extend the functionality of NEURON by writing definitions in NMODL, and then having them automatically compiled and linked into NEURON. This provides some advantages over the GENESIS approach, such as allowing a high-level scripting language to specify the differential equations to be solved, rather than writing lower level C modules. However, this means that most NEURON models of any complexity involve a mixture of HOC and NMODL, and require a recompilation and link step each time a simulation is run.

Parameter Search, Model Tuning, and Comparison

Upi's additions to XODUS enabled me to have pop-up help windows with scrolling text and images in my completed "Neuron" tutorial in time to use it with assigned exercises in the "Modeling and Analysis of Neural Networks" course (CS 189B) at Harvey Mudd in the spring of 1990. Jim suggested that I next write a simulation of a bursting molluscan neuron and develop a tutorial to go along with a manuscript "The Dance of the Ions," that he was writing to explain the role of the various types of ionic conductances in shaping the firing patterns in molluscan pacemaker cells.

It seemed like a simple thing to do. The principal channel types were well characterized with published voltage clamp data from the sea slugs *Tritonia* and *Aplysia californica*. GENESIS had all the features that I needed to implement a single-compartment model with six varieties of conductances and a calcium diffusion mechanism. That was my introduction to the difficulties and complexities of parameter searching and model comparison. My model was a "generic burster," loosely based on an *Aplysia* R15 neuron, with channel data taken from both *Tritonia* and *Aplysia* cells under different conditions.

I soon realized, along with many of the students that I tutored over the years in the MBL neural modeling courses and future ones in the EU Advanced Course in Computational Neuroscience and the Latin American School on Computational Neuroscience (LASCON), that parameter fitting is the most time-consuming task of single cell modeling. This requires scaling several conductance densities, shifting activation curves to account for different rest potentials, and varying time constants to account for temperature variations. Varying these many parameters in order to fit firing patterns obtained under current clamp conditions remains a difficult task today.

By the time I was happy with my model and tutorial, far better models of the *Aplysia* R15 neuron had been published (e.g., Canavier et al. 1991), but the tutorial and the one that I based on a GENESIS recreation of the Traub et al. (1991) hippocampal pyramidal cell, are still the best way I know of to get a feel for the role of the various conductances by modifying them within a GUI. The most sophisticated GENESIS single cell tutorial is the Purkinje cell tutorial, developed by Hugo Cornelis.

The experience of converting the Traub model and other published neural models to GENESIS revealed another sobering aspect of model replication and comparison. Of the many published descriptions of models that I have attempted to reimplement, I can think of very few that did not have errors or significant omissions. As discussed later, I feel that this stems from the limitations of the present system of publishing model-based research. Thus, there may be some parameter searching involved even when replicating an existing model, as well as the very important matter of making a meaningful comparison between the results from different implementations of what is ostensibly the same model.

A parameter search often involves doing the very easiest form of parallel computing: running many separate uncoupled simulations with different sets of parameters. One evening in the computer lab during the 1991 Woods Hole course, the students discovered that their simulations had suddenly slowed down to a crawl. It turned out that Erik De Shutter was doing a parameter search on the Purkinje cell model by running a background simulation on every workstation. I believe that he obligingly consented to cease, although John Uhley may have threatened actual physical violence.

To address the problem of comparing the results of dendritic cable model simulations when run on different simulators, Bhalla et al. (1992) developed the Rallpacks set of benchmarks. These demonstrated that GENESIS and NEURON had equivalent speed and accuracy for these models. Shortly later, GENESIS gained a number of parameter search commands that were used for tuning the olfactory bulb mitral and granule cell models of Bhalla and Bower (1993). Vanier and Bower (1999) performed a detailed study of a variety of automated parameter search methods, using the GENESIS parameter search library developed by Vanier. In most cases the simulated annealing algorithm gave the best performance.

There are, however, some pitfalls in performing an automated search. In addition to the time consumed by searching unproductive regions of a large parameter space, there can be multiple regions that give equivalent local minima in the error function for the fit.

In order to know where to start a search, one needs to have an understanding of the roles that many different ionic currents play in the timing of action potentials, in order to have a sense of which parameters are most relevant. For example, knowing the role that the "H-current" plays in producing an overshoot in the membrane potential after a hyperpolarizing current injection can help define the area in parameter space to be searched.

I have found it most helpful to begin with a manual search, starting with the best available data for initial values. By varying the parameters by hand, using a custom GUI scripted with GENESIS/XODUS, and plotting the results, I can find a much better set of initial parameters for an automated search. Figure 3.2 shows such an interface for tuning a simple pyramidal cell model.

The largest problem when fitting parameters to current clamp experiments is the same as the one when comparing the results of two different simulations. Simply matching the positions of action potentials is not a sufficient condition to judge when a simulation agrees with experimental results, or those of another model, unless the model is a tonically firing one, such as the axon model used in the Rallpacks (Bhalla et al. 1992) set of benchmarks. The difficult problem of

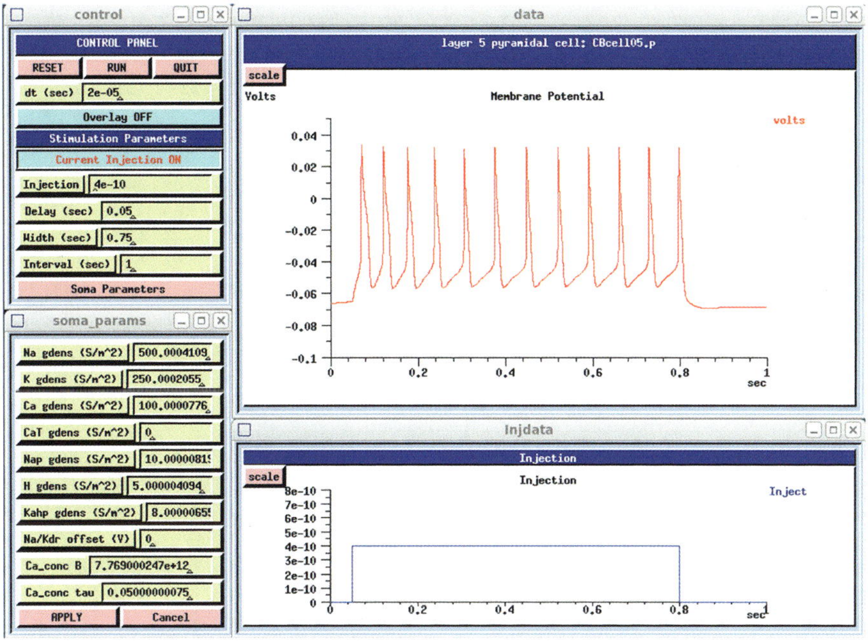

Fig. 3.2 Custom scripted GUI for adjusting channel parameters to fit response to a 0.4 nA current injection pulse to a model layer 5 pyramidal cell

reproducibility in computational neuroscience is addressed in detail in another chapter in this volume by Crook et al. (2013).

The distinction between incorrect results and those in "reasonable agreement," is particularly difficult to make in the case of cells that display spike frequency adaptation or bursting behavior. These models contain slow hyperpolarizing currents (e.g., Muscarinic or AHP) that are active near the threshold voltage for an action potential. Thus the membrane potential can hover about threshold, and these currents can have the effect of magnifying the effect of small deviations between two numerical solutions that can push the balance in one direction or the other.

Figure 3.3 shows the membrane potential for a model (Traub et al. 1994) of a burst-firing hippocampal pyramidal cell under conditions with two slightly different numerical precisions.

The upper plot, shown with a prototype Python plotting module for G-3, shows the result of a current injection over a 0.2 s interval. These small deviations eventually cause significant differences in the position of the final spike of the burst. When plotted over a 5 s period (below), the bursts have roughly the same time intervals, but drift in and out of coincidence with each other. One would call these "equivalent results," but it is difficult to quantify the differences in a meaningful way. Baldi et al. (1998) have addressed this problem by suggesting the use of Bayesian inference in the comparison of spike trains. However, making quantitative comparisons of this nature remains a largely unsolved problem.

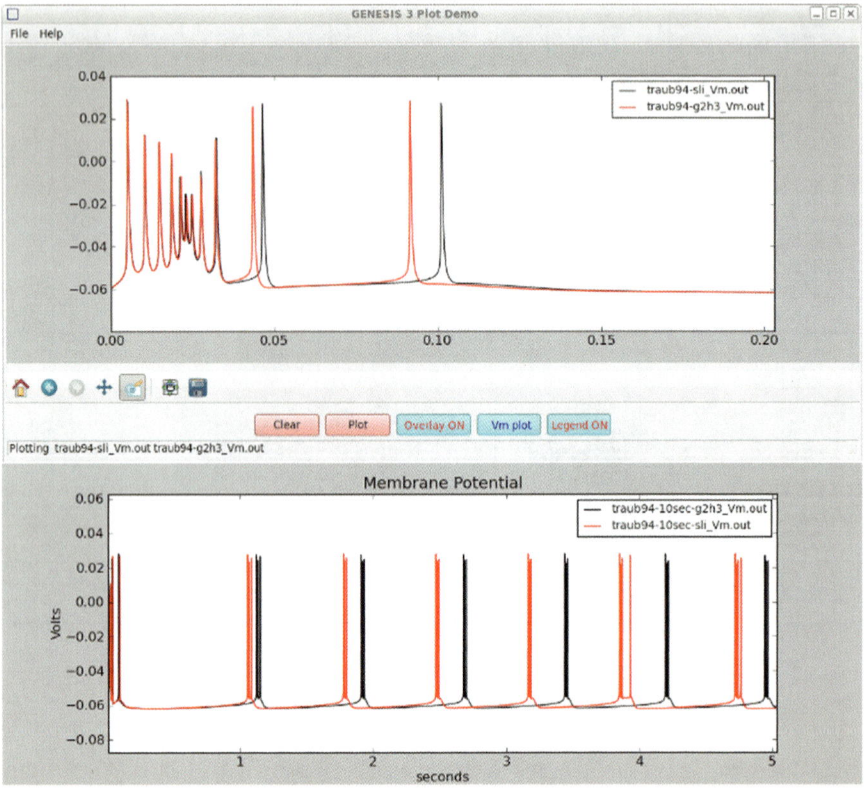

Fig. 3.3 Two simulation runs of a burst firing pyramidal cell with slightly different numerical precision. *Upper plots*: Membrane potential during the first 0.2 s. *Lower plots*: Membrane potential during 5 s of a longer run

The "Decade of the Brain" and the Human Brain Project

The US Congress established the 1990s as the "Decade of the Brain," and 16 federal agencies, including the NIH, issued program announcements in April 1993 and again in October 1995 soliciting proposals for the Human Brain Project. The research to be supported would develop informatics tools for accessing and integrating the huge amounts of data produced by neuroscience research, with an emphasis on web-based databases for data sharing (Koslow and Huerta 1997). Suddenly "neuroinformatics" became a popular word in research proposals. A great many of these proposals involved brain atlases and dealing with the huge data sets produced by neuroimaging experiments. However, there were many opportunities offered to neural modelers and simulator developers. The development of realistic neural models can benefit, not only from model sharing, but from the development of tools for managing notes and model development histories, and for linking models to experimental data and bibliographic references.

The initial goals of the SenseLab project at Yale University (Shepherd et al. 1997) were concerned with creating a comprehensive database of information about the olfactory system and tools for access. The system was to be built upon a commercial object-oriented database (OODB) called Illustra, and included an olfactory receptor database (ORDB), a database of neuron descriptions (NeuronDB), and a database of computational models of olfactory and other neurons (ModelDB).

The GENESIS group was awarded a grant in the first round to develop a "GENESIS Simulator-Based Neuronal Database" that would use an OODB with a query interface to mine the information contained within GENESIS simulations of a model neuron or network, and to link it with model descriptions, relevant data, and reference materials (Beeman et al. 1997). The initial choice of database was the commercial database UniSQL, and later prototypes were implemented with one called ObjectStore.

Two issues came up regularly at the annual Human Brain Project principal investigators meeting at the NIH in Bethesda. The question of how to best represent, store, and exchange models was a continuing theme, as well as what particular database to use. OODBs were in vogue at the time and were widely used in business software (Loomis 1995). An object-oriented representation was natural for describing neural models, and particularly for GENESIS models. However, the available OODBs were commercial products with proprietary data formats. The open-source options at that time were earlier incarnations of the relational databases PostgreSQL and MySQL. It is worth noting that none of the three commercial products mentioned above are in existence today, but PostgreSQL and MySQL are still widely used and available for a wide variety of platforms.

The other trend in the world of business software at that time was the use of SGML for representation of a large variety of data objects in a "document." A "document" could be, and often was, a description of a textual document, such a one described by a subset of SGML, HTML. The use of SGML in the publishing industry was well known. However, the use of SGML gave a powerful object-oriented description of items that could also be represented with OODBs, and it was often used as an interchange format between OODBs. An SGML document could just as well be a collection of data relating to the inventory of a business, a collection of customer contact information, or perhaps a description of a neural model.

Unfortunately, the SGML specification was needlessly complex to use or implement parsers for, and not very standard. Preliminary discussions by the World Wide Web Consortium (W3C) of a simpler standard more suitable for use with the WWW called the eXtensible Markup Language (XML) were underway at that time, but the XML 1.0 specification would not become a W3C recommendation until February 1998.

In March of 1996, Michael Arbib invited HBP participants with an interest in neural modeling to a "Workshop on Brain Models on the Web" held at the University of Southern California. His group had an HBP grant to develop a web-accessible database of neural models and to construct tools for sharing and exploring models and associated data that would be contributed by other modelers. The neural simulation system in use was NSL (Weitzenfeld 1995), a simulator for large networks of point integrate and fire neurons, and it would be advantageous to incorporate more realistic models generated with GENESIS and NEURON. These were to be stored in an Illustra OODB. Michael Hines and Matt Wilson had discussed the conversion between NEURON NMODL and GENESIS SLI scripts since 1989, and made some initial steps towards a conversion program. But, the problem was difficult because of the very different representations used in the two simulators.

The GENESIS ".p" format provides a machine-readable description of a branched compartmental cell model with active conductances. It can easily be translated to other formats. However, the channel names that appear in the file are names of elements that are created with GENESIS scripts for the SLI. Figure 3.4 is taken from a slide given at my presentation on the use of SGML for model representation, showing a fragment of an NMODL script and one of a GENESIS SLI script. Someone familiar with both simulators would recognize that both represent the same Hodgkin–Huxley model of the squid giant axon potassium channel, with one using physiological and the other SI units, but it is hard to imagine a machine translation between the two formats and their many possible variations.

The ModelDB project (Migliore et al. 2003) was one of the most successful and well known of those to come out of the HBP, primarily because of its simplicity. Rather than take the path of many others that attempted to put models into some OO format, it simply stores simulation scripts in the native simulator languages along with documentation, and is well indexed. However it suffers from the problem that the scripts are not portable, sometimes even to later versions of the simulator on which they were developed. This is because the simulators use *procedural* scripting languages that give a sequence of instructions telling the simulator how to construct a model, rather than *declarative* representations that describe the model, leaving it to the simulator to determine how it should be created in the context of the simulator implementation and its data structures.

The key to model sharing would be to translate this scripted procedural representation to a declarative object-oriented representation. I tried to argue persuasively that an SGML representation was preferable to storage in an OODB, but I don't think that anyone was convinced. My notes from the meeting say that Michael Hines was "pessimistic about the possibilities of representing a simulation outside of the structure of the simulation code." By the year 2000, when XML became widely known and open-source parsers were available, it would seem obvious that describing the objects with XML and storing XML files in a generic (and replaceable) relational database would be the best solution.

Model Sharing and Simulator Interoperability

During the final phase of the Human Brain Project, the focus of the GENESIS group turned towards creating the Modeler's Workspace (MWS). The MWS (Forss et al. 1999; Hucka et al. 2002) was a design for a graphical environment for constructing and exploring neuron and network models with simulations, experimental data, and bibliographic material. It would also allow collaborative development of models.

a NMODL example (code fragments from hh.mod)

```
COMMENT
  This is the original Hodgkin-Huxley treatment for the set of sodium,
  potassium, and leakage channels found in the squid giant axon membrane.
  ("A quantitative description of membrane current and its application
  conduction and excitation in nerve" J.Physiol. (Lond.)  117:500-544 (1952).)
ENDCOMMENT

BREAKPOINT {
        SOLVE hh METHOD runge
}

DERIVATIVE hh {
        v' = -1e3/cm * (ina + ik + il) - stim
        aux()
        rates(v)
        m' = (minf - m)/mtau
}

PROCEDURE aux() {
        gk = gkbar*n*n*n*n
        ik = gk*(v - ek)
        il = gl*(v - el)
}

PROCEDURE rates(v) {
        LOCAL  alpha, beta, sum
        .
        .
        :"n" potassium activation system
        alpha = .01*vtrap(-(v+55),10)
        beta = .125*exp(-(v+65)/80)
        sum = alpha + beta
        ntau = 1/sum
        ninf = alpha/sum
}
```

b GENESIS example (code fragments from hh_tchan.g)

```
/* FILE INFORMATION
   squid giant axon voltage-dependent channels,
   A.L.Hodgkin and A.F.Huxley, J.Physiol(Lond) 117,
   pp 500-544 (1952)

   The function setupalpha uses the form (A+B*V)/(C+exp((V+D)/F))
   for the rate variables alpha and beta to fill the channel tables
*/

// CONSTANTS
float EREST_ACT = -0.07       // resting potential (volts)
float EK        = -0.082      // potassium equil potential
float SOMA_A = len*PI*dia     // compartment area

function make_K_squid_hh
    str chanpath = "K_squid_hh"
    create tabchannel {chanpath}
    setfield ^ Ek {EK} Gbar {360.0*SOMA_A} Ik 0 Gk 0   \
            Xpower 4 Ypower 0 Zpower 0

    setupalpha {chanpath} X {10e3*(0.01 + EREST_ACT)} \
        -10.0e3 -1.0 {-1.0*(0.01 + EREST_ACT)} -0.01 \
        125.0 0.0 0.0 {-1.0*EREST_ACT} 80.0e-3
end
```

Fig. 3.4 Fragments of simulation scripts for a Hodgkin–Huxley potassium channel (**a**) NEURON NMODL script (**b**) GENESIS SLI script

The GENESIS HBP funding was to develop a prototype user interface for the examination and sharing of neural models with related metadata, rather than to enhance the functionality of the simulator core. Nevertheless, under continuing NSF funding, we were also looking forward to a major reorganization of GENESIS that would have the modularity and interoperability required to support the functionality of the MWS design.

As a key component, the MWS design contained an XML-based representation of cell and channel models. By the late 1990s, several other groups were also using XML-based descriptions of neuroscience-related data and models.

Daniel Gardner's group at Cornell had an HBP-funded effort to create an XML-based Common Data Model for the exchange of neurophysiology data and related metadata (Gardner et al. 2001). Although this was not intended for model description, it was very influential in our design of what eventually became a large part of NeuroML.

In late 1999, Michael Hucka, the main developer for the Modeler's Workspace Project, began working with what was then called the ERATO project at Caltech to develop SBML, the Systems Biology Markup Language (http://sbml.org). This team, consisting of Hamid Bolouri, Andrew Finney, and Herbert Sauro, was developing an infrastructure for computational modeling in systems biology. It faced many of the representation and design issues of the MWS project, and there was a great deal of overlap in the design of SBML and the MWS XML description of neural models (Hucka et al. 2003).

In June 2000, Hucka circulated the first draft of the notation used by the MWS for model descriptions (http://modelersworkspace.org/mws-rep/mws-rep.html) to the ERATO project and to Nigel Goddard's group in Edinburgh, who were working on model representations for their (now-defunct) simulator NEOSIM (Goddard et al. 2001a).

At about that time, Hugo Cornelis was finishing his Ph.D. thesis in Computer Science, while working in the laboratory of Erik De Schutter in Antwerp. Needing to develop a user-friendly declarative interface to the Hines method solver in GENESIS 2, he developed the Neurospaces model-container and the Neurospaces Description Format (NDF) for single neuron and network model representation in 1999 (Cornelis and De Schutter 2003). He described it in a meeting with the Goddard group in summer 2000 and, with Fred Howell, wrote a first draft of a proposal to use it as a representation for NEOSIM.

In a collaboration between these groups, a paper describing the initial specification for NeuroML (http://www.neuroml.org) was submitted in December 2000. After much discussion and further revision during a meeting in Edinburgh in early 2001, the representation was further clarified, incorporating the MWS cell and channel representations with the NEOSIM network-level representations (Goddard et al. 2001b).

In 2002, Fred Howell and Robert Cannon implemented the NeuroML Development Kit in Java, with classes corresponding to an extended version of the XML schema described in Goddard et al. (2001b), and tools for parsing NeuroML files. This was then used to implement a prototype MySQL database and Java-based GUI for to retrieve GENESIS ionic conductance models described with NeuroML, and convert them to procedural GENESIS SLI scripts (Beeman and Bower 2004).

The next significant application of NeuroML was the neuroConstruct project, initially begun in 2004 as a tool with a GUI for creating network models with

NEURON. NeuroConstruct (Gleeson et al. 2007) is implemented in Java, and the latest version (http://neuroconstruct.org/) uses the current NeuroML specification (Gleeson et al. 2010), which incorporates the MorphML (Crook et al. 2007) schema for cell morphology description. Rather than being a simulator, it provides an environment for creating large networks of biologically realistic neurons with complex connectivity patterns, using NEURON-, GENESIS-, MOOSE-, PSICS-, and PyNN-based simulators to perform the actual simulations. This is accomplished by generating simulation scripts for these simulators in their native scripting languages.

The International Neuroinformatics Coordinating Facility (INCF) has formed a program to develop another standardized description language for spiking neuronal network models, the Network Interchange for Neuroscience Modeling Language (NineML). NineML (Gorchetchnikov 2010; Raikov 2010; http://nineml.org) incorporates features of NeuroML and SBML and is based on a layered approach. An abstraction layer allows a full mathematical description of the models, including events and state transitions, while the user layer contains parameter values for specific models. There are frequent discussions between the NineML and NeuroML groups. It is likely that there will be some convergence between the two standards.

In the fall of 2005, Cornelis joined the Bower lab in San Antonio as a postdoctoral student, and began the integration of Neurospaces into G-3 as its internal data representation format and model container. Hucka is now the Team Leader and Chair of the SBML editors for the SBML project, working on all aspects of SBML development.

Choice of Programming Languages

In retrospect, the choice of C as the programming language for GENESIS and NEURON during the mid 1980s may seem like an obvious decision. But, there were other alternatives that could have been chosen, and C was a fortunate choice. At this time most scientific computations were performed in FORTRAN, running on mainframe computers without graphics. Graphical displays were available on workstations made by a variety of manufacturers including DEC, Xerox, Evans and Sutherland, Apollo, and Xerox, as well as Sun Microsystems and Silicon Graphics. These typically had their own specialized software libraries and software to take advantage of their hardware. C was the standard language for Unix-based systems, and the X Window System was just beginning to emerge as a standard hardware-independent protocol for the display of graphics.

There were a number of dialects of C, and much early software, including GENESIS, used the original informal specification by Kernighan and Ritchie (1978) that is often called "K&R C." The first standard for what became known as "ANSI C" was not adopted by the American National Standards Institute until 1989, and adopted by the International Organization for Standardization in 1990. During the 1990s, a great deal of time was spent updating and slowly "ANSIfying" the GENESIS base code in order to guarantee that it would compile under the various workstation operating system implementations of Unix, such as SunOS, Solaris, Irix, Ulttrix, and HPUX.

However, the US government was pushing strongly for the use of the language Ada. Although Pascal was originally proposed as a teaching language, it was becoming popular, and there were inexpensive Pascal compilers and interpreters widely available for personal computers. Many computer scientists favored Algol or Modula-2, or "fifth generation" languages such as Prolog or Lisp. In fact, a notable single neuron simulator, Surf-Hippo was written in Lisp (Borg-Graham 2000).

Many IBM mainframe computers ran a proprietary language PL/1/, and DEC minicomputers such as the PDP8 used FOCAL. Although object-oriented C++ had been under development since 1983 as an extension of C, standardization was slow to come and the C++ programming language standard was not ratified until 1998.

During the development of a neural simulator, it is obviously of great importance to "pick a winner" among emerging software standards for programming languages and graphical packages. The approach taken by GENESIS was a conservative one, using standard C and X libraries, at the cost of having to write much of the "middle-level" software such as the SLI and XODUS to connect low-level function calls to high-level commands in a scripting language. NEURON has tended to use available software packages (HOC, NMODL, InterViews, etc.) as an easy way to gain functionality, whereas GENESIS has tended to develop its own built-in libraries or modules. Thus, the SLI, XODUS, and the original mailing list management software were written "in-house," based on widely accepted standard low-level Unix and X libraries. Each of these approaches has both advantages and disadvantages.

Open Source vs. Proprietary Software

Many of the HBP-funded database projects elected to use proven, professionally developed commercial software, rather than less stable free open-source software. However, most of these companies and their software with proprietary data formats no longer exist. Today it is much easier to make an argument in favor of using open-source software to avoid dependence on a closed platform that may not exist in the future. However, it is certainly an advantage to have someone else do the hard work, and it is a safe bet that tools such as MATLAB and their data formats will be around for a long time.

However, the choice of an open-source package also presents a dilemma. Is it best to incorporate the needed code into the simulator, or rely on loadable libraries that are maintained elsewhere and distributed from an Internet-accessible repository? Obviously the latter is the easiest route, if only one can guarantee that the package will not become like so many orphaned software projects on Sourceforge.net. If the package is not large, at least the source code will be available to be incorporated and maintained by the simulator developers.

The GENESIS experience with the NetCDF package from Unidata is an example of a decision that wasn't entirely optimal. GENESIS has a very fast and efficient binary file format (FMT1) for outputting the state (virtually any variable of interest) of the neurons in a large network at specified time intervals. However, it is platform-dependent and cannot be reliably used for exchanging files between different computers. The NetCDF package provided an easy way to give GENESIS a platform-independent output format with many other features. However, it is a very large package with a great deal that is not relevant to GENESIS. The NetCDF license allowed the incorporation of parts of the code into GENESIS with proper attribution, so it was decided to make a GENESIS "netcdflib" from parts of NetCDF. During the early years of patching GENESIS in order to compile on the many variants of the Unix operating system, maintaining netcdflib became a tiresome task. Although the compilation of netcdflib is optional in GENESIS 2, it now runs (more slowly than the default FMT1) on all the major Unix variants. NetCDF is currently still maintained by Unidata, and updated packages are available. In retrospect, life would be simpler if netcdflib were a library maintained by Unidata, or if we had written our own software with a NetCDF-compatible format, or picked another widely used and supported format.

NEURON relied heavily on open-source software packages that might now be considered orphans, but with few ill effects. HOC and MOD (from which NMODL was developed) are used nowhere else but in NEURON. The Unix InterViews package upon which the NEURON GUI is based had its last major release in 1993. As long as these are part of NEURON and can be maintained with the rest of the code, it does not matter if they are otherwise unsupported.

In 2011, there are very powerful and complex graphical libraries available that are popular and being extensively developed. Today, no one would think of writing a programmable GUI such as XODUS from scratch, using basic X Window System function calls. For example, G-3 makes use of the platform-independent libraries for wxWidgets and the Python tool Matplotlib. Compiled binary libraries for nearly any operating system are maintained and may be updated from Internet repositories. Using these is a great advantage, but there is always the gamble of betting on a loser that will fall from favor and no longer be maintained.

Some Identified Problems with Past Simulators

Scripting a simulation with SLI or HOC is difficult because they each have an idiosyncratic syntax that must be learned. They lack the generality and the data structures available in more general purpose programming languages, and have no supported external libraries that can be used to easily gain additional functionality. Creating a custom GUI for a simulation or a tutorial with SLI or HOC is even harder, because of the lack of built-in tools for designing GUIs. Modelers also require that a simulator provide a large variety of tools for visualization, analysis, or model construction, in addition to fast and accurate simulation of the model.

Increasingly, neural simulations need to cover many scales of modeling, ranging from the subcellular level of biochemical kinetics and diffusion modeling, through single cell, network, and system modeling. Specialized simulators exist that are an excellent solution at a particular level. For example, MOOSE provides very good capabilities for biochemical kinetics modeling, based on the GENESIS 2 kinetics library component. MCell (Stiles and Bartol 2001) uses Monte Carlo methods to model diffusion and stochastic activation of synapses. Due to the lack of interoperability between simulators, it has generally required extending the simulator in order to include similar capabilities.

As mentioned in the section on parameter search, model tuning, and comparison, publication of the results of a simulation in a way that allows the results to be reproduced or compared with the results from other models is a nearly impossible task. This is true even if the source code for the simulation is made available. This is because there is no way to track either the micro-evolution of a model during a single published research project or the macro-evolution of a model across longer time periods and multiple publications by multiple researchers. There is no guarantee that the model parameters that were used to generate a particular figure are the same as those described in the paper or used in the provided simulation scripts. In the traditional scientific process the actual records of day-to-day activity are kept in research notebooks or log books. In the case of computational modeling, a model may be incrementally modified with an incomplete record kept of all changes. Most neural simulators provide facilities for keeping "Notes" files, but not a comprehensive system for tracking model evolution using a verifiable digital description of the model and the exact conditions for the simulated experiments.

The monolithic architecture of simulators that were developed during the last century makes it difficult to add "plug-in" software components to the simulator without making major changes to the core simulator code. For example, GENESIS 2 has various components such as the cell reader, the hsolve compartmental solver, the SLI parser, and XODUS. However these were not implemented in a modular fashion and cannot be used as stand-alone software components.

By the end of the twentieth century, the limitations of the 1980s software that was the basis for GENESIS, NEURON, and other simulators were becoming apparent. Although emerging standards for declarative model descriptions showed promise for simulator-independent model sharing, the fundamental barrier to simulator interoperability and collaborative model development was the monolithic architecture of the simulators. The only way to communicate with GENESIS or NEURON was via an exchange of files. Not only does this limit the ability of two simulations to closely interact in real time, but it forces an "all-or-nothing" approach to the use of tools to control the simulation, provide stimuli, or analyze and display the results. Although GENESIS has many built-in tools for spike train analysis of the output of single cell models, this can easily be done with external tools such as Matlab or custom tools written in Python, if spike times are sent to a file for post-run analysis. Modelers have long used tools such as Matlab or various plotting programs to analyze the results of simulations. New tools are being developed specifically to aid neural simulation. However, they presently communicate with simulators indirectly by loading simulation scripts and communicating via files.

The problem comes if one wants to connect them more directly to a simulation during the run, without resorting to communication via data files. When calculating extracellular potentials or simulated EEG and MEG recordings from a large network model, it is necessary to sum channel currents from many compartments of very many cells. Unless the analysis application and the simulator have been designed to communicate with each other, the only other alternative is to generate enormous data files that give meaningful results only when the simulation is over, and no feedback while it is running. Although long "production runs" may be performed non-interactively, it is useful during the exploratory phase, and essential for educational tutorials, to perform these operations while the simulation is running. Thus, GENESIS 2 and NEURON each have a great deal of code that is devoted to built-in tools that are not concerned with specifying and simulating a model. Ideally, there should be no need to have a morphology file converter or spike train analyzer encapsulated within a simulator. It would be much better if these tools were implemented as external simulator-independent plug-in modules.

The Twenty-First Century: Next Generation Neural Simulators

By the beginning of the twenty-first century, the basic simulation capabilities of GENESIS and NEURON had reached maturity, although there were continuing improvements, leading up to the latest releases of GENESIS 2.3 (May 2006) and NEURON 7.1 (October 2009). Recent development efforts have shifted to providing other capabilities for the analysis or construction of models. For NEURON, the focus has been on better integrated graphical tools, such as Cell Builder, Channel Builder, Kinetic Scheme Builder, and built-in tools, for parameter fitting and importing cell morphology files, as well as model import functions for NeuroML and other standard declarative formats. As described below, the approach taken by G-3 has been somewhat different.

I believe that the twentieth century simulators described above and their architecture have reached the end of their life cycles, and it is time for a new generation of modular, interoperable realistic neural simulators. These should be built upon modern software designs, with care to pick standards, formats, and externally developed packages that will be in existence 20 years from now.

It may be useful to briefly describe here some simulators that have been developed or extended during the present century. Brette et al. (2007) reviewed eight currently available simulators that are capable of modeling networks of spiking neurons. In addition to GENESIS and NEURON, this capability is present to one degree or another in NEST (Eppler et al. 2008), NCS (Drewes et al. 2009), CSIM (http://www.lsm.tugraz.at/csim/), SPLIT (Hammarlund and Ekeberg 1998), Mvaspike (Rochel and Martinez 2003), and XPPAUT (Ermentrout 2006).

NEST specializes in very large networks of neurons having one or a small number of compartments. It uses parallelism and has its own interpreted scripting language with no GUI. NCS, the NeoCortical Simulator, has an inherently parallel implementation and is used for large networks of multi-compartmental integrate and fire neurons.

CSIM and the Python version PCSIM (Pecevski et al. 2009) model large networks of point neurons that are typically integrate and fire, but may also include Hodgkin–Huxley channels. Rather than having its own GUI, it is controlled by Matlab, and more recently with Python.

SPLIT is designed for massively parallel simulations of networks of multi-compartmental neurons with Hodgkin–Huxley dynamics. It was recently used in a neocortical simulation with eight million neurons and four billion synapses performed on the Blue Gene/3 supercomputer (Djurfeldt et al. 2005). The user specifies the model in a C++ program, rather than using an interpreted simulation language. The program is linked to the SPLIT library, compiled, and run. It has a minimal GUI and no analysis tools.

Mvaspike is based on an event-based modeling and simulation strategy, mainly using pulse-coupled integrate-and-fire point neurons.

XPPAUT is in a class by itself. It is not so much a neural simulator, but an analysis tool for understanding the equations that are used in simulations of cells and small networks. The equations used are completely specifiable by the user and can be analyzed with bifurcation diagrams and similar phase plane representations.

Another new simulator, Brian (Goodman and Brette 2008) models networks of integrate-and-fire or Hodgkin–Huxley single or few compartment neurons. It is written entirely in Python, making it highly portable, easy to learn and use, and suitable for rapid prototyping of models. However, this prevents it from being as fast as other simulators that make use of compiled C or C++ libraries to perform most of the numerical calculations in a simulation.

PyNN (Davison et al. 2009) and MUSIC (Djurfeldt et al. 2010) are not simulators, but provide interfaces for existing simulators. PyNN, discussed below, provides a common Python-based scripting interface for many simulators. MUSIC is a C++ library implementing an API which allows large-scale neuronal network simulators to exchange data during run time. NeuroConstruct (Gleeson et al. 2007), described previously, is an environment for creating simulations that run under several simulators, using NeuroML as a declarative model description language.

The GPU-SNN simulator (Richert et al. 2011) models large networks of spiking Izhikevich model neurons, having spike-timing-dependent plasticity and short-term plasticity. It can run on an "off-the-shelf" Graphical Processing Unit (GPU) such as the Nvidia GTX-280 at speeds of up to 26 times faster than a CPU version for a simulation of 100,000 neurons with 50 million synaptic connections. It is written in C and C++ and has a Python-based user interface similar to PyNN.

MOOSE (http://moose.ncbs.res.in) is the Multiscale Object-Oriented Simulation Environment for large, detailed simulations including computational neuroscience and systems biology. It was developed by Upi Bhalla as a reimplementation of GENESIS 2 in a cleaner, more modular manner. Although it retains the GENESIS

2 use of objects that pass messages between them, it does so in a more efficient manner. It is completely rewritten in C++ and has a different architecture, with no old GENESIS 2 code except for the SLI parser definition. The SLI parser allows it to maintain a high degree of backwards compatibility with GENESIS 2 scripts, and the new Python interface (Ray and Bhalla 2008), allows scripting of simulations in Python.

The development of G-3 has taken a different path from that of the simulators described above. Rather than adding new features and capabilities to GENESIS, the functionality of its monolithic architecture has been completely reimplemented as a collection of independent software components. These, and other independently developed components may be used individually or in combination with others to perform the functions desired for running a particular simulation.

The modular CBI architecture used by G-3 (Cornelis et al. 2012a) is based on plug-ins and has multiple interfaces. This modularization provides a number of advantages for simulator development and for interoperability with other simulators across scales ranging from subcellular to systems level.

The clean separation of modules allows developers and users to choose to contribute to only a single component, instead of being exposed to the complexity of the entire simulator. Decomposition of an application into multiple software components not only allows reuse and extension of individual modules, facilitating both simulator and model development, but individual components can be independently updated, enhanced, or replaced when needed. The use of multiple parsers for scripting simulations allows G-3 users to maintain backwards compatibility with GENESIS 2, while making use of scripts written in Python or other new scripting languages. Modules can be run separately on different machines. For example, the GUI and modeling environment might be run locally, while the simulation is run remotely on more powerful, possibly parallel, machines.

Some of the more relevant G-3 components for creating and running a simulation are:

- The Neurospaces Model Container (NMC) contains the biological model description and separates it from the details of the implementation.
- Multiple solvers perform numerical calculations and allow highly efficient solvers to be implemented for particular model objects.
- The Experiment component provides experimental protocols for applying stimuli, or for recording and analyzing the model behavior.
- A scheduler (SSP in Perl or SSPy in Python) binds the contents of the NMC with needed solvers and experimental protocols, and runs the simulation.
- The G-shell (or the new Python shell) provides a console for issuing interactive commands.
- G-Tube provides a GUI for running G-3 simulations.
- Studio allows the visualization of models in the Model Container.
- NS-SLI provides backwards compatibility with the GENESIS 2 SLI.
- The Exchange component provides model exchange using common standards such as NeuroML and NineML.

- The G-3 Documentation System not only provides user and developer documentation on all aspects of G-3, but is the basis for the model publication system.

Heccer is the default fast implicit numerical solver for compartmental models. It transparently incorporates the hsolve object of GENESIS 2. The Discrete Event System (DES) component is a separate solver used for delivering spike events in network simulations. The Chemesis-3 solver is a numerical solver optimized for the solution of reaction–diffusion equations (Blackwell 2000). The use of these separate numerical solvers for different types of models allows improved optimization over that obtained by the generic solver used by GENESIS 2. It also allows the use of multiple simulation engines to perform the numerical calculations of the simulation. In principle, the solvers of simulators such as NEURON and MOOSE could be used along with the G-3 solvers to perform parts of a simulation. Keeping the declarative model description separate from the simulator-dependent solver facilitates the exchange and reuse of models. The use of separate parsers for simulator commands allows simulations to be constructed with the G-shell, NS-SLI for GENESIS 2 scripts, or SSPy for scripts written in Python.

From the standpoint of a modeler constructing a simulation, G-3 preserves the GENESIS 2 paradigm of creating simulation objects that exchange messages during a simulation. However, unlike the approach taken with MOOSE, G-3 does not internally use objects with messages. This allows highly efficient numerical methods to be used without resort to "hacks" such as the GENESIS 2 hsolve object.

The G-3 model-based publication system (Cornelis et al. 2010) addresses the limitations of current paper and digital publications, by providing model comparison tools, model lineage inspection tools, and model verification tools. This is intended to lay the ground work for making models, rather than, as at present, the written description of models, the basis for scientific publication in neuroscience. The Publication System is designed to be platform independent as it adheres to the CBI federated software architecture (Cornelis et al. 2012a).

Choice of Python as a Scripting Language

The term *scripting language* is often used for a programming language that is used for control over applications or as a "glue" that links compiled libraries that are written in other languages designed for efficient numerical computations. Scripting languages are usually interpreted, for run-time interaction with the user, and are designed to be easily writeable and modifiable by the user. This is in contrast to the programming language that is used for the implementation of the core simulator functionality. In the past, simulators have used simulator-specific scripting languages, such as SLI or HOC.

As early as 1999, Michael Vanier was working on PyGenesis to provide a better object-oriented declarative scripting language than the SLI syntax used with GENESIS 2. However, the monolithic architecture of GENESIS 2 prevented this from being an easily interfaced plug-in module, and it was never officially released. Further development of Python interfaces to GENESIS was postponed until recently, when it became possible to use them as plug-in components of G-3 (Cornelis et al. 2012b).

Python has recently become very popular as a scripting language for neurosimulators because it has a far simpler syntax than other languages such as Perl, with very flexible object-oriented capabilities and powerful built-in data structures. It also has a wide variety of well-supported open-source libraries for scientific computing and graphical display. For example, NumPy (http://numpy.scipy.org) provides arrays and fast matrix manipulation tools, and matplotlib (http://matplotlib.sourceforge.net) can duplicate most of the functionality of Matlab. These modules are widely used for neuroscience data acquisition and analysis (Spacek et al. 2009). Although it may be considered a procedural language, the ability to create purely declarative representations of models using Python objects makes it a good choice as a scripting language for OO model descriptions.

The PyNN project (Davison et al. 2009) attempts to provide a common Python-based scripting interface for nearly any neural simulator, allowing a mixture of Python and native simulator code. Many of the neural simulators described in the previous section have developed Python interfaces that aim for some degree of compatibility with PyNN, including G-3 (Cornelis et al. 2012b), NEURON (Hines et al. 2009), MOOSE (Ray and Bhalla 2008), and NEST (Eppler et al. 2008).

Conclusion: What Have We Learned?

In the preceding narrative history of neural simulator development, certain issues arose repeatedly. A summary of these may provide some guidance for future simulator development.

The advantages of using a well-supported simulator rather than dedicated simulation-specific code have now been widely recognized. However, the choice of a particular simulator usually means a commitment to spending time learning the details of using that simulator. That tends to lock the user into that choice and discourage the use of another tool more appropriate for the task. Modern simulator development has attempted to avoid this problem with efforts towards simulator interoperability and model sharing. New simulator architectures allow the use of standard, well-supported external modules, or specialized tools for neural modeling, that are implemented independently from the means of running the model simulation. This allows not only sharing of models, but sharing of research tools.

The OO paradigm of constructing models from basic reusable "objects" has now become nearly standard in current neural simulator interfaces, whether through a scripting language or a GUI. This trend has been encouraged by the development of standard declarative representations for models such as NeuroML and NineML, which are based on an inherently OO structure. However, there is more to be done in the way of standardization, and of representation of models that lie beyond the current capabilities of GENESIS 2 and NEURON.

Parameter search is a necessary part of modeling. Some simulators have implemented parameter search algorithms within the simulator. A more modular simulator architecture and the use of standard scripting languages can allow the use of more general purpose external simulator-independent search tools. Parameter searching is a very appropriate and simple use of parallelism, and simulator architectures should be designed for easy and transparent division of a simulation or many simulation runs over multiple computers or processors. It is also important to have the ability to easily add numerical solvers for new hardware devices such as powerful GPUs that were developed for display rendering and are now being used for high performance scientific computing.

As extensively discussed in the chapter by Crook et al. (2013) and earlier in this one, reproducibility of simulation results is hampered not only by insufficient, ambiguous, or inaccurate descriptions of the model in the original publication, but by sensitivity to implementation details, and dependencies on the computing environment. As with other identified problems with past simulators discussed earlier, the key to these problems seems to be a combination of standard declarative model descriptions, and modular simulator architectures that permit the use of external tools to perform the ancillary tasks of model tracking, publication, comparison, and parameter fitting.

The choice of a programming or scripting language and whether to use code developed in-house, open-source code, or proprietary commercial software has become simpler with the increased availability of well-supported open-source software packages. Their use has been made easier by attempts to increase the modularity of new simulator architectures and through the use of standard scripting languages such as Python. I have seen languages come and go in popularity. Perl is a powerful scripting language with very good string handling capabilities. It has long been a favorite scripting language of software developers and system administrators, but is difficult for the novice or occasional "script hacker." Not long ago, Java was everyone's favorite bet for a programming or scripting language with great promise to run on all platforms. Now Python is the favorite and is being challenged by other alternatives such as Ruby or C#, which also have OO capabilities, and can be used as a "glue" to provide access to compiled libraries within a script. C# (by Microsoft) and other new languages such as Go (by Google) are also "Internet aware," with built-in security and run-time execution distribution support, favoring modular architectures to prevent vendor lock-in. For these reasons, it is important to design a simulator so that its operations can easily be bound to a user's choice of scripting language. For example, the use of separate modules for a declarative model description, numerical solver, and command parser can reduce the dependency on a particular scripting language, as well as facilitate the exchange of models.

Fortunately for the future of computational neuroscience, the "lessons" mentioned above, and throughout this chapter, appear to be taken seriously by today's simulator developers. Annual workshops held at the Computational Neuroscience conference (http://cnsorg.org), by the NeuroML developers (http://neuroml.org), and the INCF (http://incf.org) bring together participants from all the major neural simulator and database projects. Standards and designs are vigorously debated, and progress continues.

Acknowledgment The author acknowledges support from the National Institutes of Health under grant R01 NS049288-06S1.

References

Alben R, Kirkpatrick S, Beeman D (1977) Spin waves in random ferromagnets. Phys Rev B15:346

Baldi P, Vanier MC, Bower JM (1998) On the use of Bayesian methods for evaluating compartmental neural models. J Comput Neurosci 5:285–314

Beeman D (1994) Simulation-based tutorials for education in computational neuroscience. In: Eeckman FH (ed) Computation in neurons and neural systems. Kluwer Academic, Norwell, MA, pp 65–70

Beeman D, Boswell J (1977) Computer graphics and electromagnetic fields. Am J Phys 45:213

Beeman D, Bower JM (2004) Simulator-independent representation of ionic conductance models with ChannelDB. Neurocomputing 58–60:1085–1090

Beeman D, Bower JM, De Schutter E, Efthimiadis EN, Goddard N, Leigh J (1997) The GENESIS simulator-based neuronal database (chap 4). In: Koslow SH, Huerta MF (eds) Neuroinformatics: an overview of the human brain project. Lawrence Erlbaum Associates, Mahwah, NJ, pp 57–80

Bhalla US (1998) Advanced XODUS techniques (chap 22). In: Bower JM, Beeman D (eds) The book of GENESIS: exploring realistic neural models with the GEneral NEural SImulation System, 2nd edn. Springer, New York, pp 381–405

Bhalla US (2000) Modeling networks of signaling pathways (chap 2). In: De Schutter E (ed) Computational neuroscience: realistic modeling for experimentalists. CRC Press, Boca Raton, FL, pp 25–48

Bhalla US (2003) Managing models of signalling networks. Neurocomputing 52–54:215–220

Bhalla US, Bower JM (1993) Exploring parameter space in detailed single neuron models: simulations of the mitral and granule cells of the olfactory bulb. J Neurophysiol 69:1948–1965

Bhalla US, Iyengar R (1999) Emergent properties of networks of biological signaling pathways. Science 283:381–387

Bhalla US, Bilitch DH, Bower JM (1992) Rallpacks: a set of benchmarks for neuronal simulators. Trends Neurosci 15:453–458

Blackwell KT (2000) Evidence for a distinct light-induced calcium-dependent potassium current in Hermissenda crassicornis. J Comput Neurosci 9:149–170

Borg-Graham LJ (2000) Additional efficient computation of branched nerve equations: adaptive time step and ideal voltage clamp. J Comput Neurosci 8:209–226

Bower JM (1991) Relations between the dynamical properties of single cells and their networks in piriform (olfactory) cortex. In: McKenna T, Davis J, Zornetzer S (eds) Single neuron computation. Academic, San Diego, pp 437–462

Bower JM (1992) Modeling the nervous system. Trends Neurosci 15:411–412

Bower JM (2005) Looking for Newton: realistic modeling in modern biology. Brains Minds Media 1:bmm217 (urn:nbn:de:0009-3-2177)

Bower JM, Beeman D (1998) The book of GENESIS: exploring realistic neural models with the GEneral NEural SImulation System, 2nd edn. Springer, New York, http://www.genesis-sim.org/GENESIS/bog/bog.html

Brette R, Rudolph M, Carnevale T, Hines M, Beeman D, Bower JM, Diesmann M, Morrison A, Goodman PH, Harris FC, Zirpe M, Natschläger T, Pecevski D, Ermentrout B, Djurfeldt M, Lansner A, Rochel O, Vieville T, Muller E, Davison AP, El Boustani S, Destexhe A (2007) Simulation of networks of spiking neurons: a review of tools and strategies. J Comput Neurosci 23:349–398. doi:10.1007/s10827-007-0038-6

Canavier CC, Clark JW, Byrne JH (1991) Simulation of the bursting activity of neuron R15 in *Aplysia*: role of ionic currents, calcium balance, and modulatory transmitters. J Neurophysiol 66:2107–2124

Carenvale NT, Woolfe TB, Shepherd GM (1990) Neuron simulations with SABER. J Neurosci Methods 33:135–148

Cole K (1968) Membranes, ions, and impulses: a chapter of classical biophysics. University of California Press, Berkeley

Connor JA, Stevens CF (1971) Prediction of repetitive firing behavior from voltage clamp data on an isolated neurone soma. J Physiol 213:31–53

Cornelis H, De Schutter E (2003) Neurospaces: separating modeling and simulation. Neurocomputing 52–54:227–231. doi:10.1016/S0925-2312(02)00750-6

Cornelis H, Coop AD, Bower JM (2010) Development of model-based publication for scientific communication. BMC Neurosci 11(suppl 1):P69. doi:10.1186/1471-2202-11-S1-P69

Cornelis H, Coop AD, Bower JM (2012a) A federated design for a neurobiological simulation engine: the CBI federated software architecture. PLoS One 7:e28956. doi:10.1371/journal.pone.0028956

Cornelis H, Rodriguez AL, Coop AD, Bower JM (2012b) Python as a federation tool for GENESIS 3.0. PLoS One 2:e29018

Crook S, Gleeson P, Howell F, Svitak J, Silver R (2007) MorphML: Level 1 of the NeuroML standards for neuronal morphology data and model specification. Neuroinformatics 5:96–104. doi:10.1007/s12021-007-0003-6

Crook S, Davison AP, Plesser HE (2013) Learning from the past: approaches for reproducibility in computational neuroscience. In: Bower JM (ed) 20 Years of computational neuroscience. Springer, New York

Davison AP, Brüderle D, Eppler JM, Kremkow J, Muller E, Pecevski D, Perrinet L, Yger P (2009) PyNN: a common interface for neuronal network simulators. Front Neuroinform 2:11. doi:10.3389/neuro.11.011.2008

De Schutter E, Smolen P (1998) Calcium dynamics in large neuronal models. In: Koch C, Segev I (eds) Methods in neuronal modeling: from ions to networks, 2nd edn. MIT Press, Boston, pp 211–250

Djurfeldt M, Johansson C, Ekeberg Ö, Rehn M, Lundqvist M, Lansner A (2005) Massively parallel simulation of brain-scale neuronal network models. Tech. Rep. QC 20100709. KTH, School of Computer Science and Communication (CSC), oai:DiVA.org:kth-10606

Djurfeldt M, Hjorth J, Eppler J, Dudani N, Helias M, Potjans T, Bhalla U, Diesmann M, Hellgren Kotaleski J, Ekeberg Ö (2010) Run-time interoperability between neuronal network simulators based on the MUSIC framework. Neuroinformatics 8:43–60. doi:10.1007/s12021-010-9064-z

Dodge FA, Cooley JW (1973) Action potential of the motor neuron. IBM J Res Dev 17:219–229

Drewes RP, Zou Q, Goodman PH (2009) Brainlab: a Python toolkit to aid in the design, simulation, and analysis of spiking neural networks with the neocortical simulator. Front Neuroinform 3:16. doi:10.3389/neuro.11.016.2009

Eppler JM, Helias M, Muller E, Diesmann M, Gewaltig MO (2008) PyNEST: a convenient interface to the NEST simulator. Front Neuroinform 2:12. doi:10.3389/neuro.11.012.2008

Ermentrout B (2006) XPPAUT. Scholarpedia 1(10):1399. doi:10.4249/scholarpedia.1399

Forss J, Beeman D, Bower JM, Eichler West RM (1999) The modeler's workspace: a distributed digital library for neuroscience. Future Gener Comp Syst 16:111–121

Gardner D, Knuth KH, Abato M, Erde SM, White T, DeBellis R, Gardner E (2001) Common data model for neuroscience data and data model interchange. J Am Med Inform Assoc 8:17–33

Getting PA (1989) Reconstruction of small neural networks (chap 6). In: Koch C, Segev I (eds) Methods in neuronal modeling. MIT Press, Cambridge, MA, pp 171–194

Gleeson P, Steuber V, Silver RA (2007) neuroconstruct: a tool for modeling networks of neurons in 3d space. Neuron 54:219–235

Gleeson P, Crook S, Cannon RC, Hines ML, Billings GO, Farinella M, Morse TM, Davison AP, Ray S, Bhalla US, Barnes SR, Dimitrova YD, Silver RA (2010) NeuroML: a language for describing data driven models of neurons and networks. PLoS Comput Biol 6(6):e1000–e1815. doi:10.1371/journal.pcbi.1000815

Goddard NH, Lynne KJ, Mintz T (1987) Rochester connectionist simulator. Tech. Rep. ADA191483. Department of Computer Science, University of Rochester

Goddard NH, Hood G, Howell FW, Hines ML, De Schutter E (2001a) NEOSIM: portable large-scale plug and play modelling. Neurocomputing 38–40:1657–1661. doi:10.1016/S0925-2312(01)00528-8

Goddard NH, Hucha M, Howell F, Cornelis H, Shankar K, Beeman D (2001b) Towards NeuroML: model description methods for collaborative modelling in neuroscience. Philos Trans R Soc Lond B Biol Sci 356:1209–1228. doi:10.1098/rstb.2001.0910

Goodman DFM, Brette R (2008) Brian: a simulator for spiking neural networks in Python. Front Neuroinform 2:5. doi:10.3389/neuro.11.005.2008

Gorchetchnikov A, The INCF Multiscale Modeling Taskforce (2010) Nineml: a description language for spiking neuron network modeling: the user layer. BMC Neurosci 11(suppl 1):P71. doi:10.1186/1471-2202-11-S1-P71

Gray CM, Konig P, Engel AK, Singer W (1989) Oscillatory responses in cat visual-cortex exhibit inter-columnar synchronization which reflects global stimulus properties. Nature 338:334–337

Hammarlund P, Ekeberg Ö (1998) Large neural network simulations on multiple hardware platforms. J Comput Neurosci 5:443–459. doi:10.1023/A:1008893429695

Hartree DR (1932) A practical method for the numerical solution of differential equations. Mem Manchester Lit Phil Soc 77:91–107

Hines M (1984) Efficient computation of branched nerve equations. Int J Biomed Comput 15:69–79

Hines M (1989) A program for the simulation of nerve equations with branching geometries. Int J Biomed Comput 24:55–68

Hines ML, Carnevale NT (2000) Expanding NEURON's repertoire of mechanisms with NMODL. Neural Comput 12:995–1007

Hines M, Davison AP, Muller E (2009) NEURON and Python. Front Neuroinform 3:1. doi:10.3389/neuro.11.001.2009

Hodgkin A, Huxley A (1952) A quantitative description of membrane current and its application to conduction and excitation in nerve. J Physiol (London) 117:500–544

Hopfield JJ (1982) Neural networks and physical systems with emergent collective computational abilities. Proc Natl Acad Sci USA 79:2554

Hucka M, Shankar K, Beeman D, Bower JM (2002) The Modeler's workspace: making model-based studies of the nervous system more accessible (chap 5). In: Ascoli G (ed) Computational neuroanatomy: principles and methods. Humana Press, Totowa, NJ, pp 83–115

Hucka M, Finney A, Sauro H, Bolouri H, Doyle J, Kitano H, Arkin A (2003) The systems biology markup language (SBML): a medium for representation and exchange of biochemical network models. Bioinformatics 19:524–531

Kernigan BW, Pike R (1984) The Unix programming environment. Prentice-Hall, Englewood Cliffs, NJ

Kernighan B, Ritchie D (1978) The C programming language. Prentice-Hall, Englewood Cliffs, NJ

Kohn MC, Hines ML, Kootsey JM, Feezor MD (1989) A block organized model builder. Math Comp Mod 19:75–97

Koslow SH, Huerta MF (eds) (1997) Neuroinformatics: an overview of the human brain project. Vol: Progress in neuroinformatics research series. Lawrence Erlbaum Associates, Mahwah, NJ

Loomis ME (1995) Object databases—the essentials. Addison-Wesley, Reading, MA

Maley N, Beeman D, Lannin JS (1988) Dynamics of tetrahedral networks: amorphous Si and Ge. Phys Rev B38:10,611

Mcullough WS, Pitts WH (1943) A logical calculus of the ideas immanent in nervous activity. Bull Math Biophys 5:115–133

Migliore M, Morse TM, Davison AP, Marenco L, Shepherd GM, Hines ML (2003) ModelDB: making models publicly accessible to support computational neuroscience. Neuroinformatics 1:135–139. doi:10.1385/NI:1:1:135

Nelson M, Rinzel J (1998) The Hodgkin-Huxley model (chap 4). In: Bower JM, Beeman D (eds) The book of GENESIS: exploring realistic neural models with the GEneral NEural SImulation System, 2nd edn. Springer, New York, pp 29–49

Nelson M, Furmanski W, Bower JM (1989) Simulating neurons and neuronal networks on parallel computers (chap 12). In: Koch C, Segev I (eds) Methods in neuronal modeling. MIT Press, Cambridge, MA, pp 397–438

Pecevski D, Natschläger T, Schuch K (2009) PCSIM: a parallel simulation environment for neural circuits fully integrated with Python. Front Neuroinform 3:11. doi:10.3389/neuro.11.011.200

Pellionisz A, Llinás R, Perkel DH (1977) A computer model of the cerebellar cortex of the frog. Neuroscience 2:19–35

Perkel DH, Watt JH (1981) A manual for MANUEL. Stanford University Press, Stanford CA

Raikov I, INCF Multiscale Modeling Taskforce (2010) NineML: a description language for spiking neuron network modeling: the abstraction layer. BMC Neurosci 11(suppl 1):P66. doi:10.1186/1471-2202-11-S1-P66

Rall W (1959) Branching dendritic trees and motoneuron membrane resistivity. Exp Neurol 1:491–527

Rall W (1962a) Electrophysiology of a dendritic neuron model. Biophys J 2:145–167

Rall W (1962b) Theory of physiological properties of dendrites. Ann N Y Acad Sci 96:1071–1092

Rall W (1964) Theoretical significance of dendritic tress for neuronal input–output relations. In: Reiss RF (ed) Neural theory and modeling. Stanford University Press, Stanford CA, pp 73–97

Rall W (1967) Distinguishing theoretical synaptic potentials computed for different soma-dendritic distributions of synaptic input. J Neurophysiol 30:1138–1168

Rall W, Agmon-Smir H (1998) Cable theory for dendritic neurons (chap 2). In: Koch C, Segev I (eds) Methods in neuronal modeling: from ions to networks, 2nd edn. MIT Press, Boston, pp 27–92

Rall W, Shepherd GM (1968) Theoretical reconstruction of field potentials and dendrodendritic synaptic interaction in olfactory bulb. J Neurophysiol 31:884–915

Ray S, Bhalla US (2008) PyMOOSE: interoperable scripting in Python for MOOSE. Front Neuroinform 2:6. doi:10.3389/neuro.11.006.2008

Richert M, Nageswaran JM, Dutt N, Krichmar JL (2011) An efficient simulation environment for modeling large-scale cortical processing. Front Neuroinform 5:19

Rinzel J (1990) Electrical excitability of cells, theory and experiment: review of the Hodgkin-Huxley foundation and an update. Bull Math Biol 52:5–23

Rochel O, Martinez D (2003) An event-driven framework for the simulation of networks of spiking neurons. In: ESANN-2003, Bruges, Belgium, pp 295–300

Santamaria F, Tripp PG, Bower JM (2007) Feedforward inhibition controls the spread of granule cell: induced Purkinje cell activity in the cerebellar cortex. J Neurophysiol 97:248–263. doi:10.1152/jn.01098.2005, http://jn.physiology.org/content/97/1/248.full.pdf+html

Sasaki K, Bower JM, Llinás R (1989) Purkinje cell recording in rodent cerebellar cortex. Eur J Neurosci 1:572–586

Segev I, Fleshman JW, Miller JP, Bunow B (1985) Modeling the electrical behaviour of anatomically complex neurons using a network analysis program: passive membrane. Biol Cybern 53:27–40

Shepherd GM, Brayton RK (1979) Computer simulation of a dendro-dendritic synapse circuit for self- and lateral-inhibition in the olfactory bulb. Brain Res 175:377–382

Shepherd GH, Healy MD, Singer MS, Peterson BE, Mirsky JS, Wright L, Smith JE, Nadkarni P, Miller PL (1997) SenseLab: a project in multidisciplinary, multilevel sensory integration (chap 3). In: Koslow SH, Huerta MF (eds) Neuroinformatics: an overview of the human brain project. Lawrence Erlbaum, Mahwah, NJ, pp 21–56

Spacek MA, Blanche T, Swindale N (2009) Python for large-scale electrophysiology. Front Neuroinform 2:1. doi:10.3389/neuro.11.009.2008

Stiles JR, Bartol TM (2001) Monte Carlo methods for simulating realistic synaptic microphysiology using MCell. In: Schutter ED (ed) Computational neuroscience: realistic modeling for experimentalists. CRC Press, Boca Raton, pp 87–127

Thorpe MF, Beeman D (1976) Thermodynamics of an Ising model with random exchange interac tions. Phys Rev B14:188

Traub R (1977) Motor neurons of different geometry and the size principle. Biol Cybern 25:163–176

Traub RD (1982) Simulation of intrinsic bursting in CA3 hippocampal neurons. Neuroscience 7:1233–1242

Traub RD, Llinás R (1979) Hippocampal pyramidal cells: significance of dendritic ionic conductances for neuronal function and epileptogenesis. J Neurophysiol 42:476–496

Traub RD, Wong RKS, Miles R, Michelson H (1991) A model of a CA3 hippocampal neuron incorporating voltage-clamp data on intrinsic conductances. J Neurophysiol 66:635–650

Traub RD, Jeffereys JGR, Miles R, Whittington MA, Tóth K (1994) A branching dendritic model of a rodent CA3 pyramidal neurone. J Physiol (London) 481:79–95

Vanier MC, Bower JM (1999) A comparative survey of automated parameter-search methods for compartmental neural models. J Comput Neurosci 7:149–171

Weitzenfeld A (1995) NSL—neural simulation language. In: Arbib MA (ed) The handbook of brain theory and neural networks, 1st edn. Bradford Books/MIT Press, Cambridge, pp 654–658

Wilson MA, Bower JM (1989) The simulation of large scale neural networks (chap 9). In: Koch C, Segev I (eds) Methods in neuronal modeling. MIT Press, Cambridge, MA, pp 291–333

Wilson M, Bower JM (1991) A computer simulation of oscillatory behavior in primary visual cortex. Neural Comput 3:498–509

Wilson M, Bower JM (1992) Cortical oscillations and temporal interactions in a computer simulation of piriform cortex. J Neurophysiol 67:981–995

Wilson MA, Bhalla US, Uhley JD, Bower JM (1989) GENESIS: a system for simulating neural networks. In: Touretzky D (ed) Advances in neural information processing systems. Morgan Kauffman, San Mateo, CA, pp 485–492

Chapter 4
Learning from the Past: Approaches for Reproducibility in Computational Neuroscience

Sharon M. Crook, Andrew P. Davison, and Hans E. Plesser

Abstract Reproducible experiments are the cornerstone of science: only observations that can be independently confirmed enter the body of scientific knowledge. Computational science should excel in reproducibility, as simulations on digital computers avoid many of the small variations that are beyond the control of the experimental biologist or physicist. However, in reality, computational science has its own challenges for reproducibility: many computational scientists find it difficult to reproduce results published in the literature, and many authors have met problems replicating even the figures in their own papers. We present a distinction between different levels of replicability and reproducibility of findings in computational neuroscience. We also demonstrate that simulations of neural models can be highly sensitive to numerical details, and conclude that often it is futile to expect exact replicability of simulation results across simulator software packages. Thus, the computational neuroscience community needs to discuss how to define successful reproduction of simulation studies. Any investigation of failures to reproduce published results will benefit significantly from the ability to track the provenance of the original results. We present tools and best practices developed over the past 2 decades that facilitate provenance tracking and model sharing.

S.M. Crook, Ph.D. (✉)
School of Mathematical and Statistical Sciences and School of Life Sciences, Center for Adaptive Neural Systems, Arizona State University, Physical Sciences Building A Room 216, 901 S. Palm Walk, Tempe, AZ 85287-1804, USA
e-mail: sharon.crook@asu.edu

A.P. Davison
Unité de Neuroscience Information et Complexité (UNIC), CNRS, Gif-sur-Yvette 91198, France

H.E. Plesser
Department of Mathematical Sciences and Technology, Norwegian University of Life Sciences, Aas, Norway

J.M. Bower (ed.), *20 Years of Computational Neuroscience*, Springer Series in Computational Neuroscience 9, DOI 10.1007/978-1-4614-1424-7_4,

Introduction

Reproducible experimental results have been the cornerstone of science since the time of Galileo (Hund 1996, p. 103): Alice's exciting finding will not become part of established scientific knowledge unless Bob and Charlie can reproduce her results independently. A related concept is provenance—the ability to track a given scientific result, such as a figure in an article, back through all analysis steps to the original raw data and the experimental protocol used to obtain it.

The quest for reproducibility raises the question of what it means to reproduce a result independently. In the experimental sciences, data will contain measurement error. Proper evaluation and judgement of such errors requires a sufficient understanding of the processes giving rise to fluctuations in measurements, as well as of the measurement process itself. An interesting historical example is the controversy surrounding Millikan's oil drop experiments for the measurement of the elementary charge, which led Schrödinger to investigate the first-passage time problem in stochastic processes (Schrödinger 1915). As experimental error can never be eliminated entirely, disciplines depending on quantitative reproducibility of results, such as analytical chemistry, have developed elaborate schemes for ascertaining the level of reproducibility that can be obtained and for detecting deviations (Funk et al. 2006). Such schemes include round robin tests in which one out of a group of laboratories prepares a test sample, which all others in the group then analyze. Results are compared across the group, and these tests are repeated regularly, with laboratories taking turns at preparing the test sample.

In computational, simulation-based science, the reproduction of previous experiments and the establishment of the provenance of results should be easy, given that computers are deterministic and do not suffer from the problems of inter-subject and trial-to-trial variability of biological experiments. However, in reality, computational science has its own challenges for reproducibility.

As early as 1992, Claerbout and Karrenbach addressed the necessity of provenance in computational science and suggested the use of electronic documentation tools as part of the scientific workflow. Some of the first computational tools that included a complete documentation of provenance were developed in the signal processing community (Donoho et al. 2009; Vandewalle et al. 2009) and other fields followed (Quirk 2005; Mesirov 2010). Generally, these important efforts are discussed as examples of *reproducible research*. However, following Drummond (2009), we find it important to distinguish between the *reproduction* of an experiment by an independent researcher and the *replication* of an experiment using the same code perhaps some months or years later.

Independent reproducibility is the gold standard of science; however, replicability is also important and provides the means to determine whether the failure of others to reproduce a result is due to errors in the original code. Replication ought to be simple—it is certainly easier than independent reproduction—but in practice replicability is often not trivial to achieve and is not without controversy. Drummond (2009) has argued that the pursuit of replicability detracts from the promotion of independent reproducibility, since the two may be confused and due to the burden that ensuring replicability places on the researcher. Indeed, when the journal *Biostatistics* recently introduced a scheme for certifying papers as replicable after a review of code and data by the associate editor for reproducibility (Peng 2009), this led to a lively debate about the relative importance of replicability of data processing steps in the context of complex scientific projects (Keiding 2010a, b; Breslow 2010; Cox and Donnelly 2010; DeAngelis and Fontanarosa 2010; Donoho 2010; Goodman 2010b; Groves 2010; Peng 2010). Generally, the risk of confusing replicability with reproducibility is an argument for education and discussion, not for neglecting replicability, and the extra workload of carefully tracking full provenance information may be alleviated or eliminated by appropriate tools. Discussions of reproducibility generally include both of these concepts, and here, we find it useful to make further distinctions as follows.

Internal replicability: The original author or someone in their group can re-create the results in a publication, essentially by rerunning the simulation software. For complete replicability within a group by someone other than the original author, especially if simulations are performed months or years later, the author must use proper bookkeeping of simulation details using version control and electronic lab journals.

External replicability: A reader is able to re-create the results of a publication using the same tools as the original author. As with internal replicability, all implicit knowledge about the simulation details must be entered into a permanent record and shared by the author. This approach also relies on code sharing, and readers should be aware that external replicability may be sensitive to the use of different hardware, operating systems, compilers, and libraries.

Cross-replicability: The use of "cross" here refers to simulating the same model with different software. This may be achieved by re-implementing a model using a different simulation platform or programming language based on the original code, or by executing a model described in a simulator-independent format on different simulation platforms. Simulator-independent formats can be divided into declarative and procedural approaches. Assuming that all simulators are free of bugs, this would expose the dependence of simulation results on simulator details, but leads to questions about how to compare results.

Reproducibility: Bob reads Alice's paper, takes note of all model properties, and then implements the model himself using a simulator of his choice. He does not download Alice's scripts. Bob's implementation thus constitutes an independent implementation of the scientific ideas in Alice's paper based on a textual description. The boundary line between cross-replicability and reproducibility is not always clear. In particular, a declarative description of a model is a structured, formalized version of what should appear in a publication, so an implementation by Charlie based on a declarative format might be considered to be just as independent as that by Bob based on reading the article.

In our terminology, the *reproducible research* approach propagated by Donoho (2010) ensures internal replicability and to quite a degree external replicability as well. As de Leeuw (2001) has pointed out, though, Donoho's specific approach depends on the commercial Matlab software together with a number of toolboxes, and thus does not aim to ensure cross-replicability and independent reproducibility as defined above. So how have approaches for replicability and reproducibility evolved in the field of computational neuroscience? As early as 1992, one group of simulation software developers realized the need for computational benchmarks as a first step toward cross-replicability (Bhalla et al. 1992); however, these benchmarks were not broadly adopted by the simulator development community. There were also early efforts to encourage simulator-independent model descriptions for complex neural models. Building on general-purpose structures proposed by Gardner et al. (2001), the first declarative description tools for models were described in 2001 by Goddard et al. Around the same time, the NMODL language developed by Michael Hines for describing biophysical mechanisms in the NEURON simulator was extended by Hines and Upinder Bhalla to work with GENESIS (GMODL; Wilson et al. 1989), making it perhaps the first programmatic simulator-independent model description language in computational neuroscience (Hines and Carnevale 2000). More recently, the activities of organizations, such as the *Organization for Computational Neuroscience* (http://www.cnsorg.org) and the *International Neuroinformatics Coordinating Facility* (http://www.incf.org), focused journals such as *Neuroinformatics* and *Frontiers in Neuroinformatics* and dedicated workshops (Cannon et al. 2007; Djurfeldt and Lansner 2007) have provided fora for an ongoing discussion of the methodological issues our field is facing in developing an infrastructure for replicability and reproducibility. The first comprehensive review of neuronal network simulation software (Brette et al. 2007) provides an example of the gains of this process.

However, there are still many improvements needed in support of reproducibility. Nordlie et al. (2009) painted a rather bleak picture of the quality of research reporting in our field. Currently, there are no established best practices for the description of models, especially neuronal network models, in scientific publications, and few papers provide all necessary information to successfully reproduce the simulation results shown. Replicability suffers from the complexity of our code and our computing environments, and the difficulty of capturing every essential piece of information about a computational experiment. These difficulties will become even more important to address as the ambition of computational neuroscience and the scrutiny placed upon science in general grow (Ioannidis 2005; Lehrer 2010).

In what follows, we will discuss further details of replicability and reproducibility in the context of computational neuroscience. In section "The Limits of Reproducibility," we examine the limits of reproducibility in the computational sciences with examples from computational neuroscience. Section "Practical Approaches to Replicability" deals with practical approaches to replicability such as code sharing, tracking the details of simulation experiments, and programmatic or procedural descriptions of complex neural models that aid in cross-replicability. In section "Structured, Declarative Descriptions of Models and Simulations," we introduce a number of efforts to formalize declarative descriptions of models and simulations and the software infrastructure to support this. Finally, in section

"Improving Research Reporting," we discuss more general efforts to improve research reporting that go beyond software development.

The Limits of Reproducibility

Independent reproduction of experimental results will necessarily entail deviations from the original experiment, as not all conditions can be fully controlled. Whether a result is considered to have been reproduced successfully thus requires careful scientific judgement, differentiating between core claims and mere detail in the original report. However, solving the same equation twice yields precisely the same result from a mathematical point of view. Consequently, it might appear that any computational study which is described in sufficient detail should be *exactly* reproducible: by solving the equations in Alice's paper using suitable software, Bob should be able to obtain figures identical to those in the paper. This implies that Alice's results are also perfectly replicable. It is obviously a prerequisite for such exact reproducibility that results can be replicated internally, externally, and across suitable software applications. In this section, we discuss a number of obstacles to external and cross-replicability of computational experiments which indicate that it is futile to expect perfect reproducibility of computational results. Rather, computational scientists need to apply learned judgement to the same degree as experimentalists in evaluating successful reproduction.

Faulty computer hardware is the principal—though not the most frequent—obstacle: digital computers are electronic devices and as such are subject to failure, which often may go undetected. For example, memory may be corrupted by radiation effects (Heijmen 2011), and as we are rapidly approaching whole-brain simulations on peta-scale and soon exa-scale computers, component failure will become a routine issue. Consider a computer with one million cores. Even if each core has a mean time between failure of a million hours (roughly 115 years), one would expect on average one core failure per hour of operation. It seems questionable whether all such errors will be detected reliably—a certain amount of undetected data corruption appears unavoidable.

Even if hardware performs flawlessly, computer simulations are not necessarily deterministic. In parallel simulations, performance of the individual parallel processes will generally depend on other activity on the computer, so that the order in which processes reach synchronization points is unpredictable. This type of unpredictability should not affect the results of correct programs, but subtle mistakes may introduce nondeterministic errors that are hard to detect. Even if we limit ourselves to serial programs running on perfect hardware, a number of pitfalls await those trying to reproduce neuronal network modeling results from the literature, which fall into the following categories:

1 Insufficient, ambiguous, or inaccurate descriptions of the model in the original publication.

2 Models that are mathematically well-defined, but numerically sensitive to implementation details.
3 Model specifications that are unambiguous and complete from a neuroscience point of view, but underspecified from a computational point of view.
4 Dependencies on the computing environment.

First, consider the first and last points. An insufficient model specification in a publication can be a major stumbling block when trying to replicate a model. Thus, ambiguous model descriptions provide a significant impediment to science that can only be avoided if authors, referees, and publishers adhere to strict standards for model specification (Nordlie et al. 2009) or rigorous, resource-intensive curation efforts (Lloyd et al. 2008); we will return to this point in section "Improving Research Reporting." Dependencies on the computing environment, such as the versions of compilers and external libraries used, will be discussed in section "Is Code Sharing Enough to Ensure Replicability?" Here, we consider model descriptions that are mathematically ambiguous or sensitive to the implementation details before discussing the consequences for computational neuroscience in section "Defining Successful Reproduction."

Ambiguous Model Numerics

The model equation for the subthreshold membrane potential V of a leaky integrate-and-fire neuron with constant input is a simple, linear first-order ordinary differential equation (Lapicque 1907)

$$\frac{dV}{dt} = -\frac{V}{\tau} + \frac{I_E}{C}. \tag{4.1}$$

Here, membrane potential V (in mV) is defined relative to the resting potential of the neuron, τ is the membrane time constant (in ms), C the membrane capacitance (in pF), and $I(t)$ the input current to the neuron (in pA). As far as differential equations go, this equation is about as simple as possible, and solutions are well-defined and well-behaved. For the initial condition $V(t=0)=V_0$, (4.1) has the analytical solution

$$V(t) = V_0 \mathrm{e}^{-t/\tau} + \frac{I_E \tau}{C}(1 - \mathrm{e}^{-(t/\tau)}). \tag{4.2}$$

Abbreviating $V_k = V(kh)$ and $a = I_E\tau/C$ gives the following iteration rule for time step h:

$$V_{k+1} = V_k \mathrm{e}^{-(h/\tau)} + a(1 - \mathrm{e}^{-(h/\tau)}). \tag{4.3}$$

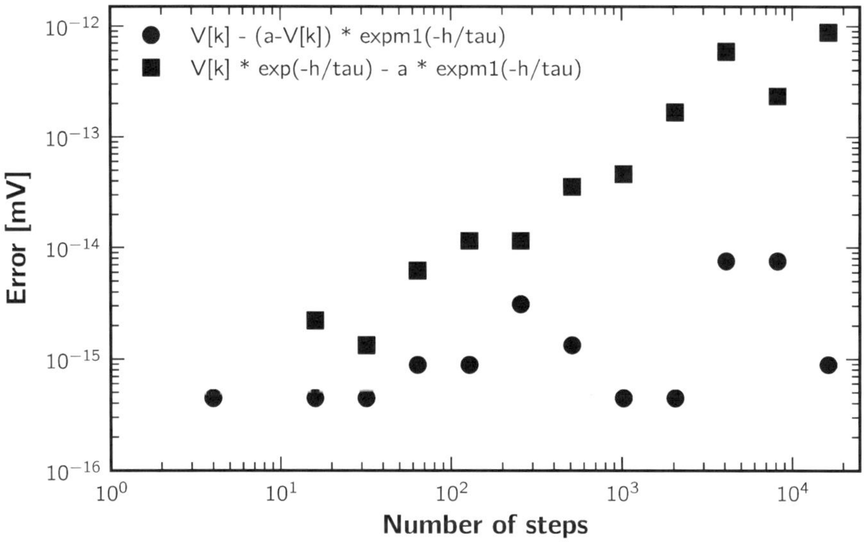

Fig. 4.1 Error of the membrane potential $V(T)$ for $T = 1$ ms computed with two different implementations of (4.3) using step sizes from $h = 2^{-1}$ to 2^{-14} ms, corresponding to 2 to 16384 steps. *Circles* show errors for the implementation given in (4.4), *squares* for the implementation given in (4.5); see text for details

This updating rule is mathematically exact, and similar rules can be found for any system of ordinary linear differential equations (Rotter and Diesmann 1999).

Now consider the following two implementations of (4.3)[1]:

$$V[k+1] = V[k] - (a - V[k]) * \mathrm{expm1}(-h/\tau) \tag{4.4}$$

$$V[k+1] = V[k] * \exp(-h/\tau) - a * \mathrm{expm1}(-h/\tau). \tag{4.5}$$

Both implementations are mathematically equivalent, but differ significantly numerically, as can be seen by computing the evolution of the membrane potential for $T = 1$ ms using different step sizes, starting from $V(t=0) = 0$ mV. We obtain the reference solution $V^* = V(T)$ directly from (4.2) as $a * \mathrm{expm1}(T/\tau)$ with the following parameter values: $I_E = 1{,}000$ pA, $C = 250$ pF, $\tau = 10$ ms, so that $a = 40$ mV. We then compute $V(T)$ using update steps from $h = 2^{-1}$ ms down to $h = 2^{-14}$ ms using both implementations and compute the difference from the reference solution; using step sizes that are powers of 2 avoids any unnecessary round-off error (Morrison et al. 2007). Results obtained with the update rule provided by (4.5) are several orders of magnitude larger than those obtained with the rule in (4.4), as shown in Fig. 4.1. Data were obtained with a custom C++ program compiled with the g++ compiler version 4.5.2 and default compiler settings on a Apple Mac Book Pro with an Intel

[1] expm1(x) is a library function computing exp(x)-1 with high precision for small x.

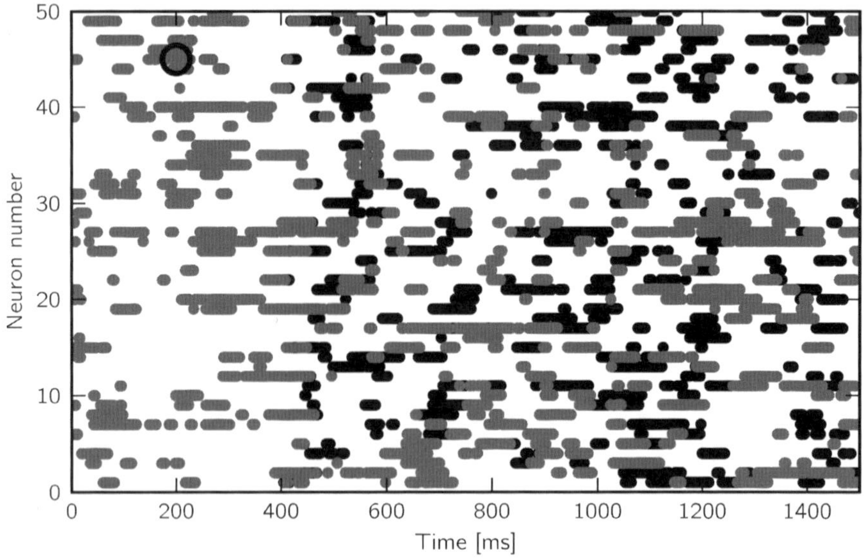

Fig. 4.2 Raster plots of spike trains of 50 excitatory neurons in a balanced network of 1,250 neurons exhibiting self-sustained irregular activity after Poisson stimulation during the first 50 ms and no external input thereafter. The network is based on Brunel (2000), but with significantly stronger synapses (Gewaltig and Koerner 2008). The first simulation (*black raster*) runs for 1,500 ms. The simulation is then repeated with identical initial conditions (*gray raster*), but after 200 ms, the membrane potential of a single neuron (*black circle*) is increased by 10^{-13} mV. From roughly 400 ms onwards, spike trains in both simulations differ significantly. Simulations were performed with NEST 2.0.0-RC2 using the iaf_psc_alpha_canon model neuron with precise spike timing (Morrison et al. 2007)

i5 CPU under Mac OSX 10.6.6. Source code for this and the other examples in this chapter is available from http://www.nest-initiative.org.

This result implies that even a model containing only equations as simple as (4.1) is not cross-replicable even if the (exact) method of iteration of (4.3) is specified as part of the model definition. Precise results depend on which of several mathematically equivalent numerical implementations is used. Clearly, different simulators should be allowed to implement exact solvers for (4.1) in different ways, and even if the solver of (4.3) were prescribed by the model, its implementation should not be. Indeed, when using modern simulators based on automatic code generation (Goodman 2010a), the computational scientist may not have any control over the precise implementation. Thus, even the simplest models of neural dynamics are numerically ambiguous, and it would be futile to expect exact model replicability across different simulation software even if detailed model specifications are provided.

One may raise the question, though, whether the errors illustrated in Fig. 4.1 are so minuscule that they may safely be ignored. Generally, the answer is no: many neuronal network models are exquisitely sensitive to extremely small errors, as illustrated in Fig. 4.2. This figure shows results from two simulations of 1,500 ms of activity in a balanced network of 1,250 neurons (80 % excitatory) based on Brunel

(2000), but with significantly stronger synapses (Gewaltig and Koerner 2008); see Fig. 4.4 for a summary overview of the model. In the second simulation run, the membrane potential of a single neuron is changed by 10^{-13} mV after 200 ms. Soon after, the spike activity in the two simulation runs has diverged entirely. In practice, this means that a scientist trying to replicate a model may obtain very different spike train results than those in the original publication. This holds both for cross-replication using different simulator software and for external replication in a different computational environment; the latter may even affect internal replication, cf. section "Is Code Sharing Enough to Ensure Replicability?" Gleeson et al. (2010) recently provided an example of the difficulties of cross-replication. They defined a network of 56 multicompartment neurons with 8,100 synapses using the descriptive NeuroML language and then simulated the model using the NEURON (Hines 1989; Hines and Carnevale 1997; Carnevale and Hines 2006), GENESIS (Bower and Beeman 1997), and MOOSE (http://moose.ncbs.res.in) software packages. These simulators generated different numbers of spikes in spite of the use of a very small time step (0.001 ms). Gleeson et al. concluded that "[t]hese results show that the way models are implemented on different simulators can have a significant impact on their behavior."

Computationally Underspecified Models

In the previous section we saw that even if the mathematics of a model are fully specified, numerical differences between implementations can lead to deviating model behavior. We shall now turn to models which may appear to be fully specified, but in fact leave important aspects to the simulation software. We refer to these models as *computationally underspecified*, as their specifications can be considered complete from a neuroscience point of view. We shall consider two cases in particular, spike timing and connection generation.

Most publications based on integrate-and-fire neurons contain a statement such as the following: "A spike is recorded when the membrane potential crosses the threshold, then the potential is reset to the reset potential." In many publications, though, it is not further specified at precisely which time the spike is recorded and the potential reset. As many network simulations are simulated on a fixed time grid, one can only assume that both events happen at the end of the time step during which the membrane potential crossed the threshold. Hansel et al. (1998) were the first to point out that tying membrane potential resets to a grid introduces a spurious regularity into network simulations that may, for example, lead to synchronization of firing activity. This observation spurred a quest for efficient methods for simulating networks with precisely timed spikes and resets (Hansel et al. 1998; Shelley and Tao 2001; Brette 2006, 2007; Morrison et al. 2007; Hanuschkin et al. 2010; D'Haene 2010).

Let us now consider how connections within neuronal network models are specified. Nordlie et al. (2009) demonstrated that connectivity information is often poorly defined in the literature, but even seemingly complete connectivity descriptions, such as given by Brunel (2000), will not ensure replicability across simulators. Specifically, Brunel states that each excitatory neuron receives input from 10 % of the excitatory neurons, chosen at random from a uniform distribution. Although different simulators may establish the correct number of connections using uniformly chosen sources, generally no two simulators will create the same network since the process of randomly drawing sources is implemented differently across simulators. Even if one prescribes which random number generators and seed values to use, little is gained unless the simulators use the random numbers in identical ways.

The only way to ensure that different software will create identical networks is to specify how the simulator iterates across nodes while creating connections. For example, the PyNN package (Davison et al. 2009) ensures that identical networks are generated in different simulators by iterating across nodes within PyNN and issuing explicit connection commands to the simulators, which will in general be slower than allowing the simulators to use their own internal routines. One might argue that such detail is beyond the scope of models created to explain brain function—after all, there is no biological counterpart to the arbitrary neuron enumeration schemes found in simulators, which naturally leads to a discussion of how one should define successful reproduction of modeling results.

Defining Successful Reproduction

Using proper documentation tools (see section "Practical Approaches to Replicability"), we can in principle achieve internal and external replicability in the short and long term. But this guarantees no more than that the same script on the same simulator generates the same results. As we have seen, in many cases we cannot take a simulation from one simulator to another and hope to obtain identical spike trains or voltage traces. Thus, there is no easy way to test the correctness of our simulations.

For many physical systems, a scientist can rely on a conservation law to provide checks and balances in simulation studies. As an example, Fig. 4.3 shows the movements of three point masses according to Newton's gravitational law where energy should be conserved in the system. Integrating the equations of motion using a forward Euler method yields incorrect results; when simulating with forward Euler, the total energy jumps to a much higher level when two planets pass close by each other. Fortunately, it is straightforward to compute the energy of the three-body system at any time, and a simulation using the LSODA algorithm (Petzold 1983) shows only a brief glitch in energy, demonstrating a better choice of numerical method. In the same manner, Ferrenberg et al. (1992) discovered important weaknesses in random number generators that were thought to be reliable when they observed implausible values for the specific heat of crystals in simulation experiments. The specific heat, a macroscopic quantity, was independently known from

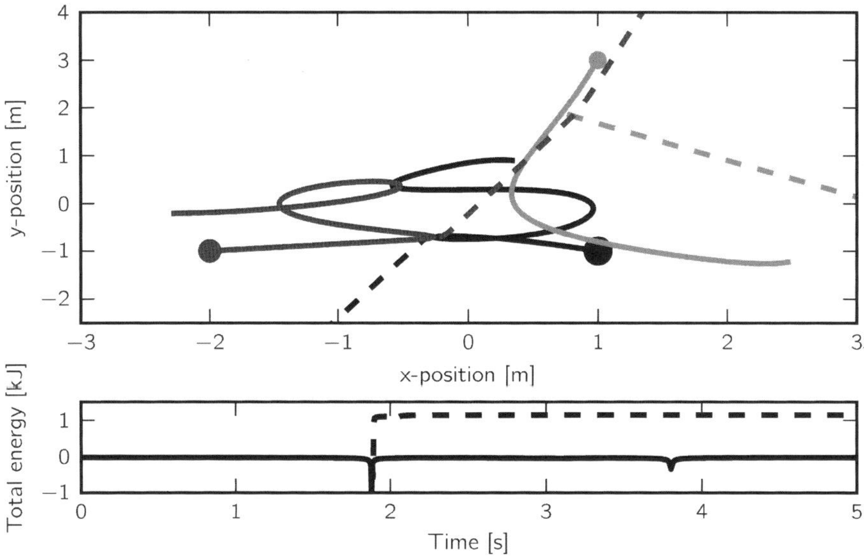

Fig. 4.3 Numerical solution of the Pythagorean planar three-body problem (Gruntz and Waldvogel 2004): Three point masses of 3 kg (*light gray*), 4 kg (*dark gray*), and 5 kg (*black*) are placed at rest at the locations marked by *circles* and move under the influence of their mutual gravitational attraction with gravitational constant $G = 1\ m^3\ kg^{-1}\ s^{-2}$. *Top*: *Solid lines* show trajectories for the three bodies up to $T = 5$ s obtained with the LSODA algorithm (Petzold 1983) provided by the SciPy Python package (Jones et al. 2001), using a step size of 0.005 s; these agree with the trajectories depicted in Gruntz and Waldvogel (2004). *Dashed lines* show trajectories obtained using a custom forward Euler algorithm using the same step size. The trajectories coincide initially but diverge entirely as the black and dark gray planets pass each other closely. *Bottom*: Total energy for the solutions obtained using LSODA (*solid*) and forward Euler (*dashed*). While total energy remains constant in the LSODA solution except for short glitches around near encounters, the forward Euler solution "creates" a large amount of energy at the first close encounter, when trajectories begin to diverge

thermodynamical theory, thus providing a method for testing the simulations. Unfortunately, there are no known macroscopic laws governing neuronal dynamics (e.g., for the number of spikes in a network, for the resonance frequencies).

Twenty years ago Grebogi et al. (1990) posed a question for simulations of chaotic systems: "For a physical system which exhibits chaos, in what sense does a numerical study reflect the true dynamics of the actual system?" An interesting point in this respect is that the model which is based most closely on the physical system may not yield the best solution. Early models of atmospheric circulations were beset with numerical instabilities, severely limiting the time horizon of simulations. This changed only when Arakawa introduced some nonphysical aspects in atmospheric models which ensured long-term stability and produced results in keeping with meteorological observations (Küppers and Lenhard 2005). Similarly, the computational neuroscience community should embark on a careful discussion of criteria for evaluating the results of neuronal network simulations. Perhaps the same criteria for evaluating whether a model network provides a good model of neural activity should be used to compare the results of two different neuronal network simulations. Regardless, computational scientists must be able to distinguish between numerical or programming errors and differences in simulator implementations.

Practical Approaches to Replicability

As noted in the "Introduction," there are some intermediate steps between pure internal replication of a result (the original author rerunning the original code) and fully independent reproduction, namely external replication (someone else rerunning the original code) and cross-replication (running cross-platform code on a different simulation platform or re-implementing a model with knowledge of the original code). Both of these intermediate steps rely on sharing code, the simplest form of model sharing (Morse 2007). Code sharing allows other researchers to rerun simulations and easily extend models, facilitating a modular, incremental approach to computational science.

Approaches to Code Sharing

There are a number of possible methods for sharing code: by e-mail, on request; as supplementary material on a publisher's web-site; on a personal web-site; on a public source-code repository such as SourceForge (http://sourceforge.net), GitHub (https://github.com), BitBucket (https://bitbucket.org), or Launchpad (https://launchpad.net); or in a curated database such as ModelDB (Peterson et al. 1996; Davison et al. 2002; Migliore et al. 2003; Hines et al. 2004, http://senselab.med.yale.edu/modeldb), the Visiome platform (Usui 2003, http://visiome.neuroinf.jp), the BioModels database (Le Novere et al. 2006, http://www.biomodels.net), or the CellML Model Repository (http://models.cellml.org).

Sharing on request is the simplest option for a model author at the time of publication, but does not provide a public reference to which any extensions or derived models can be compared, and has the risk that contacting an author may not be straightforward if his/her e-mail address changes, he/she leaves science, etc. This option can also lead to future problems for the author, if the code cannot be found when requested, or "suddenly" produces different results.

Many journals offer the possibility of making model code available as supplementary material attached to a journal article, which makes it easy to find the model associated with a particular paper. The main disadvantages of this option are (1) lack of standardization in the format of the code archive or the associated metadata; (2) difficulty in updating the code archive and lack of versioning information if bugs are found, improvements are made, or contact details are changed; and (3) quality control—article referees may check that the code runs and is correct, but this is not universal. Also, not all journals offer the option of supplementary material; and some are moving away from offering it (Maunsell 2010).

A personal or institutional web-site allows the authors to more easily make updates to the code and to maintain a list of versions. The same lack of standardization of archive formats and metadata as for publisher sites exists. The major disadvantage is discoverability: it may not always be easy to find the author's web-site in the first place, for example, in case of a change of affiliation when hosting on an institutional site, expiration of domain name registrations for personal sites, or internal site reorganizations that break links. A further disadvantage is even less quality control than for supplementary material. An attempt to address the discoverability and metadata standardization problems was made by Cannon et al. (2002), who developed an infrastructure for distributed databases of models and data in which each group maintains its own local database and a central catalogue server federates them into a single virtual resource. This idea did not take off at the time, but is similar to the approach now being taken by the Neuroscience Information Framework (Gardner et al. 2008; Marenco et al. 2010).

Public general-purpose source-code repositories have many nice features for code sharing: versioning using mainstream version control tools, standard archive formats for downloading, some standardization of metadata (e.g., authors, version, programming language), issue tracking, wikis for providing documentation, and the stability of URLs. The disadvantages are possible problems with discoverability (SourceForge, for example, hosts over 260,000 projects. Finding which if any of these is a neuroscience model could be challenging), lack of standardization of neuroscience-specific metadata (which cell types, brain regions, etc.), and lack of external quality control.

Curated model repositories, in which a curator verifies that the code reproduces one or more figures from the published article, and which often have standardized metadata which make it easier to find models of a certain type (e.g., models of cortical pyramidal cells), address the issues of quality control, discoverability, and standardization (Lloyd et al. 2008). They perhaps lack some of the features available with public general-purpose code repositories, such as easy version tracking, issue tracking, and documentation editing, although the CellML Model Repository has fully integrated the Mercurial version control system (http://mercurial.selenic.com) into its site through the concept of a workspace for each model, and ModelDB has begun experimenting with integration of Mercurial.

Currently, the best solution for an author who wishes to share the code for their published model is probably to maintain the code in a public general-purpose repository such as GitHub or BitBucket (for the very latest version of the model and to maintain a record of previous versions) and also to have an entry for the model in a curated database such as ModelDB (for the latest version to have been tested by the curators, together with neuroscience-specific metadata). This recommendation may change as curated repositories become more feature-rich over time.

When sharing code, intellectual property issues should be considered. Presently, most researchers in computational neuroscience do not provide an explicit license when sharing their code, perhaps assuming that they are placing it in the public domain or that general scientific principles of attribution will apply when others reuse their code. For much more information on legal issues related to reproducible research see Stodden (2009a, b).

Steps from Replicability to Reproducibility

One approach to cross-replicability is the use of simulator-independent formats for describing models. These may be divided into declarative and programmatic—although here again the distinction is not always clear cut, since programming languages can be used in a declarative way, and not all declarative descriptions are really simulator-independent. Declarative model and simulation experiment specification is discussed in the next section. Here we consider programmatic simulator-independent formats.

A simulator-independent/simulator-agnostic programming interface allows the code for a simulation to be written once and then run on different simulator engines. Unlike declarative specifications, such a description is immediately executable without an intermediate translation step, which gives a more direct link between description and results. The use of a programming language also provides the full power of such a language, with loops, conditionals, subroutines, and other programming constructs. The great flexibility and extensibility this gives can be a strong advantage, especially in an exploratory phase of model building. It may also be a disadvantage if misused, leading to unnecessary complexity, bugs, and difficulty in understanding the essential components of the model, which are less common with declarative specifications.

In neuroscience, we are aware of only one such simulator-independent interface, PyNN (Davison et al. 2009), which provides an API in the Python programming language, and supports computational studies using the software simulators NEURON (Hines 1989; Hines and Carnevale 1997; Carnevale and Hines 2006), NEST (Gewaltig and Diesmann 2007; Eppler et al. 2008), PCSIM (Pecevski et al. 2009), and Brian (Goodman and Brette 2008), as well as a number of neuromorphic hardware systems (Brüderle et al. 2009; Galluppi et al. 2010). This is not a recent idea: as mentioned in the "Introduction," at one time the NMODL language developed by Michael Hines for describing biophysical mechanisms in the NEURON simulator was extended by Hines and Upinder Bhalla to work with GENESIS (GMODL; Wilson et al. 1989), making it perhaps the first simulator-independent description in computational neuroscience (Hines and Carnevale 2000). To the best of our knowledge, however, GMODL no longer exists and it is certainly not compatible with the most recent evolutions of NMODL.

Is Code Sharing Enough to Ensure Replicability?

Sharing code is not a panacea for ensuring replicability. It is often the case that a given result from a published paper cannot be re-created with code that has been made available. The reasons for this may include: differences in the version of the simulator, the compiler, or of shared libraries that are used by either the simulator or the code; differences in the computing platform (e.g., 32-bit vs. 64-bit systems, changes in the operating system); or simply poor record-keeping on the part of the researcher. It is our experience that often the set of parameters used to obtain a particular figure is different from that stated in the published article, sometimes due to typographical errors. Finally, for older publications, the model may have been run originally on a platform which is no longer available.

A more systematic approach to record-keeping is essential for improving the replicability of simulation studies. An important first step is the use of version control systems so that handling the problem of tracking which version of a model and which parameters are used to produce a particular result or figure is as simple as making a note of the version number. Making the version number part of a filename or embedded comments is even better.

The problem of changing versions of simulators and their dependencies, and of a changing computing environment, may be addressed by, first, making note of the software version(s) used to produce a particular result (including compiler versions and options and library versions, when the software has been compiled locally). It may be possible to automate this process to a certain extent (see below). A second step may be to capture a snapshot of the computing environment, for example, using virtual machines. For models that were originally simulated on now-obsolete hardware, software emulators (see, for example, http://www.pdp11.org) are a possible solution. Another is for the original authors or curators to port the code to a newer system or to a declarative description when the original system nears the end of its life.

A further step would be to automate the record-keeping process as much as possible, using, for example, an electronic lab notebook to automatically record the version of all software components and dependencies, and automatically check that all code changes have indeed been committed to a version control system prior to running a simulation. One of the authors (APD) has recently initiated an open-source project to develop such an automated lab notebook for computational experiments. Sumatra (http://neuralensemble.org/sumatra) consists of a core library implemented in Python, together with a command-line interface and a web-interface that builds on the library; a desktop graphical interface is planned. Each of these interfaces enables (1) launching simulations with automated recording of provenance information (versions, parameters, dependencies, input data files, and output files) and (2) managing a simulation project (browsing, viewing, annotating, and deleting simulations). The command-line and web-interface are independent of a particular simulator, although some information (e.g., model code dependencies) can only be tracked if a plug-in is written for the simulation language of interest; plug-ins for NEURON, GENESIS, and Python are currently available.

A number of tools exist for enabling reproducible data analysis and modeling workflows in a visual environment, for example, Kepler (Ludäscher et al. 2006), Taverna (Oinn et al. 2006), and VisTrails (Silva et al. 2007). VisTrails is of particular interest since it focuses on tracking changes to workflows over time in a way that is particularly well suited to the often exploratory process of neuronal modeling. The main disadvantage of using such a workflow framework is the extra development burden of wrapping simulation software to fit within the framework.

Structured, Declarative Descriptions of Models and Simulations

As described above, the intermediate steps between replicability and reproducibility for computational models include the expression of the model in a simulator-independent format that can be used on a different computational platform from the original model. One approach is to use a declarative description of the model that is simulator-independent. Software and database developers in many fields, including neuroscience, have enthusiastically adopted EXtensible Markup Language (XML) technology (Bray et al. 1998) as an ideal representation for complex structures such as models and data, due to its flexibility and its relation to the HTML standard for web pages. Like HTML, XML is composed of text and tags that explicitly describe the structure and semantics of the content of the document. Unlike HTML, developers are free to define the tags and develop a specific XML-based markup language that is appropriate for their application. A major advantage of XML is that it provides a machine-readable language that is independent of any particular programming language or software encoding, which is ideal for a structured, declarative description that can provide a standard for the entire community.

A representation of a model in a specific markup language is essentially a text document that consists of XML descriptions of the components of the model. Usually, the structure of a valid XML document is defined using a number of XML Schema Definition (XSD) files. Using these, standard XML handling libraries can be used to check the validity of an XML document against the language elements. Once an XML file is known to be valid, the contents of the file can be transformed into other formats in a number of different ways. For example, an application can read the XML natively using one of the commonly used parsing frameworks such as SAX (Simple API for XML, http://sax.sourceforge.net) or DOM (Document Object Model, http://en.wikipedia.org/wiki/Document_Object_Model). An alternative approach is to transform the XML description into another text format that can be natively read by an application, which can be done using Extensible Stylesheet Language (XSL) files. For more details regarding the use of XML technology for declarative model descriptions, see Crook and Howell (2007).

A number of ongoing projects focus on the development of these self-documenting markup languages that are extensible and can form the basis for specific implementations covering a wide range of modeling scales in neuroscience. The Systems Biology Markup Language, SBML (Hucka et al. 2003), and CellML (Hedley et al. 2000; Lloyd et al. 2004) are two popular languages for describing systems of interacting biomolecules that comprise models often used in systems biology, and both languages can be used for describing more generic dynamical models, including neural models. NeuroML (Goddard et al. 2001; Crook et al. 2007; Gleeson et al. 2010) differs from these languages in that it is a domain-specific model description language, and neuroscience concepts such as cells, ion channels, and synaptic connections are an integral part of the language. The International Neuroinformatics Coordinating Facility aims to facilitate the development of markup language standards for model descriptions in neuroscience, and is providing support for the development of NineML (Network Interchange format for NEuroscience, http://nineml.org), which focuses on descriptions of spiking networks. Additionally, the Simulation Experiment Description Markup Language (SED-ML) (Köhn and Le Novère 2008) is a language for encoding the details of simulation experiments, which follows the requirements defined in the MIASE (Minimal Information about Simulation Experiments) guidelines (http://biomodels.net/miase). These markup languages are complementary and, taken together, they cover the scales for the majority of neuroscience models. The use of namespaces allows for unambiguous mixing of several XML languages; thus, it is possible to use multiple languages for describing different modules of a multiscale model.

Here we provide more details about these languages and their contexts for declarative descriptions of models and simulations. We also provide an introduction to how these markup languages can provide an infrastructure for model sharing, tool development and interoperability, and reproducibility.

SBML and CellML

The main focus of SBML is the encoding of models consisting of biochemical entities, or species, and the reactions among these species that form biochemical networks. In particular, models described in SBML are decomposed into their explicitly labeled constituent elements, where the SBML document resembles a verbose rendition of chemical reaction equations. The representation deliberately avoids providing a set of differential equations or other specific mathematical frameworks for the model, which makes it easier for different software tools to interpret the model and translate the SBML document into the internal representation used by that tool.

In contrast, CellML is built around an approach of constructing systems of equations by linking together the variables in those equations. This equation-based approach is augmented by features for declaring biochemical reactions explicitly, as well as grouping components into modules. The component-based architecture facilitates the reuse of models, parts of models, and their mathematical descriptions.

Note that SBML provides constructs that are more similar to the internal data objects used in many software packages for simulating and analyzing biochemical networks, but SBML and CellML have much in common and represent different approaches for solving the same general problems. Although they were initially developed independently, the developers of the two languages are engaged in exchanges of ideas and are seeking ways of making the languages more interoperable (Finney et al. 2006).

Both of these model description efforts are associated with model repositories that allow authors to share simulator-independent model descriptions. Currently, the BioModels database (Le Novere et al. 2006, http://www.biomodels.net) contains several hundred curated models, and even more non-curated models, that are available as SBML documents as well as in other formats. The CellML Model Repository (Lloyd et al. 2008, http://models.cellml.org) also contains hundreds of models that are available as CellML documents. In addition, both efforts have associated simulation tools and modeling environments, tools that validate XML documents for models against the language specifications, and translation utilities that are described in detail on their web-sites.

NeuroML

The declarative approach of the NeuroML standards project focuses on the key objects that need to be exchanged among existing applications with some anticipation of the future needs of the community. These objects include descriptions of neuronal morphologies, voltage-gated ion channels, synaptic mechanisms, and network structure. The descriptions are arranged into levels that are related to different biological scales, with higher levels adding extra concepts. This modular, object-oriented structure makes it easier to add additional concepts and reuse parts of models. As models of single neurons are at the core of most of the systems being described, neuroanatomical information about the structure of individual cells forms the core of Level 1, which also includes the specification for metadata. The focus of Level 2 is the electrical properties of these neurons which allows for descriptions of cell models with realistic channel and synaptic mechanisms distributed on their membranes. Level 3 describes networks of these cells in three dimensions including cell locations and synaptic connectivity. Networks can be described with an explicit list of instances of cell positions and connections, or with an algorithmic template for describing how the instances should be generated.

While there is overlap in the types of models that NeuroML and SBML/CellML can describe, such as a single compartment conductance-based model, NeuroML provides a concise format for neuronal model elements that can be readily understood by software applications that use the same concepts. NeuroML version 2, which is under development, will have greater interaction with SBML and CellML, with SBML being an initial focus of the work. This will allow, for example, complex signaling pathways to be expressed in one of these formats with the rest of the cell and network model specified in NeuroML. Since NeuroML is completely compatible with the structure of the user layer of NineML (see below), the goal is to be able to represent a multiscale neuroscience model that includes processes from the molecular to the network levels with a combination of SBML, NeuroML, and NineML.

The NeuroML project (http://neuroml.org) provides a validator for NeuroML documents. The project also provides XSL files for mapping NeuroML documents to a HTML format that provides detailed user-friendly documentation of the model details and for mapping NeuroML documents to a number of simulator scripting formats, including NEURON (Hines 1989; Hines and Carnevale 1997; Carnevale and Hines 2006), GENESIS (Bower and Beeman 1997), PSICS (http://www.psics.org), and additional simulators through PyNN (Davison et al. 2009). This approach has the advantage that in the short term, applications need not be extended to natively support NeuroML, but can still have access to NeuroML models.

NineML

With an increasing number of studies related to large-scale neuronal network modeling, there is a need for a common standardized description language for spiking network models. The Network Interchange for Neuroscience Modeling Language (NineML) is designed to describe large networks of spiking neurons using a layered approach. The abstraction layer provides the core concepts, mathematics, and syntax for explicitly describing model variables and state update rules, and the user layer provides a syntax to specify the instantiation and parameterization of the network model in biological terms. In particular, the abstraction layer is built around a block diagram notation for continuous and discrete variables, their evolution according to a set of rules such as a system of ordinary differential equations, and the conditions that induce a regime change, such as a transition from subthreshold mode to spiking and refractory modes. In addition, the abstraction layer provides the notation for describing a variety of topographical arrangements of neurons and populations of neurons (Raikov and The INCF Multiscale Modeling Taskforce 2010). In contrast, the user layer provides the syntax for specifying the model and the parameters for instantiating the network, which includes descriptions of individual elements such as cells, synapses, and synaptic inputs, as well as the constructs for describing the grouping of these entities into networks (Gorchetchnikov and The INCF Multiscale Modeling Taskforce 2010). Like NeuroML, the user layer of NineML defines the syntax for specifying a large range of connectivity patterns. One goal of NineML is to be self-consistent and flexible, allowing addition of new models and mathematical descriptions without modification of the previous structure and organization of the language. To achieve this, the language is being iteratively designed using several representative models with various levels of complexity as test cases (Raikov and The INCF Multiscale Modeling Taskforce 2010).

SED-ML

A simulation experiment description using SED-ML (Köhn and Le Novère 2008) is independent of the model encoding that is used in the computational experiment; the model itself is only referenced using an unambiguous identifier. Then SED-ML is used to describe the algorithm used for the execution of the simulation and the simulation settings such as step size and duration. SED-ML also can be used to describe changes to the model, such as changes to the value of an observable or general changes to any XML element of the model representation. The simulation result sometimes does not correspond to the desired output of the simulation. For this reason, SED-ML also includes descriptions of postsimulation processing that should be applied to the simulation result before output such as normalization, mean-value calculations, or any other new expression that can be specified using MathML (Miner 2005). SED-ML further allows for descriptions of the form of the output such as a 2D-plot, 3D-plot, or data table. So far, SED-ML has been used in conjunction with several different model description languages including SBML, CellML, and NeuroML, and the BioModels database supports the sharing of simulation descriptions in SED-ML.

Other Tools Based on Declarative Descriptions

neuroConstruct is an example of a successful software application that uses declarative descriptions to its advantage (Gleeson et al. 2007). This software facilitates the creation, visualization, and analysis of networks of multicompartmental neurons in 3D space, where a graphical user interface allows model generation and modification without programming. Models within neuroConstruct are based on the simulator-independent NeuroML standards, allowing automatic generation of code for multiple simulators. This has facilitated the testing of neuroConstruct and the verification of its simulator independence, through a process where published models were re-implemented using neuroConstruct and run on multiple simulators as described in section "Ambiguous Model Numerics" (Gleeson et al. 2010).

The ConnPlotter package (Nordlie and Plesser 2010) allows modelers to visualize connectivity patterns in large networks in a compact fashion. It thus aids in communicating model structures, but is also a useful debugging tool. Unfortunately, it is at present tightly bound to the NEST Topology Library (Plesser and Austvoll 2009).

Improving Research Reporting

The past 2 decades have brought a significant growth in the number of specialized journals and conferences, sustaining an ever growing volume of scientific communication. Search engines such as Google Scholar (http://scholar.google.com) and

Thomson Reuters Web of Knowledge (http://wokinfo.com) have revolutionized literature search, while the internet has accelerated the access to even arcane publications from weeks to seconds. Electronic publication permits authors to complement terse papers with comprehensive supplementary material, and some journals even encourage authors to post video clips in which they walk their audience through the key points of the paper.

But have these developments improved the communication of scientific ideas between researchers? Recently, Nordlie et al. (2009) surveyed neuronal network model descriptions in the literature and concluded that current practice in this area is diverse and inadequate. Many computational neuroscientists have experienced difficulties in reproducing results from the literature due to insufficient model descriptions. Donoho et al. (2009) propose as a cure that all scientists in a field should use the same software, where the software is carefully crafted to cover the complete modeling process from simulation to publication figure. While this approach successfully addresses software quality and replicability issues, it falls short of contributing to the independent reproduction of results, which by definition requires re-implementation of a model based on the underlying concepts, preferably using a different simulator.

To facilitate independent reproduction of neural modeling studies, a systematic approach is needed for reporting models, akin to the ARRIVE Guidelines for Reporting Animal Research (Kilkenny et al. 2010). Such guidelines can serve as checklists for authors as well as for referees during manuscript review. For neuronal network models, Nordlie et al. (2009) have proposed a good model description practice, recommending that publications on computational modeling studies should provide:

- Hypothesis: a concrete description of the question or problem that the model addresses.
- Model derivation: a presentation of experimental data that support the hypothesis, model, or both.
- Model description: a description of the model, its inputs (stimuli) and its outputs (measured quantities), and all free parameters.
- Implementation: a concise description of the methods used to implement and simulate the model (e.g., details of spike threshold detection, assignment of spike times, time resolution), as well as a description of all third party tools used, such as simulation software or mathematical packages.
- Model analysis: a description of all analytical and numerical experiments performed on the model, and the results obtained.
- Model justification: a presentation of all empirical or theoretical results from the literature that support the results obtained from the model and that were not used to derive the model.

Nordlie et al. also provide a checklist for model descriptions, requiring information on the following aspects of a model: (1) model composition, (2) coordinate systems and topology, (3) connectivity, (4) neurons, synapses, and channels, (5) model input, output, and free parameters, (6) model validation, and (7) model

A	Model Summary
Populations	Three: excitatory, inhibitory, external input
Topology	—
Connectivity	Random convergent connections
Neuron model	Leaky integrate-and-fire, fixed voltage threshold, fixed absolute refractory time (voltage clamp)
Channel models	—
Synapse model	α-shaped current inputs
Plasticity	—
Input	Independent fixed-rate Poisson spike trains to all neurons (during initial stimulation period)
Measurements	Spike activity

B		Populations
Name	**Elements**	**Size**
E	Iaf neuron	$N_E = 4N_I$
I	Iaf neuron	N_I
E_{ext}	Poisson generator	$C_E(N_E + N_I)$

C			Connectivity
Name	**Source**	**Target**	**Pattern**
EE	E	E	Random convergent $C_E \to 1$, weight J, delay D
IE	E	I	Random convergent $C_E \to 1$, weight J, delay D
EI	I	E	Random convergent $C_I \to 1$, weight $-gJ$, delay D
II	I	I	Random convergent $C_I \to 1$, weight $-gJ$, delay D
Ext	E_{ext}	$E \cup I$	Non-overlapping $C_E \to 1$, weight J, delay D

D	Neuron and Synapse Model
Name	Iaf neuron
Type	Leaky integrate-and-fire, α-current input
Subthreshold dynamics	$\tau\dot{V}(t) = -V(t) + RI(t)$ if $t > t^* + \tau_{rp}$ $V(t) = V_r$ else $I(t) = \frac{\tau}{R} \sum_{\tilde{t}} w\alpha(t - (\tilde{t} + \Delta))\Theta(t - (\tilde{t} + \Delta))$
Spiking	If $V(t-) < \theta \wedge V(t+) \geq \theta$ 1. set $t^* = t$ 2. emit spike with time-stamp t^*

E	Input
Type	**Description**
Poisson generators	Fixed rate ν_{ext}, C_E generators per neuron, each generator projects to one neuron; active only during initial stimulation period

F Measurements
Spike activity as raster plots for subset of excitatory neurons

Fig. 4.4 Concise tabular presentation of the network model introduced in section "Ambiguous Model Numerics" with spike trains shown in Fig. 4.2, using the template proposed by Nordlie et al. (2009)

implementation. They further propose a concise tabular format for summarizing network models in publications; Fig. 4.4 provides an example. These guidelines and tables present information about a model that referees can check for completeness and consistency, and also allow referees to judge whether the employed simulation methods appear adequate for the model, as well as the plausibility of the results obtained.

Publication standards such as those discussed in Nordlie et al. ensure that all possible, relevant model details are provided. However, it is also important to note what such guidelines and tables cannot provide: sufficient detail to allow exact replication of the simulation and of the figures presented in the publication for the reasons discussed in the sections above. Since it is futile to strive for details that would lead to exact replication of scientific publications, and since referees cannot confirm correctness without replicating all of the work in a manuscript, some have advocated that authors should focus only on main concepts that can be evaluated by referees, neglecting model details. Note that the Journal of Neuroscience recently ceased to publish supplementary material for the precise reason that it provides unreviewed (and essentially unreviewable) detail (Maunsell 2010).

Where do all of these issues leave reproducibility? It seems that improved reproducibility requires two important measures for reporting results. First, on the technical side, replicability of simulation studies should be ensured by requiring that authors use proper code sharing techniques and automated practices for recording provenance as described in section "Practical Approaches to Replicability." If code is not reviewed, then this is best done through a curated database rather than in the supplementary material. This model deposition provides a reference for readers if they encounter difficulties in reproducing the results of a publication. Note that the increasing use of a limited set of simulator software packages (Brette et al. 2007) facilitates this type of model archeology due to the widespread expertise with these packages in the computational neuroscience community—no need to decipher Fortran code left behind by the Ph.D.-student of yesteryear.

Second, publications in computational neuroscience should provide much more information about why a particular model formulation was chosen and how model parameters were selected. Neural models commonly require significant parameter tuning to demonstrate robust, stable, and interesting dynamics. In some cases, the selection of models for synapses and excitable membranes may be shaped by neurophysiological evidence, while in others, they are selected based on ease of implementation or mathematical analysis. Such choices should be clearly articulated, and much would be gained by details about (1) which aspects of the model proved essential to obtaining "good" simulation results and (2) which quantities and properties constrained parameter tuning. In discussing these aspects, it might be helpful to consider recent advances in the theory of science with regard to both the role of simulations in the scientific process (Humphreys 2004) and the role of explanatory models that are common in neuroscience (Craver 2007).

Discussion

In this chapter we have exposed the distinction between replicability and reproducibility, and explored the continuum between the two. We also have seen that there are scientific benefits to promoting each point on the continuum. Replication based on reuse of code enables more rapid progress of computational studies by promoting modularization, improved code quality, and incremental development, while independent reproduction of an important result (whether manually, through reading an article's Methods section, or (semi-)automatically, using a declarative, machine-readable version of the methods) remains as the gold standard and foundation of science.

In our terminology, replication of results implies reusing the original code in some way, either by rerunning it directly or by studying it when developing a new implementation of a model. If this replication is done by someone other than the original developers, the code must therefore be shared with others. This raises many issues including licensing, version tracking, discoverability, documentation, software/hardware configuration tracking, and obsolescence.

Independent reproduction of a computational experiment without using the original code is necessary to confirm that a model's results are general and do not depend on a particular implementation. Traditionally, this has been done by reading the published article(s) describing the models, and most often corresponding with the authors to clarify incomplete descriptions. In this chapter we have discussed in some depth the difficulties usually encountered in this process, and ways in which published model descriptions can be improved. We summarize these recommendations below. More recently, several efforts have been made to produce structured, declarative, and machine-readable model descriptions, mostly based on XML. Such structured descriptions allow the completeness and internal consistency of a description to be verified, and allow for automated reproduction of simulation experiments in different simulation environments.

In considering how to improve the reproducibility of computational neuroscience experiments, it is important to be aware of the limits of reproducibility, due to component failure, environmental influences on hardware, floating point numerics, and the amplification of small errors by sensitive model systems. The question of how best to determine whether differences between two simulations are due to unavoidable computational effects or whether they reflect either errors in the code or important algorithmic differences has not been satisfactorily answered in computational neuroscience.

Based on the issues identified and discussed in this chapter, we propose a number of steps that can be taken to improve replication and reproduction of computational neuroscience experiments:

Modelers

- Use version control tools.
- Keep very careful records of the computational environment, including details of hardware, operating system versions, versions of key tools, and software libraries. Use automated tools where available.
- Use best practices in model publications; see Nordlie et al. (2009).
- Plan to release your code from the beginning of development to aid in code sharing.
- Make code available through ModelDB, BioModels, or other appropriate curated databases.

- Make models available using simulator-independent model descriptions if possible.
- Evaluate your career by downloads and users in addition to citations.

Tool developers

- Incorporate version control tools and tools for automated environment tracking into your software.
- Collaborate with model description language efforts.

Reviewers and editors

- Demand clear model descriptions following Nordlie et al. (2009).
- Demand and verify code and/or model availability.
- Make sure publications include details of model choices and behavior.

Computational neuroscience has made enormous progress in the past 20 years. Can the same be said for the reproducibility of our models and our results? The range and quality of the available tools for ensuring replicability and reproducibility has certainly improved, from better version control systems to structured, declarative model description languages and model databases. At the same time, the typical complexity of our models has also increased, as our experimental colleagues reveal more and more biological detail and Moore's Law continues to put more and more computing power in our laboratories. It is this complexity which is perhaps the major barrier to reproducibility. As the importance of computational science in scientific discovery and public policy continues to grow, demonstrable reproducibility will become increasingly important. Therefore, it is critical to continue the development of tools and best practices for managing model complexity and facilitating reproducibility and replicability. We must also attempt to change the culture of our computational community so that more researchers consider whether their reported results can be reproduced and understand what tools are available to aid in reproducibility. These changes are needed so that reproducibility can be front and center in the thinking of modelers, reviewers, and editors throughout the computational neuroscience community.

Acknowledgments S.C. acknowledges support from the National Institute of Mental Health under grant R01MH061905. A.P.D. acknowledges support from the Centre National de la Recherche Scientifique and from the European Union under grant FP7-ICT-FET-269921 (BrainScaleS). H.E.P. acknowledges support from the Research Council of Norway under grant 178892/V30 eNeuro.

References

Bower JM, Beeman D (1997) The book of GENESIS: exploring realistic neural models with the GEneral NEural SImulation System. Springer, New York

Bhalla U, Bilitch DH, Bower J (1992) Rallpacks: a set of benchmarks for neural simulators. Trends Neurosci 15:453–458

Bray T, Paoli J, Sperberg-McQueen C (1998) Extensible markup language (XML) 1.0. http://www.w3.org/TR/REC-xml

Breslow NE (2010) Commentary. Biostatistics 11(3):379–380. doi:10.1093/biostatistics/kxq025

Brette R (2006) Exact simulation of integrate-and-fire models with synaptic conductances. Neural Comput 18:2004–2027

Brette R (2007) Exact simulation of integrate-and-fire models with exponential currents. Neural Comput 19:2604–2609

Brette R, Rudolph M, Carnevale T, Hines M, Beeman D, Bower JM, Diesmann M, Morrison A, Goodman PH, Harris FC Jr, Zirpe M, Natschläger T, Pecevski D, Ermentrout B, Djurfeldt M, Lansner A, Rochel O, Vieville T, Muller E, Davison AP, Boustani SE, Destexhe A (2007) Simulation of networks of spiking neurons: a review of tools and strategies. J Comput Neurosci 23:349–398

Brüderle D, Muller E, Davison A, Muller E, Schemmel J, Meier K (2009) Establishing a novel modeling tool: a Python-based interface for a neuromorphic hardware system. Front Neuroinform 3:17. doi:10.3389/neuro.11.017.2009

Brunel N (2000) Dynamics of sparsely connected networks of excitatory and inhibitory spiking neurons. J Comput Neurosci 8:183–208

Cannon R, Howell F, Goddard N, De Schutter E (2002) Non-curated distributed databases for experimental data and models in neuroscience. Network 13:415–428

Cannon RC, Gewaltig MO, Gleeson P, Bhalla US, Cornelis H, Hines ML, Howell FW, Muller E, Stiles JR, Wils S, Schutter ED (2007) Interoperability of neuroscience modeling software: current status and future directions. Neuroinformatics 5:127–138

Carnevale N, Hines M (2006) The NEURON book. Cambridge University Press, Cambridge, UK

Claerbout JF, Karrenbach M (1992) Electronic documents give reproducible research a new meaning. In: SEG expanded abstracts, Society of Exploration Geophysicists, vol 11, pp 601–604

Cox DR, Donnelly C (2010) Commentary. Biostatistics 11(3):381–382. doi:10.1093/biostatistics/kxq026

Craver CF (2007) Explaining the brain: mechanisms and the Mosaic Unity of Neuroscience. Oxford University Press, New York

Crook S, Howell F (2007) XML for data representation and model specification. In: Crasto C (ed) Methods in molecular biology book series: neuroinformatics. Humana Press, Totowa, NJ, pp 53–66

Crook S, Gleeson P, Howell F, Svitak J, Silver R (2007) MorphML: Level 1 of the NeuroML standards for neuronal morphology data and model specification. Neuroinformatics 5:96–104

D'Haene M (2010) Efficient simulation strategies for spiking neural networks. PhD thesis, Universiteit Gent

Davison A, Morse T, Migliore M, Marenco L, Shepherd G, Hines M (2002) ModelDB: a resource for neuronal and network modeling. In: Kötter R (ed) Neuroscience databases: a practical guide. Kluwer Academic, Norwell, MA, pp 99–122

Davison A, Brüderle D, Eppler J, Kremkow J, Muller E, Pecevski D, Perrinet L, Yger P (2009) PyNN: a common interface for neuronal network simulators. Front Neuroinform 2:11. doi:10.3389/neuro.11.011.2008

de Leeuw J (2001) Reproducible research: the bottom line. Tech. Rep., Department of Statistics, UCLA, UC Los Angeles. http://escholarship.org/uc/item/9050x4r4

DeAngelis CD, Fontanarosa PB (2010) The importance of independent academic statistical analysis. Biostatistics 11(3):383–384. doi:10.1093/biostatistics/kxq027

Djurfeldt M, Lansner A (2007) Workshop report: 1st INCF Workshop on Large-scale Modeling of the Nervous System. Available from Nature Precedings http://dx.doi.org/10.1038/npre.2007.262.1

Donoho DL (2010) An invitation to reproducible computational research. Biostatistics 11(3):385–388. doi:10.1093/biostatistics/kxq028

Donoho DL, Maleki A, Rahman IU, Shahram M, Stodden V (2009) 15 years of reproducible research in computational harmonic analysis. Comput Sci Eng 11:8–18. doi:10.1109/MCSE.2009.15

Drummond C (2009) Replicability is not reproducibility: nor is it good science. In: Proceedings of the evaluation methods for machine learning workshop at the 26th ICML, Montreal, CA
Eppler JM, Helias M, Muller E, Diesmann M, Gewaltig MO (2008) PyNEST: a convenient interface to the NEST simulator. Front Neuroinform 2:12. doi:10.3389/neuro.11.012.2008
Ferrenberg AM, Landau DP, Wong YJ (1992) Monte Carlo simulations: hidden errors from "good" random number generators. Phys Rev Lett 69(23):3382–3384. doi:10.1103/PhysRevLett.69.3382
Finney A, Hucka M, Bornstein B, Keating S, Shapiro B, Matthews J, Kovitz B, Schilstra M, Funahashi A, Doyle J, Kitano H (2006) Software infrastructure for effective communication and reuse of computational models. In: Szallasi Z, Stelling J, Periwal V (eds) Systems modeling in cell biology: from concepts to nuts and bolts. MIT Press, Boston, pp 369–378
Funk W, Dammann V, Donnevert G (2006) Quality assurance in analytical chemistry, 2nd edn. Wiley-VCH, Weinheim
Galluppi F, Rast A, Davies S, Furber S (2010) A general-purpose model translation system for a universal neural chip. In: Wong K, Mendis B, Bouzerdoum A (eds) Neural information processing. Theory and algorithms, Lecture notes in computer science, vol 6443. Springer, Berlin, pp 58–65
Gardner D, Knuth KH, Abato M, Erde SM, White T, DeBellis R, Gardner EP (2001) Common data model for neuroscience data and data model exchange. J Am Med Inform Assoc 8:17–33
Gardner D, Akil H, Ascoli G, Bowden D, Bug W, Donohue D, Goldberg D, Grafstein B, Grethe J, Gupta A, Halavi M, Kennedy D, Marenco L, Martone M, Miller P, Müller H, Robert A, Shepherd G, Sternberg P, Van Essen D, Williams R (2008) The neuroscience information framework: a data and knowledge environment for neuroscience. Neuroinformatics 6(3):149–160
Gewaltig MO, Diesmann M (2007) NEST (NEural Simulation Tool). Scholarpedia 2(4):1430
Gewaltig MO, Koerner EE (2008) Self-sustained dynamics of sparsely connected networks without external noise. In: 2008 Neuroscience meeting planner, Society for Neuroscience, Washington, DC, program No. 220.1
Gleeson P, Steuber V, Silver RA (2007) neuroConstruct: a tool for modeling networks in 3D space. Neuron 54:219–235
Gleeson P, Crook S, Cannon RC, Hines ML, Billings GO, Farinella M, Morse TM, Davison AP, Ray S, Bhalla US, Barnes SR, Dimitrova YD, Silver RA (2010) NeuroML: a language for describing data driven models of neurons and networks with a high degree of biological detail. PLoS Comput Biol 6(6):e1000815. doi:10.1371/journal.pcbi.1000815
Goddard N, Hucka M, Howell F, Cornelis H, Shankar K, Beeman D (2001) NeuroML: model description methods for collaborative modelling in neuroscience. Philos Trans R Soc Lond B Biol Sci 356:1209–1228
Goodman D (2010a) Code generation: a strategy for neural network simulators. Neuroinformatics 8:183–196. doi:10.1007/s12021-010-9082-x
Goodman SN (2010b) Commentary. Biostatistics 11(3):389–390. doi:10.1093/biostatistics/kxq030
Goodman D, Brette R (2008) Brian: a simulator for spiking neural networks in Python. Front Neuroinform 2:5. doi:10.3389/neuro.11.005.2008
Gorchetchnikov A, The INCF Multiscale Modeling Taskforce (2010) NineML—a description language for spiking neuron network modeling: the user layer. BMC Neurosci 11(suppl 1):P71
Grebogi C, Hammel SM, Yorke JA, Sauer T (1990) Shadowing of physical trajectories in chaotic dynamics: containment and refinement. Phys Rev Lett 65(13):1527–1530. doi:10.1103/PhysRevLett.65.1527
Groves T (2010) The wider concept of data sharing: view from the BMJ. Biostatistics 11(3):391–392. doi:10.1093/biostatistics/kxq031
Gruntz D, Waldvogel J (2004) Orbits in the planar three-body problem. In: Gander W, Hřebíček J (eds) Solving problems in scientific computing using Maple and MATLAB, 4th edn. Springer, Berlin, pp 51–72

Hansel D, Mato G, Meunier C, Neltner L (1998) On numerical simulations of integrate-and-fire neural networks. Neural Comput 10:467–483
Hanuschkin A, Kunkel S, Helias M, Morrison A, Diesmann M (2010) A general and efficient method for incorporating exact spike times in globally time-driven simulations. Front Neuroinform 4:113. doi:10.3389/fninf.2010.00113
Hedley W, Nelson M, Nielsen P, Bullivant D, Hunter P (2000) XML languages for describing biological models. In: Proceedings of the Physiological Society of New Zealand, vol 19
Heijmen T (2011) Soft errors from space to ground: historical overview, empirical evidence, and future trends (chap 1). In: Nicolaidis M (ed) Soft errors in modern electronic systems, Frontiers in electronic testing, vol 41. Springer, New York, pp 1–25
Hines M (1989) A program for simulation of nerve equations with branching geometries. Int J Biomed Comput 24:55–68
Hines ML, Carnevale NT (1997) The NEURON simulation environment. Neural Comput 9(6):1179–1209
Hines ML, Carnevale NT (2000) Expanding NEURON's repertoire of mechanisms with NMODL. Neural Comput 12:995–1007
Hines ML, Morse T, Migliore M, Carnevale NT, Shepherd GM (2004) ModelDB: a database to support computational neuroscience. J Comput Neurosci 17(1):7–11. doi:10.1023/B:JCNS.0000023869.22017.2e
Hucka M, Finney A, Sauro H, Bolouri H, Doyle J, Kitano H, Arkin A (2003) The systems biology markup language (SBML): a medium for representation and exchange of biochemical network models. Bioinformatics 19:524–531
Humphreys P (2004) Extending ourselves: computational science, empiricism, and scientific method. Oxford University Press, Oxford
Hund F (1996) Geschichte der physikalischen Begriffe. Spektrum Akademischer Verlag, Heidelberg
Ioannidis JPA (2005) Why most published research findings are false. PLoS Med 2:e124. doi:10.1371/journal.pmed.0020124
Jones E, Oliphant T, Peterson P et al (2001) SciPy: open source scientific tools for Python. http://www.scipy.org/
Keiding N (2010a) Reproducible research and the substantive context. Biostatistics 11(3):376–378. doi:10.1093/biostatistics/kxq033
Keiding N (2010b) Reproducible research and the substantive context: response to comments. Biostatistics 11(3):395–396. doi:10.1093/biostatistics/kxq034
Kilkenny C, Browne WJ, Cuthill IC, Emerson M, Altman DG (2010) Improving bioscience research reporting: the ARRIVE guidelines for reporting animal research. PLoS Biol 8(6):e1000412. doi:10.1371/journal.pbio.1000412
Köhn D, Le Novère N (2008) SED-ML—an XML format for the implementation of the MIASE guidelines. In: Heiner M, Uhrmacher A (eds) Computational methods in systems biology, Lecture notes in computer science, vol 5307. Springer, Berlin, pp 176–190
Küppers G, Lenhard J (2005) Validation of simulation: patterns in the social and natural sciences. J Artif Soc Soc Simul 8(4):3
Lapicque L (1907) Recherches quantitatives sur l'excitation électrique des nerfs traitée comme une polarization. J Physiol Pathol Gen 9:620–635
Le Novere N, Bornstein B, Broicher A, Courtot M, Donizelli M, Dharuri H, Li L, Sauro H, Schilstra M, Shapiro B, Snoep J, Hucka M (2006) BioModels database: a free, centralized database of curated, published, quantitative kinetic models of biochemical and cellular systems. Nucleic Acids Res 34(Database issue):D689–D691
Lehrer J (2010) The truth wears off. The New Yorker, 13 Dec 2010:52
Lloyd C, Halstead M, Nielsen P (2004) CellML: its future, present and past. Prog Biophys Mol Biol 85:433–450
Lloyd CM, Lawson JR, Hunter PJ, Nielsen PF (2008) The CellML model repository. Bioinformatics 24(18):2122–2123. doi:10.1093/bioinformatics/btn390

Ludäscher B, Altintas I, Berkley C, Higgins D, Jaeger-Frank E, Jones M, Lee E, Tao J, Zhao Y (2006) Scientific workflow management and the Kepler system. Concurrency Comput Pract Exp 18(10):1039–1065

Marenco L, Wang R, Shepherd G, Miller P (2010) The NIF DISCO framework: facilitating automated integration of neuroscience content on the web. Neuroinformatics 8(2):101–112

Maunsell J (2010) Announcement regarding supplemental material. J Neurosci 30:10599–10600

Mesirov JP (2010) Accessible reproducible research. Science 327(5964):415–416. doi:10.1126/science.1179653

Migliore M, Morse TM, Davison AP, Marenco L, Shepherd GM, Hines ML (2003) ModelDB: making models publicly accessible to support computational neuroscience. Neuroinformatics 1(1):135–139

Miner R (2005) The importance of MathML to mathematical communication. Notices AMS 52:532–538

Morrison A, Straube S, Plesser HE, Diesmann M (2007) Exact subthreshold integration with continuous spike times in discrete time neural network simulations. Neural Comput 19:47–79

Morse TM (2007) Model sharing in computational neuroscience. Scholarpedia 2(4):3036

Nordlie E, Plesser HE (2010) Visualizing neuronal network connectivity with connectivity pattern tables. Front Neuroinform 3:39. doi:10.3389/neuro.11.039.2009

Nordlie E, Gewaltig MO, Plesser HE (2009) Towards reproducible descriptions of neuronal network models. PLoS Comput Biol 5(8):e1000456. doi:10.1371/journal.pcbi.1000456

Oinn T, Greenwood M, Addis M, Alpdemir M, Ferris J, Glover K, Goble C, Goderis A, Hull D, Marvin D, Li P, Lord P, Pocock MR, Senger M, Stevens R, Wipat A, Wroe C (2006) Taverna: lessons in creating a workflow environment for the life sciences. Concurr Comput Pract Exp 18(10):1067–1100. doi:10.1002/cpe.993

Pecevski D, Natschläger T, Schuch K (2009) PCSIM: a parallel simulation environment for neural circuits fully integrated with Python. Front Neuroinform 3:11. doi:10.3389/neuro.11.011.2009

Peng RD (2009) Reproducible research and biostatistics. Biostatistics 10(3):405–408. doi:10.1093/biostatistics/kxp014

Peng RD (2010) Discussion of Keiding. Biostatistics 11(3):393–394. doi:10.1093/biostatistics/kxq032

Peterson B, Healy M, Nadkarni P, Miller P, Shepherd G (1996) ModelDB: an environment for running and storing computational models and their results applied to neuroscience. J Am Med Inform Assoc 3:389–398

Petzold L (1983) Automatic selection of methods for solving stiff and nonstiff systems of ordinary differential equations. SIAM J Sci Comput 4:136–148. doi:10.1137/0904010

Plesser HE, Austvoll K (2009) Specification and generation of structured neuronal network models with the NEST topology module. BMC Neurosci 10(suppl 1):P56. doi:10.1186/1471-2202-10-S1-P56

Quirk JJ (2005) Computational science: "same old silence, same old mistakes; something more is needed". In: Plewa T, Linde T, Weirs VG (eds) Adaptive mesh refinement—theory and applications. Springer, Berlin, pp 3–28

Raikov I, The INCF Multiscale Modeling Taskforce (2010) NineML—a description language for spiking neuron network modeling: the abstraction layer. BMC Neurosci 11(suppl 1):P66

Rotter S, Diesmann M (1999) Exact digital simulation of time-invariant linear systems with applications to neuronal modeling. Biol Cybern 81:381–402

Schrödinger E (1915) Zur Theorie der Fall-und Steigversuche an Teilchen mit Brownscher Bewegung. Phys Zeit 16:289–295

Shelley MJ, Tao L (2001) Efficient and accurate time-stepping schemes for integrate-and-fire neuronal networks. J Comput Neurosci 11:111–119

Silva CT, Freire J, Callahan S (2007) Provenance for visualizations: reproducibility and beyond. Comput Sci Eng 9(5):82–90

Stodden V (2009a) Enabling reproducible research: open licensing for scientific innovation. Int J Commun Law Policy 13:2–25, http://www.ijclp.net/files/ijclp_web-doc_1-13-2009.pdf

Stodden V (2009b) The legal framework for reproducible research in the sciences: licensing and copyright. IEEE Comput Sci Eng 11(1):35–40

Usui S (2003) Visiome: neuroinformatics research in vision project. Neural Netw 16:1293–1300

Vandewalle P, Kovačević J, Vetterli M (2009) Reproducible research in signal processing. IEEE Signal Proc Mag 26(3):37–47. doi:10.1109/MSP.2009.932122

Wilson MA, Bhalla US, Uhley JD, Bower JM (1989) GENESIS: a system for simulating neural networks. In: Touretzky D (ed) Advances in neural information processing systems. Morgan Kaufmann, San Mateo, CA, pp 485–492

Chapter 5
The Emergence of Community Models in Computational Neuroscience: The 40-Year History of the Cerebellar Purkinje Cell

James M. Bower

Abstract The previous chapter in this volume considers the 20-year development of technology supporting the reuse and reproducibility of computational models. This chapter considers the specific case of the 40-year history of modeling cerebellar Purkinje cells, resulting in the emergence of one of the first "community models" in computational neuroscience. The chapter traces the model-based progress in understanding the relationship between Purkinje cell structure and function, as well as the implications of those results for our understanding of the function of this cell and the cerebellum in general. Using the history of Purkinje cell modeling as an example, the chapter also identifies the importance of the development of community models as a base for the eventual establishment of a quantitative understructure for neuroscience as a whole.

Introduction

While all chapters in this book describe progress made in our computational understanding of specific neural systems over the last 20 years, this chapter, in addition, considers how the use of models can potentially change the structure and organization of neuroscience itself. Specifically, this chapter reviews the now 40+-year effort to realistically model the cerebellar Purkinje cell, focused on the evolution of a particular model first described by Rapp et al. (1992, 1994) and then extended by De Schutter and Bower (1994a, b, c). To quote from a recent review article on data sharing in neuroscience, this model "remain(s) among the most successful, cited, and re-used/updated in computational neuroscience" (Ascoli 2007, p. 156).

J.M. Bower, Ph.D. (✉)
Health Sciences Center, University of Texas, San Antonio, TX, USA
e-mail: bower@uthscsa.edu

J.M. Bower (ed.), *20 Years of Computational Neuroscience*, Springer Series in Computational Neuroscience 9, DOI 10.1007/978-1-4614-1424-7_5,

In fact, it is the overall thesis of this chapter that the Rapp et al., De Schutter/Bower model (R-DB Model) has become one of the first "community models" in neuroscience. While there are a few models of biophysical processes that have transcended individual research laboratories (e.g., Hodgkin and Huxley 1952), most models of neurobiological structures whether at the synaptic, neuronal, network, or systems levels have not migrated out of their laboratories of origin with many resulting in only a single publication. As a result, computational neuroscience is full of separate models of the same structures, many addressing the same issues, but with no clear articulation or connection between them (Manninen et al. 2010).

While isolated models can be valuable for individual researchers to explore function, in principle, community use of models can provide a mechanism supporting scientific communication, coordination, and collaboration, coming to represent the current state of our understanding of the relationship between biological structure and function. Models shared in this way can also reflect the continued evolution of the views of a field as a whole, providing a mechanism to organize and evaluate debates and opinions. While still rare, the development of such models is likely to be essential to construct a formal, quantitative underpinning for neuroscience which has been essential in other scientific fields to make measurable scientific progress (Kuhn 1962).

With these ideas in mind, this chapter begins by considering the first efforts to build models of cerebellar Purkinje cells in the 1960s and 1970s. The chapter then describes the emergence of the R-DB Model as a community model, subsequently considering how the model has changed our views of cerebellar organization and function. Finally, the chapter briefly discusses technical and structural issues related to the emergence and further development of community models. It is my hope that the story told here will encourage other modelers to adopt and support community models of their own systems.

Early Stages in Modeling the Cerebellar Purkinje Cell

The cerebellum was one of the first brain structures in which computer models were built reflecting the actual morphological structure of its networks (Bower 2012). This is in part because the overall structural organization of cortical circuitry as well as its major afferent and efferent connections has been known since the turn of the twentieth century (Cajal 1911), but is also due to the fact that the regular structure of this network allowed for a relatively early detailed analysis of its physiological properties and cellular relationships (Eccles et al. 1967). Those relationships are shown in Fig. 5.1.

The historically close link between cerebellar modeling and specific structure/function relationships is actually reflected in the first published discussions of Purkinje cell modeling, which arose in the context of a controversy involving the existence of active conductances in the Purkinje cell dendrite (Calvin and Hellerstein 1969). Specifically, Llinas et al. had proposed that the Purkinje cell dendrite was

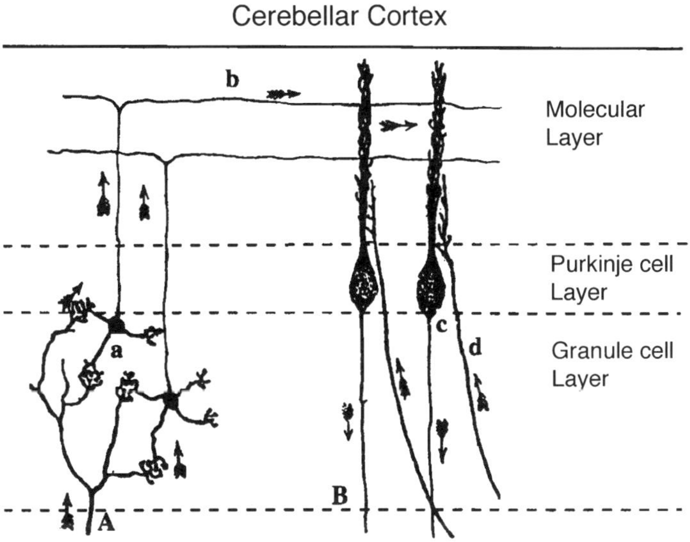

Fig. 5.1 Connections of the cerebellar cortex as drawn by Ramon y Cajal (1911). Following the *arrows* on the left side of the diagram, mossy fiber inputs (A) make excitatory connections with granule cells (a), whose axons ascend through the molecular layer, bifurcate and then all course as parallel fibers (b) through the isoplanar dendritic tree of the Purkinje cell (c). The Purkinje cell axon (B) then provides the sole output of the cerebellar cortex. Also shown in this diagram is the other major afferent input to cerebellar cortex, the climbing fiber projection (d) arising in the inferior olive. In the adult each Purkinje cell receives only one climbing fiber projection but tens of thousands of granule cell inputs. Figure used with permission from Ito (2001) modified from Cajal (1911)

capable of "active" current conduction due to differences in the latencies of field potentials recorded at different depths in the molecular layer following direct cortical stimulation (Llinas et al. 1968). This interpretation was challenged by Calvin and Hellerstein (1969), who pointed out that the recent cable theory models developed by Rall (1964) predicted similar delays from passive dendritic current conduction alone. In response, Llinas et al. asserted that models based on volume conductors rather than cables were more appropriate for the analysis of extracellular field potentials (Calvin and Hellerstein 1969). A few months later, Zucker entered the debate by actually performing calculations comparing both types of models, concluding that neither approach in its classical form could resolve the issue (Zucker 1969). However, Zucker pointed out that similarities in simulated field potential results recently obtained from cable theory models for olfactory bulb mitral cells (Rall and Shepherd 1968) supported Llinas' original interpretation. In response, Calvin suggested that Zucker's model had too many free parameters, and defended his own argument as based on "the simplest possible model consistent with our objective (to demonstrate that a) commonplace explanation for conduction velocities was as good as the more esoteric" (Calvin 1969, p. 637). It took 10 more years and the development of brain slice procedures and more sophisticated intracellular

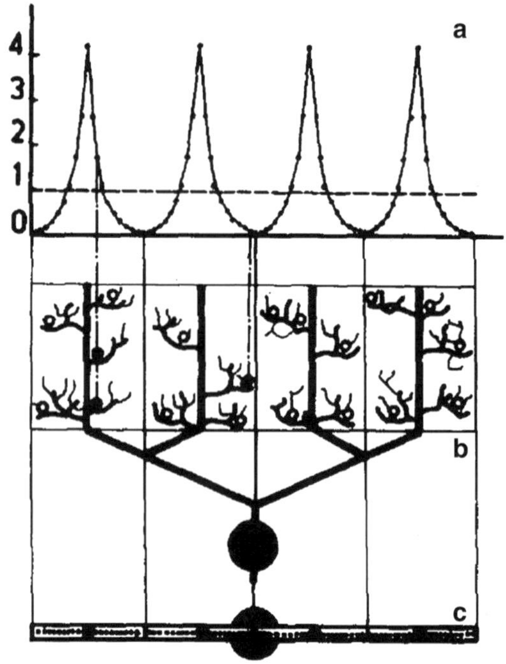

Fig. 5.2 Schematic representation of a model Purkinje cell model simulated in Pellionisz and Szentagothai (1974). The dendritic tree is divided into four nonoverlapping synaptic territories meant to represent the main Purkinje cell dendritic branches. (A) Shows the distribution of parallel fiber synapses on each dendritic branch, (B) is the modeled Purkinje cell viewed in a parasagittal plane and (C) is the Purkinje cell viewed from above. The fine structure within each branch in this figure is only for illustrative purposes and did not influence the summation of synaptic inputs. The model was programmed in FORTRAN and ran on a CDC 3300 computer. Reproduced with permission from Pellionisz and Szentagothai (1974)

recording techniques for Llinas and Sugimori to provide conclusive experimental evidence that Purkinje cell dendrites are in fact electrically active (Llinas and Sugimori 1980a). As will be clear from this chapter, the functional significance of these active dendritic properties, however, continues to be a major computational and modeling issue to this day.

It is important to point out that while this early debate concerned modeling, no effort was actually made to build a model of the Purkinje cell dendrite (Calvin and Hellerstein 1969). The first model of a Purkinje cell with dendritic structure was actually published in 1974 by Pellionisz and Szentagothai as the last of a series of cerebellar network models (Pellionisz 1970; Pellionisz and Szentagothai 1973, 1974). As shown in Fig. 5.2, in that model the Purkinje cell dendrite was represented as four branches in which synaptic influences were calculated independently using a simple algebraic summation. On reaching threshold each branch independently generated a dendrite spike which was then assumed to be summed by the soma. Because each branch generated a dendritic spike, in some sense this model

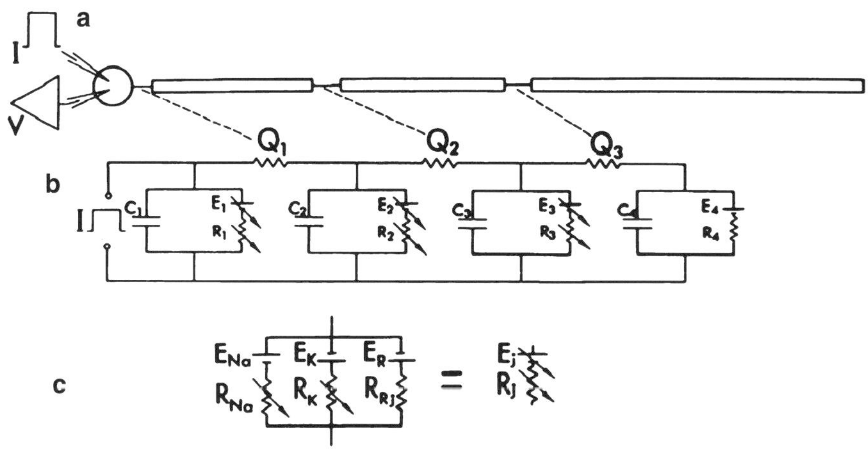

Fig. 5.3 The first published compartmental model of a Purkinje cell consisting of a soma and three dendritic compartments. As shown, the soma and first two dendrites included an element representing synaptic input in the form of a variable battery (E_j) and a variable resistor (R_j). The model was used in conjunction with experimental data to support the hypothesis that the climbing fiber made multiple synaptic inputs on the proximal Purkinje cell dendrite. Reproduced with permission from Llinas and Nicholson (1976)

was also the first with some form of active dendritic structure. Comparing results of network simulations using these four-branch Purkinje cells to previous results with no dendritic structure these authors concluded that: "the simulation experiments are giving quite strong hints in favor of the importance of dendritic geometry" (Pellionisz and Szentagothai 1974, p. 28).

Perhaps reflecting the influence of the original debate between Llinas and Calvin and Hallerstein in the 1960s (Calvin and Hellerstein 1969), Llinas and Nicholson published the first compartmental model of the Purkinje cell dendrite to test new speculations on cerebellar organization made on the basis of field potential recordings (Llinas and Nicholson 1976). In this case, the experiments involved climbing fiber-evoked responses in cat cerebellar cortex, which, Llinas and Nicholson (also it turns out again correctly) believed, were likely a result of synapses distributed widely over the Purkinje cell dendrite. As shown in Fig. 5.3, while a compartmental model that also included for the first time conductances represented with Hodgkin/Huxley model parameters (Hodgkin and Huxley 1952), the entire Purkinje cell dendrite was represented by only three dendritic compartments and despite the earlier debate regarding active dendritic properties, the only active conductances were synaptic.

One year later, Llinas, now working with Pellionisz, published the first compartmental Purkinje cell model that included a full dendritic tree (Pellionisz and Llinas 1977) as shown in Fig. 5.4. Based on extending a compartmental model of the spinal motorneuron (Dodge and Cooley 1973) this more realistic Purkinje cell model consisted of 62 compartments with the soma and initial segment and also

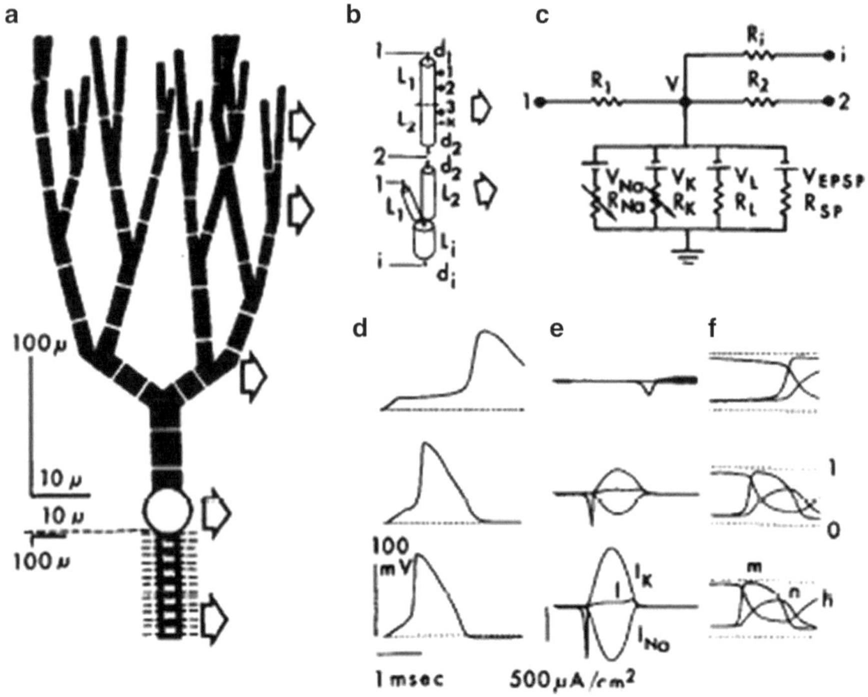

Fig. 5.4 The first full compartmental model of the Purkinje cell dendritic tree represented by 62 dendritic compartments (**a**), with each of the compartments (**b**) simulating ionic conductances using an equivalent electrical circuit (**c**). (**d–f**) show the responses of three different compartments after a simulated somatic current injection (dendritic branch point, upper row; soma middle row; node of Ranvier, lower row). The 62 compartmental model was implemented in Fortran and run on a PDP-15 computer (Digital Equipment Corp.) with the authors explicitly noting that it generating data 5×10^5 times slower than real time activity. Reproduced with permission from Pellionisz and Llinas (1977)

incorporated Hodgkin/Huxley conductance kinetics (Hodgkin and Huxley 1952). Schematically representing a frog Purkinje cell, the authors also sought, for the first time, to replicate basic characteristics of the Purkinje cell physiological responses: (1) the rapid "antidromic" decrement in action potential amplitude in the dendrite following somatic current injection (Llinas et al. 1969b; Freeman and Nicholson 1975); (2) the orthodromic activation of Purkinje cells following parallel fiber stimulation (Eccles et al. 1966a); and (3) the spike burst resulting from climbing fiber synaptic input (Eccles et al. 1966b, 1967). In fact, the authors suggested that the replication of these three physiological effects should constitute the minimum requirement for Purkinje cell modeling. While the authors state explicitly in the paper that compartmental modeling is an essential technique to: "(handle) a partially or totally active dendritic tree" (Pellionisz and Llinas, 1977, p. 37) this model still did not include active voltage-dependent dendritic conductances.

Establishing the Purkinje Cell Community Model Lineage

In the justification for building the first large scale Purkinje cell model, Llinas and Pellionisz explicitly state that: "Rigorous mathematical models of the electrical activity of central neurons (are) a powerful tool to test and interpret experimental data" (Pellionisz and Llinas 1977, p. 37). And, in fact this model was the first to represent the complete Purkinje cell dendritic tree and also consider the biophysical mechanisms responsible for generating this cell's specific and characteristic physiological responses (Pellionisz and Llinas 1977). However, while the model did include a full dendrite, the modeling effort was actually not an exploration of the influence of morphology on physiology, but instead was used to demonstrate the plausibility of mechanisms previously inferred from physiological data. Accordingly, the model was used to demonstrate mechanisms rather than discover them.

The first published Purkinje cell model that explicitly set out to deduce function from structure was published by Shelton 8 years later (Shelton 1985). As shown in Fig. 5.5, this was also the first model based on an actual anatomical reconstruction of a real Purkinje cell. Like the earlier Purkinje cell models, this model also did not include active dendritic properties, an omission justified by the author's assertion that: "the part of the dendritic tree of the Purkinje cell which is thought to be essentially passive forms a very large fraction of the total membrane surface area of the cell" (Shelton 1985, p. 111), although the author later notes that dendritic passivity is an assumption of the model, rather than a conclusion. Instead the model was used to provide a description of the expected passive electrical properties of the Purkinje cell given the morphology of its dendrite. This was accomplished by tuning the model to replicate experimentally observed differences in dendritic and somatic input conductances. It should be noted that while the model was built on an actual anatomical reconstruction of a rat Purkinje cell, the physiological data was actually obtained from Guinea Pigs. Accordingly in the model, the dendritic morphology was actually "stretched" to better resemble a Guinea Pig Purkinje cell.

In addition to being the first Purkinje cell model used to deduce function from structure, the Shelton model is also the first model whose components have been reused by other modelers (Blum and Wang 1990; Bush and Sejnowski 1990; Genet et al. 2010; Brown et al. 2011). This use was in keeping with Shelton's intent that his exploration of the passive properties of the dendrite "form the substrate for extensions which would treat more complex properties" (Shelton 1985, p. 111). As such, the Shelton model was the first constructed with the explicit intent of reuse.

The model, however, that seeded the R-DB Model lineage was published 7 years later by Rapp et al. (1992, 1994) and was based on reconstructions of three Guinea Pig Purkinje cells (see Fig. 5.6). Like the Shelton model, the Rapp model was used to study the passive electrical properties of the dendrite on the similar assumption that this was "an essential step—a skeleton—for constructing biologically more realistic models of PC dendrites" (Rapp et al. 1994, p. 114). For the first time, these publications also included new experimental data obtained by the authors themselves specifically to parameterize the model. While Shelton had speculated on the possible influence of active synaptic conductances on passive membrane properties,

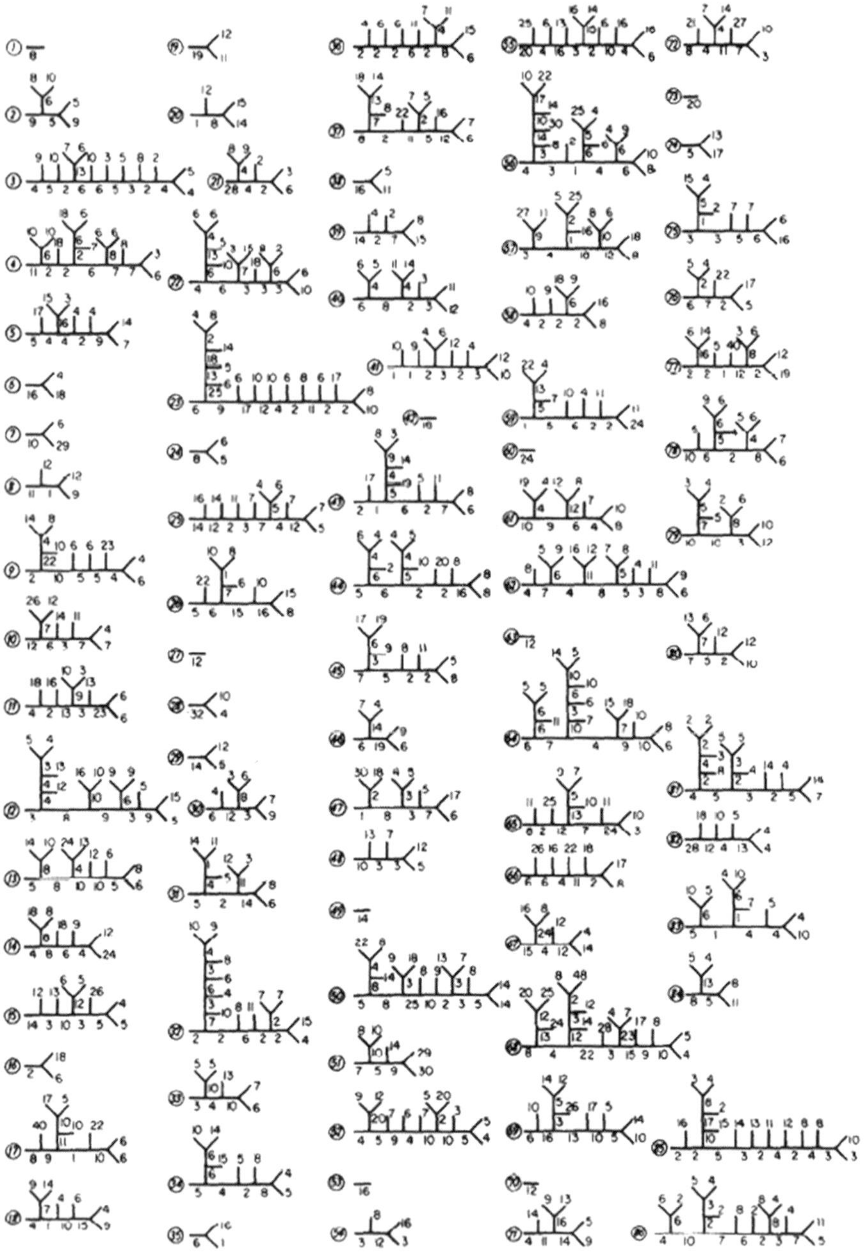

Fig. 5.5 From Shelton (1985) showing details of each of the modeled Purkinje spiny dendritic branches. Simulations were again written in FORTRAN and run on a VAX 11/780 computer (Digital Equipment Corporation) with 10 μs of real time requiring 1 s of compute time. Used with permission from Shelton (1985)

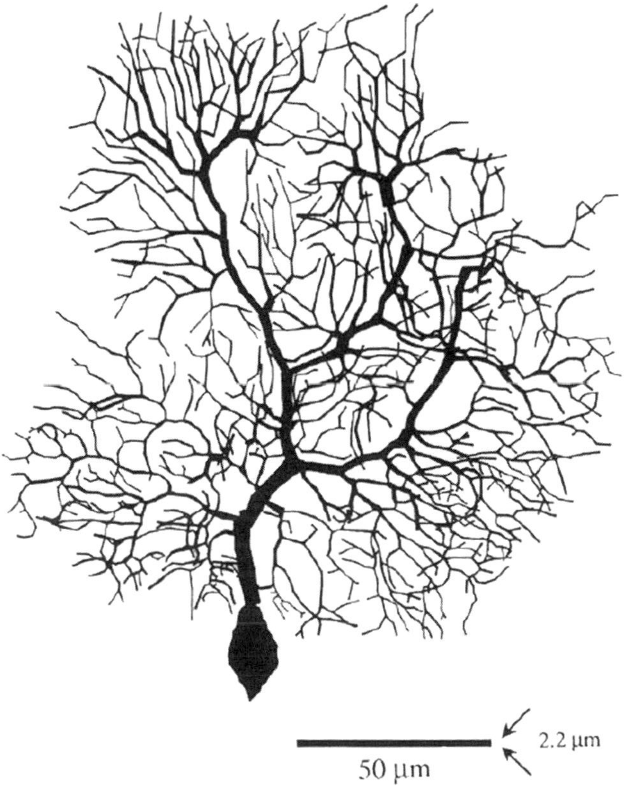

Fig. 5.6 The original Rapp et al. Purkinje cell model, reconstructed from a Guinea Pig. The model was built in SPICE (Vladimirescu et al. 1981), and ran on a VAX/VMS 6830 computer (Digital Equipment Corp), with 10 min of processor time necessary to simulate a 200 ms of real time. Reproduced with permission from Rapp et al. (1992)

Rapp et al. actually applied the first synaptic inputs to a fully realistic dendritic model (Rapp et al. 1992). These authors also considered in some detail the application of newly developed parameter estimation methods for large compartmental models (Holmes and Rall 1992), and tested their results using all three reconstructed morphologies (Rapp et al. 1994). The modeling effort also explicitly compared compartmental modeling results to analytical cable model solutions (Rall 1964; Segev et al. 1985).

Modeling the Active Properties of the Purkinje Cell Dendrite

As just described, the first two Purkinje cell models based on actual anatomical reconstructions both considered only the passive electrical properties of the dendrite (Shelton 1985; Rapp et al. 1992, 1994). However, by the 1990s, the significance of the active properties of the Purkinje cell dendrite had been a subject of debate and discussion for almost 30 years. In fact, as already described, the very first published discussion of Purkinje cell modeling involved speculations involving the active properties of the Purkinje cell dendrite (Calvin and Hellerstein 1969), and the construction of the first compartmental Purkinje cell model was justified as necessary to model those active properties (Pellionisz and Llinas 1977).

By the 1990s, due largely to the pioneering experimental studies of Llinas and Sugimori (1980a, b), there was little doubt that not only the Purkinje cell dendrites but also the Purkinje cell soma included many more voltage-dependent conductances than those traditionally associated with the action potential and dendritic synapses. As already mentioned, both Rapp and Shelton considered their modeling efforts as a first step towards the study of active dendritic properties with Rapp et al. explicitly declaring that it was now essential that Purkinje cell models, "incorporate a variety of non-linear voltage- and ligand-gated channels that we know exist in the Purkinje cell dendrite" (Rapp et al. 1994, p. 114), and two Purkinje cell models with some active dendritic components were actually published in the proceedings of the 1990 Conference on Analysis and Modeling of Neural Systems from which the CNS meetings evolved (Blum and Wang 1990; Travis 1990). However, the first realistic large scale compartmental Purkinje cell model with full Hodgkin/Huxley active voltage-dependent conductances was presented at the first CNS meeting in San Francisco in 1992 and then published for the first time in that meeting's first proceedings volume (De Schutter and Bower 1993; Jaeger et al. 1993). These initial conference reports were followed by three full length papers published in the following year (De Schutter and Bower 1994a, b, c). As shown in Fig. 5.7, this model was based on the Rall et al. dendritic morphology and included all ten then known active conductances differentially distributed in the soma and dendrite as suggested by data from in vitro voltage clamp experiments (Gähwiler and Llano 1989; Hirano and Hagiwara 1989; Kaneda et al. 1990; Regan 1991; Wang et al. 1991).

The first of the 1994 papers (De Schutter and Bower 1994a) explicitly extended the work of Shelton (1985) and Rapp et al. (1992, 1994) with an analysis of the electrical structure of the Purkinje cell dendrite now including active voltage-dependent conductances (Fig. 5.7). The second paper (De Schutter and Bower 1994b) explored dendritic responses to climbing fiber input and then extended the study of background excitatory synaptic inputs first introduced by Rapp et al. (1992, 1994) but for the first time including inhibitory synapses. The third paper (De Schutter and Bower 1994c) considered for the first time the response of Purkinje cells to the type of synaptic activity expected to result from stimulus-driven input. As the first neuronal model to use concurrent supercomputers (De Schutter and Bower 1992), these simulations involved a much more extensive test of parameter space than previously possible, demonstrating that modeled responses were quite robust to changes in its primary parameters. Importantly for the reuse of this model by others, this was also the first model published online using a simulation system specifically developed for realistic neurobiological modeling (Bower and Beeman 1995). It should be noted that the wholesale migration of a complex multi-compartment single neuron model from one set of investigators (Rapp et al. 1994) to another (De Schutter and Bower 1994a, b, c) was also one of the first in computational neuroscience.

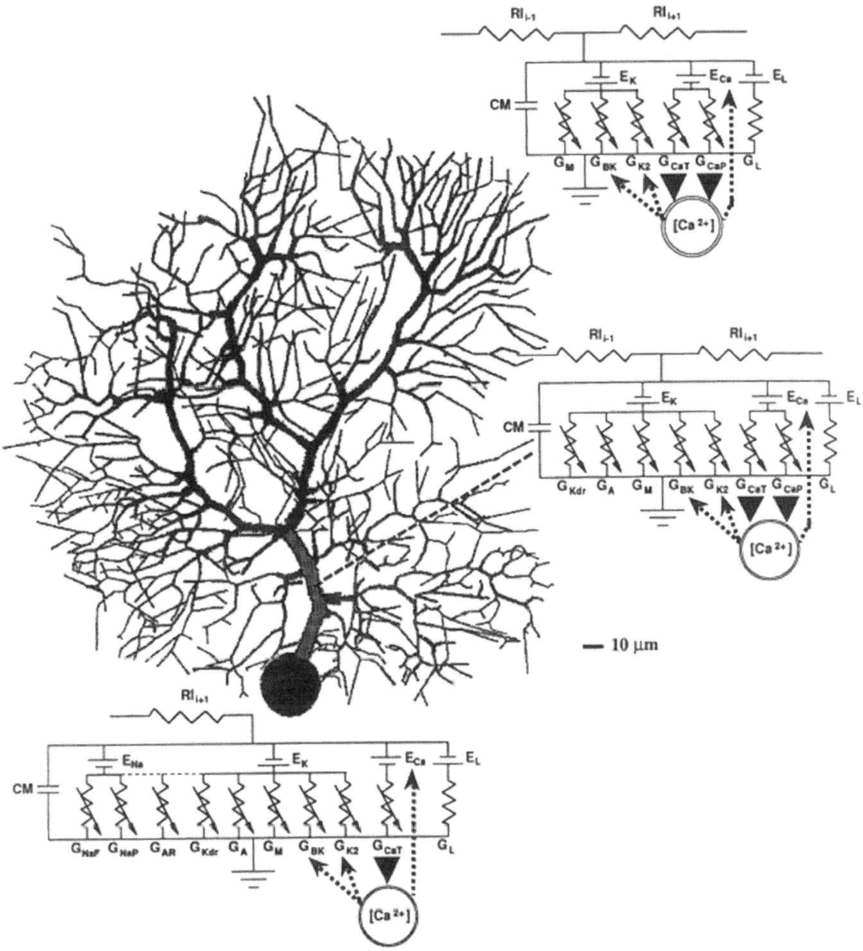

Fig. 5.7 Schematic description of the De Schutter and Bower Purkinje cell model with equivalent circuit diagrams for the modeled ionic conductance included in each section of the cell. Reproduced with permission from De Schutter (1999)

Emergence of a Community Model

The papers by Rapp et al. (1992, 1994) and De Schutter and Bower (1994a, b, c) have collectively been cited more than 500 times, with the first description of the active Purkinje cell model (De Schutter and Bower 1994a) responsible for almost half those citations. Perhaps more importantly for the definition of a "community model" the R-DB Model has also been used by a growing number of authors as a base for further modeling work outside the laboratories of its origin. Thus, as described in the next section, over the last 20 years, the model has been contributed to, evaluated, and adopted by the larger scientific community. It also continues to be used to test and generate new hypothesis about the functional organization of the Purkinje cell as well as cerebellar cortical circuitry.

Model Reuse

The adoption of the R-DB Model as a community model can be explicitly represented by an expanding lineage of model-dependent publications. As already described, the core lineage root for the R-DB Model was the original Rapp et al. (1992, 1994) model, enhanced with active properties by De Schutter and Bower (1994a, b, c). This model subsequently formed the basis for a series of modeling and experimental publications over the next 20 years originating from the Bower laboratory and reflecting the efforts of a sequence of students (De Schutter 1994; Jaeger et al. 1996; Baldi et al. 1998; Sultan and Bower 1998; Jaeger and Bower 1999; Mocanu et al. 2000; Santamaria et al. 2002, 2007; Santamaria and Bower 2004; Lu et al. 2005; Cornelis et al. 2010). Further, the model has continued to be used by students of these original students in their own independent laboratories (De Schutter 1998; Vos et al. 1999; Howell et al. 2000; Steuber and De Schutter 2001, 2002; Gauck and Jaeger 2003; Solinas et al. 2003, 2006; Kreiner and Jaeger 2004; Shin and De Schutter 2006; Shin et al. 2007; Steuber et al. 2007; Achard and De Schutter 2006, 2008; De Schutter and Steuber 2009; Anwar et al. 2010; Coop et al. 2010; Santamaria et al. 2011; Tahon et al. 2011). Finally, and perhaps most importantly from the point of view of a community model, the R-DB Model has become the basis for a growing number of publications in laboratories not directly related to my own (Staub et al. 1994; Coop and Reeke 2001; Mandelblat et al. 2001; Miyasho et al. 2001; Roth and Häusser 2001; Chono et al. 2003; Khaliq et al. 2003; Ogasawara et al. 2007; Yamazaki and Tanaka 2007; Kulagina et al. 2008; Traub et al. 2008; Brown et al. 2011; Brown and Loew 2012; Forrest et al. 2012) and several of these modeling efforts have now initiated their own lineage sequences with, for example, the adaptation of the original R-DB Model by Miyasho et al. (2001), being further extended by Chono et al. (2003), Kulagina (2008), and Brown et al. (2011).

Testing Model Parameters Against New Experimental Data

One of the first uses of the R-DB Model outside of my own laboratory's lineage explicitly tested the model's ability to replicate PC physiological responses obtained using a new set of ion channel blockers (Miyasho et al. 2001). Using dendritic morphology from the rat (Shelton 1985) parameterized with data from the R-DB Model, these authors modified channel descriptions and conductance densities to reproduce the repetitive Ca^{2+} spike firing they had found after the application of TTX in vitro. These authors also refined the kinetics of the K^{+}-delayed rectifier current, applying a new mechanism for calculating intracellular Ca^{2+} concentration while also changing the Ca^{2+} sensitivity of the calcium-activated dendritic K^+ conductance. With these changes, the model was extended to replicate physiological responses including: (1) characteristic Ca^{2+} dendritic spikes in the presence of TTX; (2) repetitive Ca^{2+} spiking patterns resulting from the presence of TTX; (3) the lack of Ca^{2+} spikes found after application of a P-type Ca^{2+} channel blocker; (4) the slow onset of the Ca^{2+} spikes in response to depolarizing current steps; and (5) the marked shortening of the Ca^{2+} spike onset seen in the presence of 4-AP. These authors also added a Ni^{2+}-sensitive Ca^{2+} (class-E type) channel to the model's dendrites, allowing it to replicate the longer onset Ca^{2+} spikes found in the presence of Ni^{2+}. Two years later, Chono et al. (2003) further refined the Miyasho et al. (2001) model by adding new channel descriptions as well as refinements in the conductance values for the simulated Ca^{2+} and Ca^{2+}-dependent K^+ channels. These enhancements have since been incorporated into subsequent Purkinje cell modeling efforts by other groups (Traub et al. 2008; Brown et al. 2011).

Additional Analysis

Equally important to changes in the structure of a community model is the use of that model to explore new forms of behavior or perform new forms of analysis not considered by the original model's authors. Several authors have used the R DB Model in a reduced form to more closely examine Purkinje cell neuronal dynamics (Mandelblat et al. 2001; Fernandez et al. 2007). In a series of recent publications, Brown et al. have adapted the original R-DB Model to explore how mechanisms at the subcellular (biochemical) levels might be linked to somatic output (Brown et al. 2011; Brown and Loew 2012). At the subcellular level as well, the R-DB Model has provided a larger context for studies of calcium diffusion (Santamaria et al. 2006, 2011; Anwar et al. 2010) as well as biophysical mechanisms of synaptic plasticity (Vladimirescu et al. 1981; Antunes and De Schutter 2012; De Schutter 2012). At more network levels, Traub et al. recently built a new model based in large part on R-DB Model parameters to explore the possible role of gap junctions between the initial axon segments of Purkinje cells in cerebellar cortical oscillations (Traub et al. 2008). To do so these authors reduced overall dendritic complexity while maintaining a "realistic" path from the distal dendrite to the soma (see Fig. 5.8). The model has also been used as a base to build network level simulations in reduced (Yuen et al. 1995; Coop and Reeke 2001; Sarro 2004) and full (Howell et al. 2000; Solinas et al. 2003; Santamaria et al. 2007) forms. The R-DB Model has also been applied to new analytical studies, including, for example, questions involving the information processing potential of dendrites (Coop et al. 2010) as well as possible spike coding strategies (Steuber and De Schutter 2001, 2002; Steuber et al. 2007; De Schutter and Steuber 2009). Efforts have also been made to link the structure of the R-DB Model to the kind of analysis involved in the field of artificial neural networks (Steuber and De Schutter 2001; Sarro 2004). Finally, the R-DB Model has also been used as a base for assessing modeling technology itself, including parameter estimation techniques (Van Geit et al. 2007) and the relationship between parameter

Fig. 5.8 Schematic representation of the cerebellar Purkinje cell model in Traub et al. (2008). Reflecting the focus of the study on putative gap junctions between the initial axon segments of Purkinje cells, the axonal region was represented by 6 compartments while the dendrite was reduced to 553 compartments with a particular emphasis on the spiny branchlets. This model was hand coded entirely in Fortran. Used with permission from Traub et al. (2008)

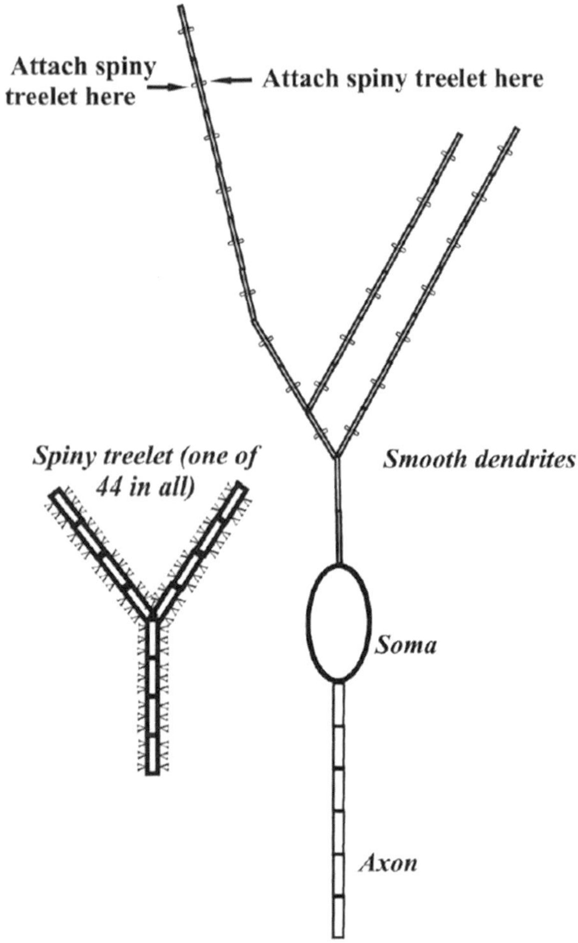

variations and modeling results (Achard and De Schutter 2008). The R-DB Model has even been used to test whether experimental techniques like the voltage clamp are appropriate for evaluating the physiological properties of Purkinje cells (Staub et al. 1994).

Understanding Purkinje Cell Physiological Responses

While the previous sections have discussed the general reuse and improvement of the R-DB Model, ultimately the utility of any model, whether used by the community or not, is its ability to generate and test hypothesis regarding physiological function. This is also the most complex use of any model, and perhaps especially a community model. As context for considering what has been learned about the

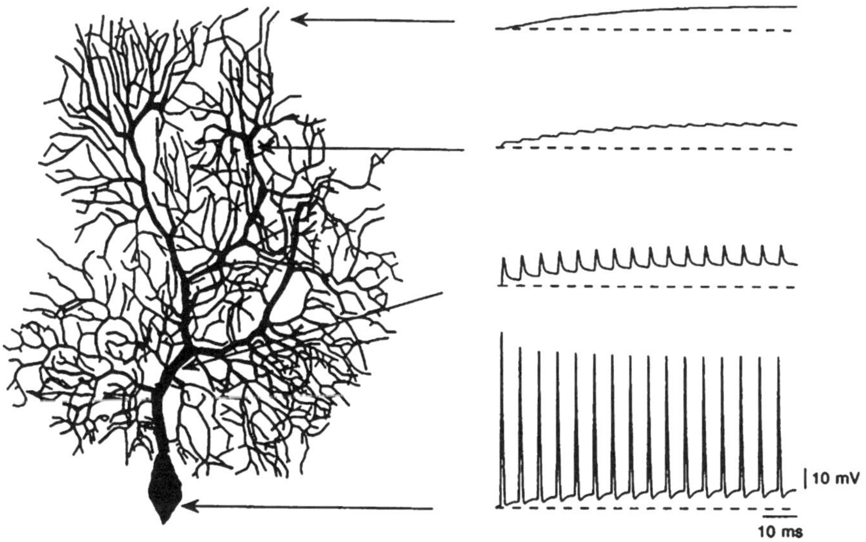

Fig. 5.9 Simulation of the lack of antidromic action potential dendritic invasion in a modeled Purkinje cell following simulated current injection in the soma. Used with permission from Rapp et al. (1994)

functional organization of Purkinje cells using the R-DB Model, the next section is organized around the R-DB Model's replication of the electrical behavior of Purkinje cell that Pellionisz and Llinas proposed in 1977 as a standard for, "any Purkinje cell model which claims to be adequate" (Pellionisz and Llinas 1977, p. 42). Some of these results are described in more detail in a review by De Schutter (1999).

Antidromic Spike Activation of the Purkinje Cell Dendrite

Perhaps the most straightforward characteristic Purkinje cell response identified by Pellionisz and Llinas (1977) as a core requirement for any model of this neuron is the fact that action potentials generated in its soma do not propagate to the dendrite (Fig. 5.9). At the time of the first Purkinje cell modeling studies, this lack of antidromic dendritic invasion had already been predicted based on field potential recordings (Llinas et al. 1969b; Freeman and Nicholson 1975), although the phenomenon was not directly observed until much later (Llinas and Sugimori 1980b). In the early passive models, the lack of back propagation was attributed to the relative surface area of the cell dendrite compared to its soma (Pellionisz and Llinas 1977; Rapp et al. 1994). This explanation was further elaborated in a recent passive modeling study using parameters obtained from the R-DB Model (although with different dendritic morphology) as due to a large cumulative impedance mismatch resulting from the high branching density of the Purkinje cell dendrite

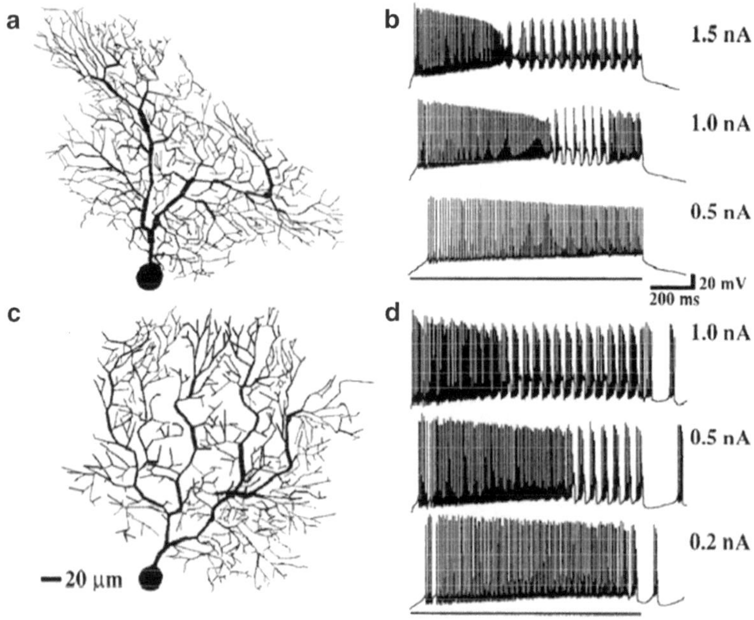

Fig. 5.10 Simulation of somatic responses to three different amplitude synaptic current injections in models with two different dendritic morphologies. Reproduced with permission from De Schutter and Bower (1994a)

(Roth and Häusser 2001). Active dendritic modeling efforts have also noted that the very low Na^+ channel density in Purkinje cell dendrites provides no mechanism to overcome these morphological effects (Kitamura and Häusser 2011) as has been found in other types of mammalian neurons (Vetter et al. 2001).

Responses to Somatic Current Injection

The results shown in Fig. 5.9 were obtained from a passive Purkinje cell dendritic model in response to current injection in the soma. In fact, as shown in Fig. 5.10, current injection in a real Purkinje cell produces a much more complex pattern of somatic and dendritic activity (Gähwiler and Llano 1989; Hirano and Hagiwara 1989; Kaneda et al. 1990; Regan 1991; Wang et al. 1991; Lev-Ram et al. 1992). In part for this reason, although not explicitly a part of the original Pellionisz and Llinas (1977) standard for Purkinje cell models, the ability to replicate the results of in vitro current injection studies has become the defacto standard for testing and tuning realistic Purkinje cell models (Bush and Sejnowski 1990; De Schutter and Bower 1994a; Coop and Reeke 2001; Mandelblat et al. 2001; Miyasho et al. 2001; Forrest et al. 2012). Accordingly, as shown in Fig. 5.10, the first test of the active

R-DB Model involved determining whether a reasonable set of model parameters could replicate this in vitro data.

While a full description of the mechanisms responsible for these in vitro response patterns is beyond the scope of this chapter, the general result from modeling studies is that this behavior is a function of a complex interaction between all the Purkinje cell's biophysical and anatomical properties (De Schutter 1999). This conclusion is somewhat in contrast with the more typical analysis from experimental studies which tend to associate different features of the in vitro response properties specifically to different kinds of afferent input (Gähwiler and Llano 1989; Hirano and Hagiwara 1989; Kaneda et al. 1990; Regan 1991; Wang et al. 1991; Lev-Ram et al. 1992). Thus, for example, typically, these complex waveforms are considered to be made up of three different types of electrical events: fast events associated with somatic action potential generation; the somewhat slower Ca^{2+}-related dendritic bursting behavior assumed to be related to climbing fiber inputs; and longer time course events assumed to be influenced by granule cell-related synaptic inputs (Isope et al. 2012; Kitamura and Kano 2012). Again, the analysis of modeling results suggests that all components of the dendrite contribute to each type of Purkinje cell activity as well as its response to afferent inputs (De Schutter 1999).

Interestingly, the interdependence of physiological properties revealed by model analysis is even evident in understanding the somewhat unusual in vitro behavior of the Purkinje cell. Specifically it has been known for many years that the spontaneous behavior of Purkinje cells in vitro is quite different from the spontaneous behavior of Purkinje cell in vivo (Llinas and Sugimori 1980b). As shown in Fig. 5.11A, in vitro behavior consists of relatively rapid (usually >100 Hz) action potentials, interrupted periodically by spontaneous dendritic calcium spikes. In contrast, as simulated in Fig. 5.11C, Purkinje cells in vivo generate spontaneous action potentials at lower frequencies (usually <80 Hz) that are quite irregular. In vivo, dendritic Ca^{2+} spikes only appear in response to climbing fiber inputs (Llinas and Nicholson 1976) while they occur spontaneously in vitro.

Given that Purkinje cells are essentially deafferented in brain slice preparations, it seemed reasonable to assume that these differences in in vivo and in vitro behavior might be due to a lack of spontaneous background input from the 150,000 excitatory parallel fiber inputs. However, when provided with background excitatory input alone, the R-DB Model produced a pattern of output that resembled neither the in vitro nor the in vivo conditions (Fig. 5.11B). Instead, replication of in vivo patterns required spontaneous input from both excitatory and inhibitory synaptic inputs (Fig. 5.11C). The model both in single cell (Jaeger et al. 1996; Watanabe et al. 1998) and network form (Howell et al. 2000) predicted that normal Purkinje cell behavior depends on the presence of constant background synaptic inputs interacting with the active Ca^{2+}- and K^{+}-dependent channels in the dendrite and soma (De Schutter 1999). Experimental studies specifically designed to test these modeling predictions are consistent with this interpretation (Jaeger and Bower 1999; Kreiner and Jaeger 2004). These results also suggest that caution is necessary in inferring too much about the natural in vivo behavior of Purkinje cells based on their in vitro response properties.

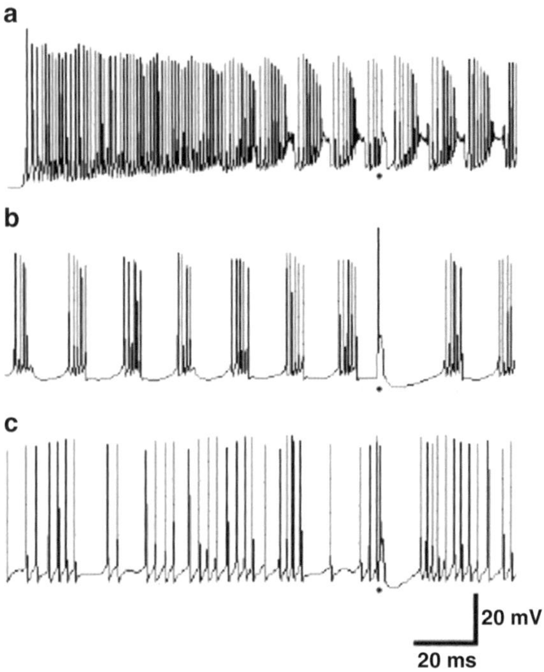

Fig. 5.11 Comparison of responses of the R-DB Model in the absence of background synaptic input to the dendrite (**a**), in the presence of only excitatory synaptic input (**b**) and both excitatory and inhibitory input (**c**). As described in the text, the firing pattern in (**a**) resembles Purkinje cell activity recorded in vitro, while (**c**) resembles in vivo activity. Figure used with permission from De Schutter (1999); Traub et al. (2008)

Purkinje Cell Responses to Climbing Fiber Activation

Returning to the properties of Purkinje cells that Pellionisz and Llinas (1977) suggested were a critical test for any Purkinje cell model, the fact that the Purkinje cell responds to climbing fiber activation in vivo with a burst of action potentials has also been known for many years (Eccles et al. 1966b). In fact as already noted, the first compartmental Purkinje cell model was specifically constructed to test the experimentally derived prediction (Llinas and Hillman 1969) that single climbing fibers made multiple synaptic contacts distributed over the Purkinje cell dendrite (Llinas and Nicholson 1976). One year later, the modeling focus shifted to a consideration of the actual biophysical mechanisms responsible for producing the "oscillatory wavelets" or "spike burst" characteristic (see Fig. 5.12F) of climbing fiber somatic Purkinje cell responses (Pellionisz and Llinas 1977). At the time, these authors concluded that the different peaks in the somatic burst response were generated by repetitive firing of the initial segment of the axon rather than by an active dendritic mechanism as had previously been proposed (Eccles et al. 1966b).

Neither Shelton (1985) nor Rapp et al. (1992, 1994) attempted to replicate Purkinje cell responses to climbing fiber activation, however, this was an important component of the initial analysis of the active dendritic and somatic model of De Schutter and Bower (1994b). In fact, after tuning model parameters to replicate responses to somatic current injection data (De Schutter and Bower 1994a), the ability of the

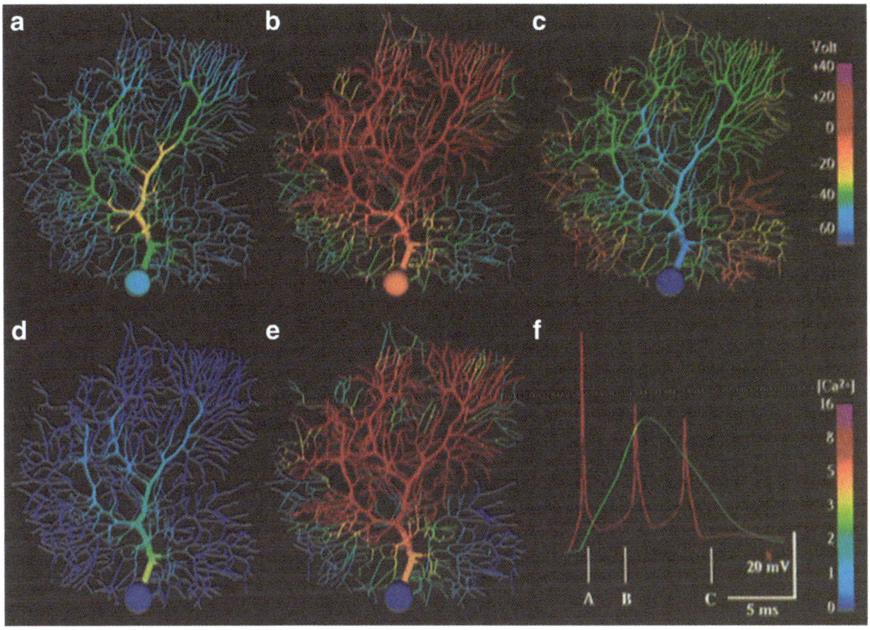

Fig. 5.12 False color representation of membrane potential and Ca^{2+} concentration during simulation of a climbing fiber input. (**a**) Membrane potential 1.4 ms after beginning of the resulting complex spike. (**b**) Membrane potential 4.0 ms after beginning of complex spike. (**c**) Membrane potential 10.0 ms after beginning of a complex spike (after the last somatic action potential). (**d** and **e**) sub membrane Ca^{2+} concentration at same times as (**a**) and (**b**), respectively. (**f**) complex spike as it appears in the soma (*red*) and distal dendrite (*green*) at the same times represented by (**a–c**) as indicated. Note the nonlinear [Ca^{2+}] scales. This model was built in GENESIS (version 1.4) and was the first simulation to be run concurrently on a large scale parallel supercomputer (De Schutter and Bower 1992), and across multiple workstations (8 Sun Sparc2). Using this technology, 550 ms of real time activity could be simulated in approximately 1 h. Figure used with permission from De Schutter and Bower (1994b)

model to generate climbing fiber burst responses was the first measure of the model's likely realism (see Fig. 5.12). As already described, the model predicted that the correct in vivo form of the climbing fiber response was dependent on background patterns of excitatory and inhibitory synaptic inputs. But analysis of the model also predicted that the dendritic response was dependent on the activation of P-type Ca^{2+} channels in both the cells' smooth and the spiny dendrites, with the duration of the dendritic spike being regulated by Ca^{2+}-activated K^+ conductances. The modeling results also suggested that the duel reversal potential for the climbing fiber input previously shown experimentally (Llinas and Hillman 1969) and attributed to the spatial distribution of climbing fiber synapses (Llinas and Nicholson 1976) was also dependent on the active properties of the Purkinje cell dendrite. Further, an unexpected but important prediction of the model was that climbing fiber activation resulted in substantial increases in intracellular calcium not only in the smooth dendrites, where climbing fiber synapses terminate, but also in the smallest spiny dendritic branches receiving granule cell synaptic inputs (Gundappa-Sulur et al. 1999; Lu et al. 2009). The involvement of the entire dendrite in the climbing fiber event was simultaneously shown experimentally (Konnerth et al. 1992; Miyakawa et al. 1992). The model also predicted that inhomogeneities in local levels of calcium activation in the dendrite did not depend on a nonuniform distribution of Ca^{2+} channels as had previously been suggested (Tank et al. 1988; Llinas and Sugimori 1992). Instead the pattern of calcium response was a consequence of the nonuniform geometry of the Purkinje cell dendrite, and likely varied from Purkinje cell to Purkinje cell. Thus, unlike Rapp et al. (1994), who reported little effect of individual dendritic variations on cellular passive properties, the active model suggested that differences in individual Purkinje cell morphologies might, in fact have functional significance.

Purkinje Cell Responses to Granule Cell Pathway Related Input

The third standard for Purkinje cell modeling proposed by Pellionisz and Llinas (1977) was the ability to replicate simple spike firing in response to granule cell (parallel fiber) input. It is, in fact, on this question that the R-DB Model has actually produced the most interesting and provocative set of predictions.

Significance of Background Synaptic Inputs

As already described, one important prediction of the R-DB Model is that the natural behavior of the Purkinje cell dendrite depends on the presence of continuous background excitatory and inhibitory synaptic input from the granule cell pathway. Having replicated the characteristic somatic response to climbing fiber input, the next test for the R-DB Model was to determine whether these background patterns of granule cell pathway synaptic inputs would generate the proper frequencies of Purkinje cell simple spike firing (De Schutter and Bower 1994b). Again, while background excitatory granule cell (parallel fiber) synaptic activity had been anticipated for some time to influence ongoing Purkinje cell firing (Llinas et al. 1969a), in order to get realistic patterns of spiking out of the active Purkinje cell model it was necessary to also add background inhibitory synaptic inputs (De Schutter and Bower 1994b). This was the first time that inhibitory inputs had been included in a Purkinje cell model. Further, the model predicted that the same frequency of Purkinje cell output could be generated by different combinations of background excitatory and inhibitory inputs.

It is again not possible to fully describe the complex dendritic and somatic interactions resulting in the ongoing patterns of simple spike activity. Those explanations can be found in the original modeling and experimental papers themselves (De Schutter and Bower 1994a, b; Jaeger et al. 1996; De Schutter 1999; Jaeger and Bower 1999), as well as subsequent R-DB Model-based investigations (Santamaria

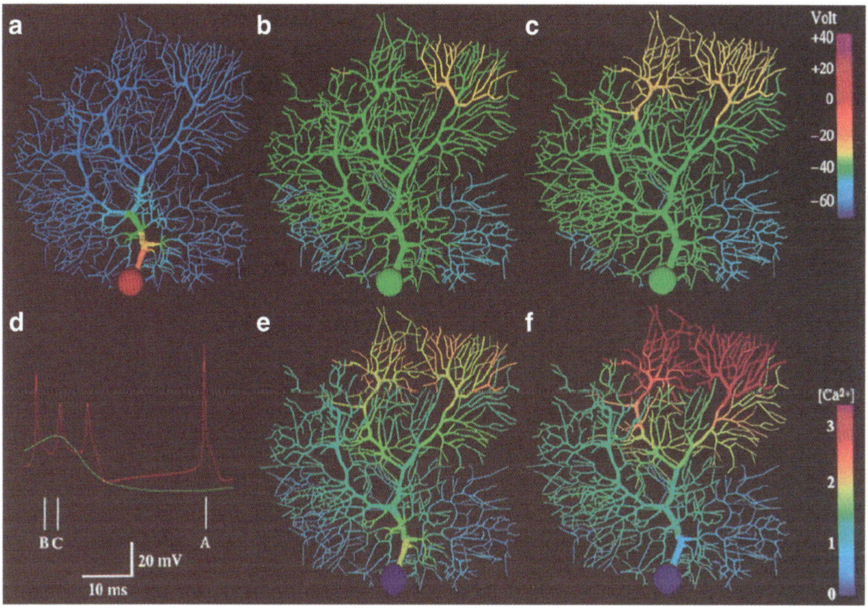

Fig. 5.13 False color representation of membrane potential and Ca^{2+} concentration during a 2.0 nA current injection in the soma of the modeled Purkinje cell. Simulated membrane potential is shown during a somatic action potential (**a**), at the beginning of a dendritic spike (**b**) and 1.6 ms later (**c**). (**d**) Shows predicted somatic (*red*) and dendritic (*Green*) membrane potential at the times indicated. (**e**) and (**f**) indicated submembrane Ca^{2+} concentration at the same time as (**b**) and (**c**) respectively. Reproduced with permission from De Schutter and Bower (1994b)

et al. 2002). However, these modeling efforts resulted in several key predictions. First the model predicted that Purkinje cell behavior was dependent on the ability of the soma, itself, to spontaneously generate action potentials. This ability now specifically demonstrated experimentally (Pugh and Raman 2009) has recently been further studied using a model derived from the R-DB line (Forrest et al. 2012). Second, as shown in Fig. 5.13, the model predicted that the large intrinsic voltage gated currents and not the relatively smaller currents associated with synaptic activation, most influenced ongoing somatic spiking (Jaeger et al. 1996; De Schutter 1998; Jaeger and Bower 1999). In fact, the model predicted that the Purkinje cell dendrite is actually dominantly a current synch rather than a source. Further, the model suggested that background spontaneous parallel fiber inputs had much less of an effect on the actual timing of Purkinje cell spikes than did inhibitory synaptic input (Jaeger et al. 1996). While a full description is again beyond the scope of this chapter, experimental (Jaeger and Bower 1999; Womack and Khodakhah 2002a, b, 2004; Womack et al. 2004; Santamaria et al. 2007) and subsequent R-DB Model related studies (Howell et al. 2000; Miyasho et al. 2001; Coop et al. 2010; Brown et al. 2011; Forrest et al. 2012) have supported these unexpected interactions between the Purkinje cell dendrite and soma.

Dendritic Democracy and Distal Synaptic Inputs

The influence of excitatory synaptic input in such a large dendrite has been a central issue for Purkinje cell modeling for many years. In fact, the publication by Llinas et al. (1968) that sparked the first consideration of modeling in Purkinje cells (Calvin 1969; Calvin and Hellerstein 1969; Zucker 1969) started by posing the following fundamental question: "In studying the anatomy of the Purkinje cell, one wonders how the distal region of (these large) dendrites can act upon the soma and axon …" (Llinas et al. 1968, p. 1132). That paper went on to identify two possibilities: "(1) by direct electrotonic spread from the distal dendrite to the soma, or (2) by the initiation of action potentials or local responses which can be conducted either in an all- or-none manner or in a decremental fashion down to the axon." (Llinas et al. 1968, p. 1132). Considering this question was also a primary objective of the modeling efforts of both Shelton (1985) and Rapp et al. (1992, 1994), who, based on their passive models both predicted that the Purkinje cell dendrite was actually electrotonically compact and therefore that distal and proximal synaptic inputs should, in principle, have an equal influence on the soma. Shelton specifically describes the functional significance of the high passive dendritic input resistance and thus the electrotonic compactness as "a specialization which optimizes the dendrites for signaling (the soma) with minimum (synaptic) attenuation" (Shelton 1985, p. 127). This apparent characteristic of the passive electrical properties of the Purkinje cell dendrite has since been described as promoting "dendritic democracy" so that: "somatic EPSP amplitude is only weakly dependent on synaptic location on Purkinje cell spiny branchlets" (Roth and Hausser 2001, p. 469).

Of course, Llinas et al. (1968); Pellionisz and Llinas (1977); Shelton (1985); and Rapp et al. (1992, 1994), all recognized that this baseline "dendritic democracy" only applied to the passive electrical properties of the dendrite, and was therefore likely to change with the addition of active conductances. Shelton specifically predicted that the addition of synaptic conductances would likely "swamp" (Shelton 1985, p. 128) the passive membrane conductivity, significantly extending the electrotonic length of the dendrite. Actual simulations by Rapp et al. (1992, 1994) supported Shelton's speculation, predicting that individual parallel fiber synapses "essentially loose their functional meaning (in the presence of large amounts of background synaptic input) and only activation of a large number of parallel fibers will significantly displace the membrane potential" (Rapp et al. 1992, p. 530).

It therefore was not surprising that adding both synaptic conductances as well as the large voltage dependent dendritic Ca^{2+} related membrane conductances further extend the electrotonic length of the dendrite (De Schutter and Bower 1994a) a modeling result subsequently tested experimentally (Staub et al. 1994; Ascoli 2007). However, as described in the third paper in the 1994 series (De Schutter and Bower 1994b; De Schutter 1999), what was surprising was that the addition of dendritic voltage dependent Ca^{2+} membrane conductances uncovered a new and unexpected biophysical mechanism in which synchronously activated granule cell inputs induced a subthreshold Ca^{2+} dependent amplification mechanism that restored "democracy" to the dendrite (Fig. 5.14). While Pellionisz and Llinas (1977) had

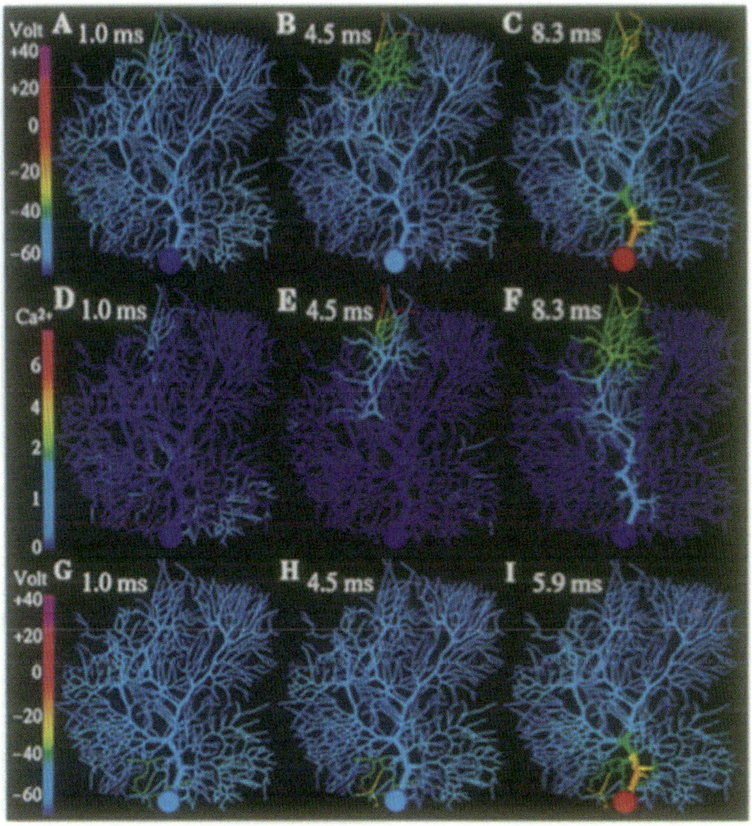

Fig. 5.14 False color images of the response of the R-DB Model to a synchronous synaptic input on a distal (**a**–**f**) and proximal (**g**–**i**) branchlet. Membrane potential is shown in (**a**–**c**) and (**g**–**i**) while (**d**–**f**) indicates sub membrane Ca^{2+} concentrations corresponding to activity in (**a**)–(**c**). Reproduced with permission from De Schutter and Bower (1994c)

suggested the general possibility that active membrane properties could facilitate the influence of synapses on the soma, and Shelton specifically speculated that "active dendritic spikes or active graded potentials may act as a booster mechanism to overcome the electrotonic lengthening of the dendrite due to synaptic activation" (Shelton 1985, p. 128), the particular mechanism that emerged from the active R-DB Model was unexpected. Instead of being dependent on a dendritic calcium spiking mechanism as previously assumed (Pellionisz and Szentagothai 1974), the mechanism involved activation of a sub-spiking threshold event (Fig. 5.14). Because of this mechanism, the model predicted that a small number of synchronously activated granule cell synaptic inputs would produce a similar level of depolarization in the soma regardless of where they were located on the dendrite (De Schutter and Bower 1994c). Similar results have now been shown using the model for synchronous inhibition (Solinas et al. 2006). Importantly, while generating a somatic spike in the passive dendritic models required the activation of large numbers of excitatory synapses (Llinas and Sugimori 1980b; Rapp et al. 1992, 1994), spike generation in the active membrane model required an order of magnitude fewer active synapses (De Schutter and Bower 1994c). Since the original model-based prediction that dendritic calcium channels can have significant effects on the transmission of synaptic input to the soma, considerable experimental work has been done to characterize the effects of dendritic voltage gated calcium channels on Purkinje cell spiking (Cavelier et al. 2002; Rancz and Häusser 2010) and experimental studies have shown that relatively few synchronous excitatory inputs are necessary to generate a somatic spike (Isope and Barbour 2002).

Purkinje Cells in Their Network Context

As is probably the case for all neurons, the functional significance of the physiological properties of any neuron must be considered and will probably only be fully understood in the context of the network in which they are embedded. For Purkinje cells, this has meant embedding the R-DB Model within realistic network simulations (Santamaria et al. 2007). As with single cell modeling, it is also important for network level modeling to have a clearly defined set of physiological behaviors it seeks to simulate, preferably behaviors that are surprising or not yet well understood (Bower 1990). As it turns out the original motivation for cerebellar modeling in my laboratory was to investigate an unexpected and counterintuitive pattern of experimentally observed Purkinje cell responses in response to peripheral sensory stimuli (Bower and Woolston 1983). Specifically, as shown in Fig. 5.15, the spatial extent of Purkinje cell responses to peripheral stimuli is far more restricted than expected from the spatial spread of the parallel fibers (Eccles et al. 1972; Bower and Woolston 1983). Further, as also shown in Fig. 5.15, when the spatial location of Purkinje cells was compared to the spatial distribution of activity in the granule cell layer the results demonstrated that only Purkinje cells recorded immediately over the region of activated granule cells were directly excited (Bower and Woolston 1983). Results consistent or directly supporting this finding have now been reported in numerous subsequent experiments (Kolb et al. 1997; Cohen and Yarom 1998; Lu et al. 2005; Holtzman et al. 2006; Heck et al. 2007; de Solages et al. 2008; Rokni et al. 2008; Brown and Ariel 2009; Walter et al. 2009; Dizon and Khodakhah 2011).

In the original experimental studies, the restricted extent of Purkinje cells activated by peripheral stimuli was interpreted to suggest that parallel fibers were less influential on Purkinje cell output than had previously been assumed (Bell and Grimm 1969; Eccles et al. 1972; Bower et al. 1980; Bower and Woolston 1983). As diagrammed in Fig. 5.16, Llinas subsequently suggested that the experimental data could be explained if Purkinje cells were driven by synchronous input from synapses made by granule cells as they ascend through the molecular layer (Mugnaini 1972), but not by a more asynchronous input from parallel fibers activated by the same stimulus (Llinas 1982). Considered now in the context of the R-DB Modeling results, this explanation seemed perfectly consistent with the relative lack of direct influence

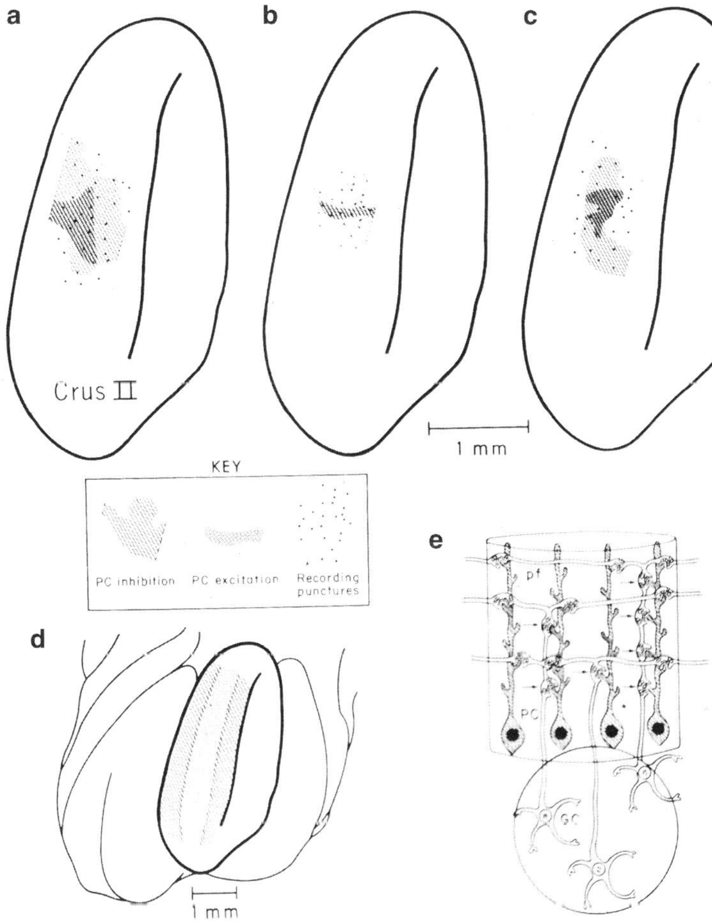

Fig. 5.15 (**a**, **b**, and **c**): show the restricted spatial pattern of excitatory (*dark stippling*) and inhibitory (*light hatching*) Purkinje cell responses following peripheral stimulation in three different experiments. The stimulus activated only granule cells beneath the region of excitatory PC responses. (**d**) Shows the expected pattern of activation if parallel fibers drove Purkinje cell responses. (**e**) Original drawing from Llinas (1982) illustrating the hypothesis that synapses associated with the ascending segment of the granule cell axon are responsible for the restricted spatial activation of Purkinje cells. Reprinted with permission from Bower and Woolston (1983)

of background parallel inputs on Purkinje cell spiking, and the existence of an amplification mechanism for synchronous excitatory inputs discovered using the model (De Schutter and Bower 1994c). Accordingly it was fully expected that the R-DB Model, when placed in a network context, would fully support the Llinas hypothesis. It was surprising therefore, that even a highly desynchronized pattern of parallel fibers following simulated peripheral stimulation, still drove Purkinje cell spiking as a result of the dendritic boosting mechanism (Santamaria et al. 2007). Instead, once again replicating the physiological data required the addition of feedforward

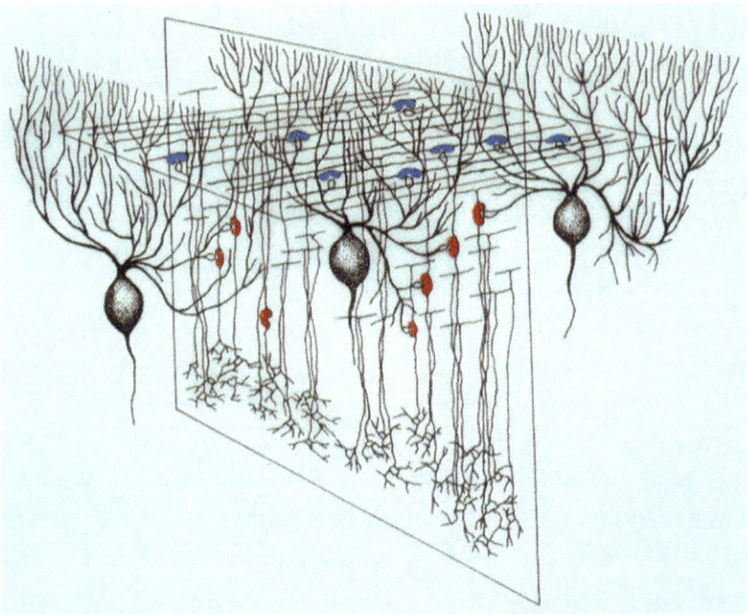

Fig. 5.16 Diagram contrasting synapses made on Purkinje cell dendrites by ascending (*red*) and parallel (*blue*) branches of the granule cell. Anatomical studies have shown that ascending segment synapses make direct projections on only the most distal fine Purkinje cell dendritic branches (Gundappa-Sulur et al. 1999; Lu et al. 2009)

inhibitory synaptic inputs. In fact, the model predicted (Santamaria et al. 2007), and subsequent experimental results confirmed (Santamaria et al. 2007; Walter et al. 2009), that parallel fiber inputs would drive Purkinje cell responses in vivo when inhibition was blocked.

Implications for Functional Structure

What has emerged from the combination of single cell and network modeling results using the R-DB Model is a very different view of the functional organization of cerebellar cortical networks (Bower 1997a, 2010). Instead of being driven by parallel fiber excitation, somatic output appears to be influenced primarily by synapses associated with the ascending segment of the granule cell axon. Parallel fibers and the inhibition they also drive through feedforward molecular layer inhibitory neurons, instead seem to indirectly influence the ongoing spiking behavior of the soma. As diagramed in Fig. 5.17, what is interesting is that the interactions between these different synaptic influences turn out to be manifest in the fine physical structure of the Purkinje cell dendrite itself. Specifically, anatomical studies have shown that the synapses associated with the ascending granule cell axon segments are found only

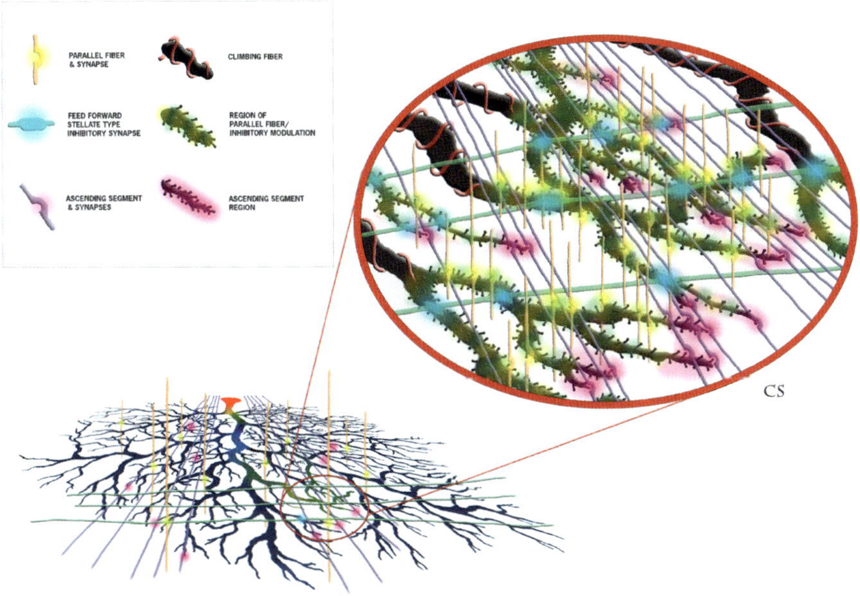

Fig. 5.17 Schematic representation of the proposed synaptic and functional structure of cerebellar Purkinje cells. Each element and region is color coded as shown in the figure legend. The diagram demonstrates that the influence of ascending segment synapses must traverse regions of the spiny dendrite influenced by parallel fibers and molecular layer inhibitory interneurons. This is predicted to form the anatomical basis for modulation of the Purkinje cell's response to peripheral input. Reproduced with permission from Bower (2012)

on the distal regions of the dendrite, while parallel fiber inputs are found dominantly on the more proximal spiny dendrites (Gundappa-Sulur et al. 1999, Lu et al. 2009). In principle what this means is that synapses from parallel fiber (and molecular layer interneurons) are in an ideal position to modulate somatic responses to ascending segment inputs. A modulatory effect is further expected by the fact that both the amplification mechanism and the influence of background parallel fiber and molecular layer inhibitory on the soma appear to be mediated through voltage dependent calcium channels (Bower 2010). It is interesting to note that the R-DB Model also predicts that climbing fiber activation would flood the entire spiny branchlet region with calcium, potentially providing a general reset function for these modulatory effects (Bower 1997c).

This R-DB Model-based reevaluation of the functional structure of cerebellar cortical circuitry has significant implications for cerebellar function (Bower 1997b, 2002, 2012; Bower and Parsons 2003; Manto et al. 2012). Instead of performing a traditional parallel fiber mediated role in timing (Braitenberg 1967; Heck and Sultan 2002), or pattern recognition (Marr 1969; Albus 1971; Ito 2006; Ohyama et al. 2010), Purkinje cells would seem to be more involved in a contextual kind of function, where specific sensory inputs via the ascending segment synapses are placed in a modulatory context provided the parallel fiber and molecular layer interneuron activity (Bower 2012). As only one resulting change in perspective, parallel fiber synaptic plasticity has generally been interpreted in the context of Purkinje cells serving a pattern recognition function. Specifically, first predicted in the context of theories of cerebellar motor learning (Marr 1969; Albus 1971), parallel fiber synaptic plasticity has been assumed for many years to be a mechanism to select which of the tens of thousands converging parallel fiber's actually drive Purkinje cell output (Ito 2006; Steuber et al. 2007). In the context of the results of the R-DB Model, however, it has been proposed instead that parallel fiber synaptic plasticity serves to keep the overall levels of excitatory input to the Purkinje cell dendrite in balance with overall levels of inhibitory input (De Schutter 1997). In other words, synaptic plasticity in cerebellar cortex may be providing what is, in effect, a more homeostatic function than one specifically related to pattern recognition and learning.

Why Build and Use Community Models?

First, before considering why it is important to build and use community models, it is important to state that one purpose of community models is NOT to force everyone to accept the same functional interpretations. In fact, having originally published the first suggestion that parallel fiber synaptic modification might not be related to learning (De Schutter 1997), De Schutter et al. have recently been using the R-DB Model to explore synaptic learning mechanisms (Achard and De Schutter 2008). However, even or maybe especially in the context of ongoing controversies and debates regarding Purkinje cell and cerebellar function, there are several distinct and clear advantages to the use of community models:

Confidence in Model Structure

As should be clearly evident in this chapter, realistic biological models are becoming more and more complex as more information is available and the behavior we seek to replicate is becoming finer grained. Some have suggested, for this reason, that modelers should build less complex, more abstract Purkinje cell models (Bush and Sejnowski 1990; Coop and Reeke 2001). These models, however, have barely been referenced in the literature and have not resulted in any further modeling studies. Accordingly, realistic models are likely to continue to be developed and propagate. Therefore, especially given their complexity a basic but important use of community models is to develop confidence in model structure through testing by multiple investigators in multiple-laboratories.

Of course, the growing complexity of realistic models as well as their common use also requires the concomitant growth in modeling infrastructure and techniques. In fact, anticipating the importance of modeling systems, the original transfer of the model from the Segev to Bower laboratories involved converting from the electronic circuit modeling system SPICE (Bunow et al. 1985) into GENESIS, a modeling system specifically designed for models of this type (Bower and Beeman 1995, 2007). More recently, the R-DB Model, has been translated into NEURON (Hines and Carnevale 1997), and been built into Neuroconstruct (Gleeson et al. 2007), further extending its reach and availability. It is also necessary to continue to develop new tools for model visualization, publishing and perhaps especially the quantitative study of parameter variations (Cornelis et al. 2010, 2011).

Reinventing the Wheel

While the R-DB Model has emerged as a community model, this does not mean that other realistic Purkinje cell models have not been developed. In the last 10 years several other multi-compartmental single cell Purkinje cell models have been published (Roth and Häusser 2001; Vetter et al. 2001; Heck et al. 2003; Steuber and Willshaw 2004; Kulagina et al. 2008; Sjostrom et al. 2008; Genet et al. 2010). In some cases, new models were developed so that experimental results could be evaluated in the context of the specific dendritic morphologies of the recorded neurons (Roth and Häusser 2001). In most others however, it is not at all clear why it was necessary to build an entirely new model. So far there is no evidence that these models have been used outside of their labs of origin and most have been described in only one publication, reflecting what is the more common general state of computational neuroscience (Manninen et al. 2010). The risk to the field is additional confusion and uncertainty as well as less clarity in measuring and understanding progress. As one recent example, the "new unifying hypothesis" on the influence of the Purkinje cell dendrite on its soma, recently proposed by Genet et al. 2010, is largely a replication of work done 20 years earlier with the R-DB Model (Jaeger et al. 1996). This earlier effort is only referenced for a technical detail. In another example, the role of intrinsic somatic spike generation in Purkinje cell behavior is the subject of a recent modeling study that actual does uses the same morphologies as Shelton and Rapp (Forrest et al. 2012), but fails to reference or take advantage of previous related work.

Establishing Modeling Standards

Perhaps one of the most important reasons to develop and adopt community models is to establish specific and agreed upon modeling standards. It is a remarkable fact that even though Pellionisz and Llinas first proposed more than 25 years ago that the "adequacy" of a Purkinje cell model should depend on its ability to replicate its characteristic response properties (Pellionisz and Llinas 1977), many Purkinje cell models, especially those providing the base for network simulations, have made no attempts to do so (Blum et al. 1993; Buonomano and Mauk 1994; Yuen et al. 1995; Barto et al. 1999; Chauvet and Chauvet 1999; Medina and Mauk 2000; Spoelstra et al. 2000; Kistler and De Zeeuw 2002; Brunel et al. 2004; Mauk and Ohyama

2004; Steuber and Willshaw 2004; Yamazaki and Tanaka 2007; Carrillo et al. 2008; Kulagina et al. 2008; de Gruijl et al. 2009; Dean et al. 2009; Abrams et al. 2010; Ohyama et al. 2010; Dean and Porrill 2011; Li et al. 2012; Yamazaki and Nagao 2012). It is entirely unclear what the value of a network model is if the properties of its principle neuron bears little resemblance to physiological reality.

Establishing an Accepted Understanding of Purkinje Cell Behavior and Function

Quoting from the summary of De Schutter and Bower (1994b): "This simulation work demonstrates that a model based on voltage clamp data and tuned entirely on the response of Purkinje cells to current injection is capable of reproducing a wide range of synaptically activated responses" (De Schutter and Bower 1994b, p. 401). By incorporating active dendritic properties, the model addressed questions regarding the influence of an active dendrite on cellular processes that had motivated Purkinje cell modeling from the previous 25 years (Calvin and Hellerstein 1969; Pellionisz and Llinas 1977; Shelton 1985; Rapp et al. 1992, 1994; Bush and Sejnowski 1990; Travis 1990). In doing so, the model revealed many new features of cerebellar Purkinje cells many of which were subsequently supported by experimental studies (De Schutter 1999; Bower 2010). Perhaps especially important, this combination of modeling and experimental studies has revealed that Purkinje cell responses to granule cell-related excitatory and inhibitory synaptic inputs are quite different from the parallel fiber dominant, integrate and fire type cellular dynamics assumed by the most dominant current theories of cerebellar function (Braitenberg 1967; Marr 1969; Albus 1971; Pellionisz and Szentagothai 1974; Medina and Mauk 2000; Vetter et al. 2001; Heck and Sultan 2002; Ito 2006; Kitamura and Kano 2012). Perhaps it is not surprising that the models built primarily to support those theories continue to assume that Purkinje cells can be represented as simple integrate and fire neurons (Buonomano and Mauk 1994; Medina and Mauk 2000; Mauk and Ohyama 2004; Yamazaki and Tanaka 2007; Carrillo et al. 2008; Hong and Optican 2008; Ohyama et al. 2010; Li et al. 2012). In principle, full adoption of a community model could help develop some uniformity in how Purkinje cells are represented and considered.

Efficient Collaborative Communication

This then leads to what is perhaps the most practical and important value of a community model: its use as a means of communication and collaboration. For the R-DB Model, this is evident in the continuing expansion of related modeling studies. In quantitative fields, like physics and chemistry, this type of shared model is the basis for publishing as well as funding results. Philosophers of science have long recognized the importance for science of an eventual transition from an observation-based story telling, to a quantitative model-based structure (Kuhn 1962). It is worth noting again, that models that misrepresent the actual physical properties of the neurons or circuits in order to demonstrate the plausibility of a preexisting theory or hypothesis are essentially an extension of the story telling tradition.

Beyond modelers, while the R-DB Model has been referenced more than 500 times in the last 20 years, the same 20 years have seen the publication of over 10,000 experimental papers on Purkinje cells. Many of those papers, even today, raise issues that modelers have considered for many years, and in some cases have even resolved years ago. Yet even review articles on subjects as central to 40 years of Purkinje cell modeling as the active properties of the Purkinje cell dendrite can quite remarkably be published with hardly any mention of these modeling results (Kitamura and Kano 2012).

This therefore, is perhaps the most important reason that over the next 20 years it will be critical for the computational neuroscience community to adopt and build community models. By committing to the use of community models we establish a common structure that can be presented to the larger neuroscience community, not as just another model, but as a model that has been built, tested, verified and accepted. Why shouldn't those models find their way into graduate training programs, or neuroscience textbooks? As long as we fail to cooperate, we will likely continue to be largely ignored, not only be experimentalists, but also by our fellow modelers. It is only through the cooperative building and testing of models that an underlying quantitative infrastructure will begin to be constructed for neuroscience.

References

Abrams ZR, Warrier A, Trauner D, Zhang X, Abrams ZR, Warrier A, Trauner D, Zhang X (2010) A signal processing analysis of Purkinje cells in vitro. Front Neural Circuits 4:13

Achard P, De Schutter E (2006) Complex parameter landscape for a complex neuron model. PLoS Comput Biol 2:794–804

Achard P, De Schutter E (2008) Calcium, synaptic plasticity and intrinsic homeostasis in Purkinje neuron models. Front Comput Neurosci 2:8

Albus JS (1971) A theory of cerebellar function. Math Biosci 10:25–61

Antunes G, De Schutter E (2012) A stochastic signaling network mediates the probabilistic induction of cerebellar long-term depression. J Neurosci 32:9288–9300

Anwar H, Hong S, Schutter E (2010) Controlling Ca2+-activated K+ channels with models of Ca2+ buffering in Purkinje cells. Cerebellum 11:681–693

Ascoli GA (2007) Successes and rewards in sharing digital reconstructions of neuronal morphology. Neuroinformatics 5:154–160

Baldi P, Vanier MC, Bower JM (1998) On the use of Bayesian methods for evaluating compartmental neural models. J Comput Neurosci 5:285–314

Barto AG, Fagg AH, Sitkoff N, Houk JC (1999) A cerebellar model of timing and prediction in the control of reaching. Neural Comput 11:565–594

Bell CC, Grimm RJ (1969) Discharge properties of Purkinje cells recorded on single and double microelectrodes. J Neurophysiol 32:1044–1055

Blum EK, Wang X (1990) Design for Purkinje cells. In: Eeck AFH (ed) Analysis and modeling of neural systems. Kluwer Academic, Boston, pp 123–129
Blum EK, Khademi PM, Thompson RF (1993) Model and simulation of a simplified cerebellar neural network for classical conditioning of the rabbit eyeblink response. In: Proceedings of analysis and modeling of neural systems, vol 2. Kluwer Academic
Bower JM (1990) Reverse engineering the nervous system: an anatomical, physiological, and computer based approach. In: Zornetzer S, Davis J, Lau C (eds) An introduction to neural and electronic networks. Academic, New York, pp 3–24
Bower JM (1997a) What do parallel fibers do? Commentary on "the detection and generation of sequences as a key to cerebellar function: experiments and theory" by: Valentino Braitenberg, Detlef Heck, and Fahad Sultan Behavioral Brain Sci 20:229–277
Bower JM (1997b) Is the cerebellum sensory for motor's sake, of motor for sensory's sake: the view from the whiskers of a rat? Prog Brain Res 114:483–516
Bower JM (1997c) The cerebellum and the control of sensory data acquisition. In: Schmahmann J (ed) International review of neurobiology vol 41: the cerebellum and cognition, 1997(null) edn. Academic, San Diego, CA, pp 489–513
Bower JM (2002) The organization of cerebellar cortical circuitry revisited: implications for function. Ann N Y Acad Sci 978:135–155
Bower JM (2010) Model-founded explorations of the roles of molecular layer inhibition in regulating Purkinje cell responses in cerebellar cortex: more trouble for the beam hypothesis. Front Cell Neurosci 4:27
Bower JM (2012) The computational structure of the cerebellar molecular layer. In: Manto M, Gruol DL, Schmahmann JD, Koibuchi N, Rossi F (eds) Handbook of cerebellum and cerebellar disorders. Springer, Berlin
Bower JM, Beeman D (1995) The book of GENESIS. Springer, New York
Bower JM, Beeman D (2007) Constructing realistic neural simulations with GENESIS. Methods Mol Biol 401:103–125
Bower JM, Parsons LM (2003) Rethinking the "lesser brain". Sci Am 289:50–57
Bower JM, Woolston DC (1983) Congruence of spatial organization of tactile projections to granule cell and Purkinje cell layers of cerebellar hemispheres of the albino rat: vertical organization of cerebellar cortex. J Neurophysiol 49:745–766
Bower JM, Woolston DC, Gibson JM (1980) Congruence of spatial patterns of receptive field projections to Purkinje cell and granule cell layers in the cerebellar hemispheres of the rat. Soc Neurosci Abstr 6:511
Braitenberg V (1967) Is the cerebellar cortex a biological clock in the millisecond range? Prog Brain Res 25:334–346
Brown ME, Ariel M (2009) Topography and response timing of intact cerebellum stained with absorbance voltage-sensitive dye. J Neurophysiol 101:474–490
Brown S-A, Loew LM (2012) Computational analysis of calcium signaling and membrane electrophysiology in cerebellar Purkinje neurons associated with ataxia. BMC Syst Biol 6:70
Brown S-A, Moraru II, Schaff JC, Loew LM (2011) Virtual NEURON: a strategy for merged biochemical and electrophysiological modeling. J Comput Neurosci 31:385–400. Available at: http://www.springerlink.com/index/10.1007/s10827-011-0317-0
Brunel N, Hakim V, Isope P, Nadal JP, Barbour B (2004) Optimal information storage and the distribution of synaptic weights: perceptron versus Purkinje cell. Neuron 43:745–757
Bunow B, Segev I, Fleshman JW (1985) Modeling the electrical behavior of anatomically complex neurons using a network analysis program: excitable membrane. Biol Cybern 53:41–56
Buonomano DV, Mauk MD (1994) Neural-network model of the cerebellum—temporal discrimination and the timing of motor-responses. Neural Comput 6:38–55
Bush PC, Sejnowski TJ (1990) Simulations of a reconstructed cerebellar Purkinje cell based on simplified channel kinetics. Neural Comput 3:321–332
Cajal S (1911) Histologie du System Nerveux de l'Homme et des Vertebres. Maloine, Paris
Calvin WH (1969) Dendritic spikes revisited. Science 166:637–638
Calvin WH, Hellerstein D (1969) Dendritic spikes versus cable properties. Science 163:96–97

Carrillo RR, Ros E, Boucheny C, Coenen OJ (2008) A real-time spiking cerebellum model for learning robot control. Biosystems 94:18–27
Cavelier P, Pouille F, Desplantez T, Beekenkamp H, Bossu JL (2002) Control of the propagation of dendritic low-threshold Ca(2+) spikes in Purkinje cells from rat cerebellar slice cultures. J Physiol 540:57–72
Chauvet P, Chauvet GA (1999) Purkinje local circuits with delays: mathematical conditions of stability for learning and retrieval. Neural Netw 12:59–77
Chono K, Takagi H, Koyma S, Suzuki H, Ito E (2003) A cell model study of calcium influx mechanism regulated by calcium-dependent potassium channels in Purkinje cell dendrites. J Neurosci Methods 129:115–127
Cohen D, Yarom Y (1998) Patches of synchronized activity in the cerebellar cortex evoked by mossy-fiber stimulation: questioning the role of parallel fibers. Proc Natl Acad Sci USA 95:15032–15036
Coop AD, Reeke GNJ (2001) The composite neuron: a realistic one-compartment Purkinje cell model suitable for large-scale neuronal network simulations. J Comput Neurosci 10:173–186
Coop AD, Cornelis H, Santamaria F (2010) Dendritic excitability modulates dendritic information processing in a Purkinje cell model. Front Comput Neurosci 4:6
Cornelis H, Coop AD, Rodriquez M, Beeman D, Bower JM (2010) Using models to collaborate, communicate, and publish: an introduction to GENESIS 3.0 and the future of computational neurobiology. In: Proceedings of CNS 2010
Cornelis H, Coop AD, Bower JM (2011) A federated design for a neurobiological simulation engine: the CBI federated software architecture. PLoS One 7(1):e28956
de Gruijl JR, van der Smagt P, De Zeeuw CI (2009) Anticipatory grip force control using a cerebellar model. Neuroscience 162:777–786
De Schutter E (1994) Modelling the cerebellar Purkinje cell: experiments in computo. Prog Brain Res 102:427–441
De Schutter E (1997) A new functional role for cerebellar long-term depression. Prog Brain Res 114:529–542
De Schutter E (1998) Dendritic voltage and calcium-gated channels amplify the variability of postsynaptic responses in a Purkinje cell model. J Neurophysiol 80:504–519
De Schutter E (1999) Using realistic models to study synaptic integration in cerebellar Purkinje cells. Rev Neurosci 10:233–245
De Schutter E (2012) The importance of stochastic signaling processes in the induction of long-term synaptic plasticity, Neural Networks, Available online 7 December 2012, ISSN 0893-6080, 10.1016/j.neunet.2012.11.015
De Schutter E, Bower JM (1992) Purkinje neuron simulation on the Intel Touchstone Delta with GENESIS. In: Mihaly T, Messina P (eds) Proceedings of the grand challenge computing fair. CCSF Publishing, Pasadena, CA, pp 268–279
De Schutter E, Bower JM (1993) Integration of synchronous and asynchronous synaptic inputs in a detailed compartmental model of the cerebellar Purkinje cell. In: Bower JM, Eeckman F (eds) Computation and neural systems 1992. Kluwer Press, Boston, pp 355–362
De Schutter E, Bower JM (1994a) An active membrane model of the cerebellar Purkinje cell. I. Simulation of current clamps in slice. J Neurophysiol 71:375–400
De Schutter E, Bower JM (1994b) An active membrane model of the cerebellar Purkinje cell II. Simulation of synaptic responses. J Neurophysiol 71:401–419
De Schutter E, Bower JM (1994c) Simulated responses of cerebellar Purkinje cells are independent of the dendritic location of granule cell synaptic inputs. Proc Natl Acad Sci USA 91:4736–4740
De Schutter E, Steuber V (2009) Patterns and pauses in Purkinje cell simple spike trains: experiments, modeling and theory. Neuroscience 162:816–826
de Solages C, Szapiro G, Brunel N, Hakim V, Isope P, Buisseret P, Rousseau C, Barbour B, Léna C (2008) High-frequency organization and synchrony of activity in the Purkinje cell layer of the cerebellum. Neuron 58:775–788

Dean P, Porrill J (2011) Evaluating the adaptive-filter model of the cerebellum. J Physiol 589:3459–3470
Dean P, Porrill J, Ekerot C-F, Jorntell H (2009) The cerebellar microcircuit as an adaptive filter: experimental and computational evidence. Nat Rev Neurosci 11:30–43
Dizon M-J, Khodakhah K (2011) The role of interneurons in shaping Purkinje cell responses in the cerebellar cortex. J Neurosci 31:10463–10473
Dodge FA, Cooley JW (1973) Action potential of the motorneuron. IBM J Res Develop 17:219–229
Eccles JC, Llinas R, Sasaki K (1966a) Parallel fibre stimulation and the responses induced thereby in the Purkinje cells of the cerebellum. Exp Brain Res 1:17–29
Eccles JC, Llinas R, Sasaki K (1966b) The excitatory synaptic action of climbing fibres on the Purkinje cells of the cerebellum. J Physiol (Lond) 182:268–296
Eccles JC, Ito M, Szentagothai J (1967) The cerebellum as a neuronal machine. Springer, Berlin
Eccles JC, Faber DS, Murphy JT, Sabah NH, Taborikova H (1972) Investigation on integration of mossy fiber inputs to Purkyne cells in the anterior lobe. Exp Brain Res 13:54–77
Fernandez FR, Engbers JDT, Turner RW (2007) Firing dynamics of cerebellar Purkinje cells. J Neurophysiol 98:278–294
Forrest MD, Wall MJ, Press DA, Feng J (2012) The sodium-potassium pump controls the intrinsic firing of the cerebellar Purkinje neuron. PLoS One 7:e51169
Freeman JA, Nicholson C (1975) Experimental optimization of current source-density technique for anuran cerebellum. J Neurophysiol 38:369–382
Gähwiler BH, Llano I (1989) Sodium and potassium conductances in somatic membranes of rat Purkinje cells from organotypic cerebellar cultures. J Physiol (Lond) 417:105–122
Gauck V, Jaeger D (2003) The contribution of NMDA and AMPA conductances to the control of spiking in neurons of the deep cerebellar nuclei. J Neurosci 23:8109–8118
Genet S, Sabarly L, Guigon E, Berry H, Delord B (2010) Dendritic signals command firing dynamics in a mathematical model of cerebellar Purkinje cells. Biophys J 99:427–436
Gleeson P, Steuber V, Silver RA (2007) neuroConstruct: a tool for modeling networks of neurons in 3D space. Neuron 54:219–235
Gundappa-Sulur G, De Schutter E, Bower JM (1999) Ascending granule cell axon: an important component of cerebellar cortical circuitry. J Comp Neurol 408:580–596
Heck D, Sultan F (2002) Cerebellar structure and function: making sense of parallel fibers. Hum Mov Sci 21:411–421
Heck D, Borst A, Antkowiak B (2003) Passive spatial and temporal integration of excitatory synaptic inputs in cerebellar Purkinje cells of young rats. Neurosci Lett 341:79–83
Heck DH, Thach WT, Keating JG (2007) On-beam synchrony in the cerebellum as the mechanism for the timing and coordination of movement. Proc Natl Acad Sci USA 104:7658–7663
Hines ML, Carnevale NT (1997) The NEURON simulation environment. Neural Comput 9:1179–1209
Hirano T, Hagiwara S (1989) Kinetics and distribution of voltage-gated Ca, Na and K channels on the somata of rat cerebellar Purkinje cells. Pflugers Arch 413:463–469
Hodgkin AL, Huxley AF (1952) A quantitative description of membrane current and its application to conduction and excitation in nerve. J Physiol 117:500–544
Holmes WR, Rall W (1992) Electrotonic length estimates in neurons with dendritic tapering or somatic shunt. J Neurophysiol 68:1421–1437
Holtzman T, Rajapaksa T, Mostofi A, Edgley SA (2006) Different responses of rat cerebellar Purkinje cells and Golgi cells evoked by widespread convergent sensory inputs. J Physiol (Lond) 574:491–507
Hong S, Optican LM (2008) Interaction between Purkinje cells and inhibitory interneurons may create adjustable output waveforms to generate timed cerebellar output. PLoS one 3:e2770
Howell FW, Dyhrfjeld-Johnsen J, Maex R, Goddard N, De Schutter E (2000) A large scale model of the cerebellar cortex using PGENESIS. Neurocomputing 32–33:1041–1046
Isope P, Barbour B (2002) Properties of unitary granule cell: Purkinje cell synapses in adult rat cerebellar slices. J Neurosci 22:9668–9678

Isope P, Hildebrand ME, Snutch TP (2012) Contributions of T-type voltage-gated calcium channels to postsynaptic calcium signaling within Purkinje neurons. Cerebellum 11:651–665

Ito M (2001) Cerebellar long-term depression: characterization, signal transduction, and functional roles. Physiol Rev 81:1143–1195

Ito M (2006) Cerebellar circuitry as a neuronal machine. Prog Neurobiol 78:272–303

Jaeger D, Bower JM (1999) Synaptic control of spiking in cerebellar Purkinje cells: dynamic current clamp based on model conductances. J Neurosci 19:6090–6101

Jaeger D, De Schutter E, Bower JM (1993) Prolonged activation following brief synaptic input in the cerebellar Purkinje cell: intracellular recording and compartmental modeling. In: Bower JM, Eeckman F (eds) Computation and neural systems 1992. Kluwer Press, Boston, pp 343–348

Jaeger D, De Schutter E, Bower JM (1996) The role of synaptic and voltage-gated currents in the control of Purkinje cell spiking: a modeling study. J Neurosci 17:91–106

Kaneda M, Wakamori M, Ito C, Akaike N (1990) Low-threshold calcium current in isolated Purkinje cell bodies of rat cerebellum. J Neurophysiol 63:1046–1051

Khaliq ZM, Gouwens NW, Raman IM (2003) The contribution of resurgent sodium current to high-frequency firing in Purkinje neurons: an experimental and modeling study. J Neurosci 23:4899–4912

Kistler WM, De Zeeuw CI (2002) Dynamical working memory and timed responses: the role of reverberating loops in the olivo-cerebellar system. Neural Comput 14:2597–2626

Kitamura K, Häusser M (2011) Dendritic calcium signaling triggered by spontaneous and sensory-evoked climbing fiber input to cerebellar Purkinje cells in vivo. J Neurosci 31:10847–10858

Kitamura K, Kano M (2012) Dendritic calcium signaling in cerebellar Purkinje cell, Neural Networks, Available online 5 September 2012, ISSN 0893-6080, 10.1016/j.neunet.2012.08.001

Kolb FP, Arnold G, Lerch R, Straka H, Buttner-Ennever J (1997) Spatial distribution of field potential profiles in the cat cerebellar cortex evoked by peripheral and central inputs. Neuroscience 81:1155–1181

Konnerth A, Dreessen J, Augustine GJ (1992) Brief dendritic calcium signals initiate long-lasting synaptic depression in cerebellar Purkinje cells. Proc Natl Acad Sci USA 89:7051–7055

Kreiner L, Jaeger D (2004) Synaptic shunting by a baseline of synaptic conductances modulates responses to inhibitory input volleys in cerebellar Purkinje cells. Cerebellum 3:112–125

Kuhn T (1962) The structure of scientific revolutions, 4(null) edn. Chicago University Press, Chicago

Kulagina IB, Korogod SM, Horcholle-Bossavitt G, Batini C, Tyc-Dumont S (2008) Phase relationships between calcium and voltage oscillations in different dendrites of Purkinje neurons. Neurophysiology 145:404–411

Kulagina IB (2008) Phase relationships between calcium and voltage oscillations in different dendrites of Purkinje Neurons. Proceedings of the International School "Problems of Experimental, Clinical and Theoretical Neurosciences". Dnepropetrovsk, Ukraine, May 2–5, 2008. Neirofiziologiya/Neurophysiology, 40(5–6):477–485

Lev-Ram V, Miyakawa H, Lasser-Ross N, Ross WN (1992) Calcium transients in cerebellar Purkinje neurons evoked by intracellelular stimulation. J Neurophysiol 68:1167–1177

Li W-K, Hausknecht MJ, Stone P, Mauk MD (2012) Using a million cell simulation of the cerebellum: Network scaling and task generality, Neural Networks, Available online 20 November 2012, ISSN 0893-6080, 10.1016/j.neunet.2012.11.005

Llinas R (1982) General discussion: radial connectivity in the cerebellar cortex: a novel view regarding the functional organization of the molecular layer. In: Palay SL, Chan-Palay V (eds) The cerebellum: new vistas, 0(null)th edn, Exp. Brain Res. Suppl. vol. 6. Springer, New York, pp 189–194

Llinas R, Hillman DE (1969) Physiological and morphological organization of cerebellar circuits in various vertebrates. In: Llinas R (ed) Neurobiology of cerebellar evolution and development. American Medical Association, Chicago, pp 43–76

Llinas R, Nicholson C (1976) Reversal properties of climbing fiber potential in cat Purkinje cells: an example of a distributed synapse. J Neurophysiol 39:311–323

Llinas R, Sugimori M (1980a) Electrophysiological properties of in vitro Purkinje cell dendrites in mammalian cerebellar slices. J Physiol 305:197–213

Llinas R, Sugimori M (1980b) Electrophysiological properties of in vitro Purkinje cell somata in mammalian cerebellar slices. J Physiol 305:171–195

Llinas R, Sugimori M (1992) The electrophysiology of the cerebellar Purkinje cell revisited. In: Llinas R, Sotelo C (eds) The cerebellum revisited. Springer Verlag, New York, pp 167–181

Llinas R, Nicholson C, Freeman JA, Hillman DE (1968) Dendritic spikes and their inhibition in alligator Purkinje cells. Science 163:1132–1135

Llinas R, Bloedel JR, Hillman DE (1969a) Functional characterization of neuronal circuitry of frog cerebellar cortex. J Neurophysiol 32:847–870

Llinas R, Bloedel JR, Roberts W (1969b) Antidromic invasion of Purkinje cells in frog cerebellum. J Neurophysiol 32:881–891

Lu H, Hartmann MJ, Bower JM (2005) Correlations between purkinje cell single-unit activity and simultaneously recorded field potentials in the immediately underlying granule cell layer. J Neurophysiol 94:1849–1860

Lu H, Esquivel AV, Bower JM (2009) 3D electron microscopic reconstruction of segments of rat cerebellar purkinje cell dendrites receiving ascending and parallel fiber granule cell synaptic inputs. J Comp Neurol 514:583–594

Mandelblat Y, Etzion Y, Grossman Y, Golomb D (2001) Period doubling of calcium spike firing in a model of a Purkinje cell dendrite. J Comput Neurosci 11:43–62

Manninen T, Hituri K, Kotaleski JH, Blackwell KT, Linne M-L (2010) Postsynaptic signal transduction models for long-term potentiation and depression. Front Comput Neurosci 4:152

Manto M, Bower JM, Conforto AB, Delgado-García JM, da Guarda SNF, Gerwig M, Habas C, Hagura N, Ivry RB, Mariën P, Molinari M, Naito E, Nowak DA, Oulad Ben Taib N, Pelisson D, Tesche CD, Tilikete C, Timmann D (2012) Consensus paper: roles of the cerebellum in motor control—the diversity of ideas on cerebellar involvement in movement. Cerebellum 11:457–487

Marr D (1969) A theory of cerebellar cortex. J Physiol (Lond) 202:437–471

Mauk MD, Ohyama T (2004) Extinction as new learning versus unlearning: considerations from a computer simulation of the cerebellum. Learn Mem 11:566–571

Medina JF, Mauk MD (2000) Computer simulation of cerebellar information processing. Nat Neurosci 3(Suppl):1205–1211

Miyakawa H, Lev-Ram V, Lasser-Ross N, Ross WN (1992) Calcium transients evoked by climbing fiber and parallel fiber synaptic inputs in guinea pig cerebellar Purkinje neurons. J Neurophysiol 68:1178–1189

Miyasho T, Takagi H, Suzuki H, Watanabe S, Inoue M, Kudo Y, Miyakawa H (2001) Low-threshold potassium channels and a low-threshold calcium channel regulate Ca2+ spike firing in the dendrites of cerebellar Purkinje neurons: a modeling study. Brain Res 891:106–115

Mocanu OD, Oliver J, Santamaria F, Bower JM (2000) Branching point effects on the passive properties of the cerebellar granule cell axon. Neurocomputing 32–33:207–212

Mugnaini E (1972) The histology and cytology of the cerebellar cortex. In: Larsell O, Jansen J (eds) The comparitive anatomy and histology of the cerebellum: the human cerebellum, cerebellar connections, and cerebellar cortex. University of Minnesota Press, Minneapolis, pp 201–262

Ogasawara H, Doi T, Doya K, Kawato M (2007) Nitric oxide regulates input specificity of long-term depression and context dependence of cerebellar learning. PLoS Comput Biol 3:e179

Ohyama T, Voicu H, Kalmbach B, Mauk MD (2010) A decrementing form of plasticity apparent in cerebellar learning. J Neurosci 30:16993–17003

Pellionisz A (1970) Computer simulation of the pattern transfer of large cerebellar neuronal fields. Acta Biochim Biophys Acad Sci Hung 5:71–79

Pellionisz A, Llinas R (1977) A computer model of cerebellar Purkinje cells. Neuroscience 2:37–48

Pellionisz A, Szentagothai J (1973) Dynamic single unit simulation of a realistic cerebellar network model. Brain Res 49:83–99

Pellionisz A, Szentagothai J (1974) Dynamic single unit simulation of a realistic cerebellar network model. II. Purkinje cell activity within the basic circuit and modified by inhibitory systems. Brain Res 68:19–40

Pugh JR, Raman IM (2009) Nothing can be coincidence: synaptic inhibition and plasticity in the cerebellar nuclei. Trends Neurosci 32:170–177

Rall W (1964) Theoretical significance of dendritic trees for neuronal input–output relations. In: Reiss R (ed) Neural theory and modeling. Stanford University Press, Stanford, CA, pp 73–97

Rall W, Shepherd GM (1968) Theoretical reconstruction of field potentials and dendrodendritic synaptic interactions in olfactory bulb. J Neurophysiol 31:884–915

Rancz EA, Häusser M (2010) Dendritic spikes mediate negative synaptic gain control in cerebellar Purkinje cells. Proc Natl Acad Sci USA 107:22284–22289

Rapp M, Yarom Y, Segev I (1992) The impact of parallel fiber background activity on the cable properties of cerebellar Purkinje cells. Neural Comput 4:518–533

Rapp M, Segev I, Yarom Y (1994) Physiology, morphology and detailed passive models of guinea-pig cerebellar Purkinje cells. J Physiol (Lond) 474:101–118

Regan LJ (1991) Voltage-dependent calcium currents in Purkinje cells from rat cerebellar vermis. J Neurosci 11:2259–2269

Rokni D, Llinas R, Yarom Y (2008) Stars and stripes in the cerebellar cortex: a voltage sensitive dye study. Front Syst Neurosci 1:1–9

Roth A, Häusser M (2001) Compartmental models of rat cerebellar Purkinje cells based on simultaneous somatic and dendritic patch-clamp recordings. J Physiol 535:445–472

Santamaria F, Bower JM (2004) Background synaptic activity modulates the response of a modeled Purkinje cell to paired afferent input. J Neurophysiol 93:237–250

Santamaria F, Jaeger D, De Schutter E, Bower JM (2002) Modulatory effects of parallel fiber and molecular layer interneuron synaptic activity on purkinje cell responses to ascending segment input: a modeling study. J Comput Neurosci 13:217–235

Santamaria F, Wils S, De Schutter E, Augustine GJ (2006) Anomalous diffusion in Purkinje cell dendrites caused by spines. Neuron 52:635–648

Santamaria F, Tripp PG, Bower JM (2007) Feedforward inhibition controls the spread of granule cell-induced Purkinje cell activity in the cerebellar cortex. J Neurophysiol 97:248–263

Santamaria F, Wils S, De Schutter E, Augustine GJ (2011) The diffusional properties of dendrites depend on the density of dendritic spines. Eur J Neurosci 34:561–568

Sarro LM (2004) Characterization of dendrites as nonlinear computation devices. Neurocomputing 58–60:581–586

Segev I, Fleshman JW, Miller JP, Bunow B (1985) Modeling the electrical behavior of anatomically complex neurons using a network analysis program: passive membrane. Biol Cybern 53:27–40

Shelton DP (1985) Membrane resistivity estimated for the Purkinje neuron by means of a passive computer model. Neuroscience 14:111–131

Shin SL, De Schutter E (2006) Dynamic synchronization of Purkinje cell simple spikes. J Neurophysiol 96:3485–3491

Shin S-L, Hoebeek FE, Schonewille M, De Zeeuw CI, Aertsen A, De Schutter E (2007) Regular patterns in cerebellar Purkinje cell simple spike trains. PLoS One 2:e485

Sjostrom PJ, Rancz EA, Roth A, Häusser M (2008) Dendritic excitability and synaptic plasticity. Physiol Rev 88:769–840

Solinas S, Maex R, De Schutter E (2003) Synchronization of Purkinje cell pairs along the parallel fiber axis: a model. Neurocomputing 52–54:97–102

Solinas SMG, Maex R, Schutter E (2006) Dendritic amplification of inhibitory postsynaptic potentials in a model Purkinje cell. Eur J Neurosci 23:1207–1218

Spoelstra J, Schweighofer N, Arbib MA (2000) Cerebellar learning of accurate predictive control for fast-reaching movements. Biol Cybern 82:321–333

Staub C, De Schutter E, Knopfel T (1994) Voltage-imaging and simulation of effects of voltage- and agonist-activated conductances on soma-dendritic voltage coupling in cerebellar Purkinje cells. J Comput Neurosci 1:301–311

Steuber V, De Schutter E (2001) Long-term depression and recognition of parallel fibre patterns in a multi-compartmental model of a cerebellar Purkinje cell. Neurocomputing 38:383–388

Steuber V, De Schutter E (2002) Rank order decoding of temporal parallel fibre input patterns in a complex Purkinje cell model. Neurocomputing 44:183–188
Steuber V, Willshaw D (2004) A biophysical model of synaptic delay learning and temporal pattern recognition in a cerebellar Purkinje cell. J Comput Neurosci 17:149–164
Steuber V, Mittmann W, Hoebeek FE, Silver RA, De Zeeuw CI, Hausser M, De Schutter E (2007) Cerebellar LTD and pattern recognition by Purkinje cells. Neuron 54:121–136
Sultan F, Bower JM (1998) Quantitative Golgi study of the rat cerebellar molecular layer interneurons using principal component analysis. J Comp Neurol 393:353–373
Tahon K, Wijnants M, De Schutter E, Maex R (2011) Current source density correlates of cerebellar Golgi and Purkinje cell responses to tactile input. J Neurophysiol 105:1327–1341
Tank DW, Sugimori M, Connor JA, Llin s RR (1988) Spatially resolved calcium dynamics of mammalian Purkinje cells in cerebellar slice. Science 242:773–777
Traub RD, Middleton SJ, Knöpfel T, Whittington MA (2008) Model of very fast (>75 Hz) network oscillations generated by electrical coupling between the proximal axons of cerebellar Purkinje cells. Eur J Neurosci 28:1603–1616
Travis BJ (1990) A computational model of the cerebellum. In: Eeckmann (ed), Analysis and Modeling of Neural Systems. Kluwer Academic Publishers, Boston, pp 131–137
Van Geit W, Achard P, De Schutter E, Van Geit W, Achard P, De Schutter E (2007) Neurofitter: a parameter tuning package for a wide range of electrophysiological neuron models. Front Neuroinform 1:1
Vetter P, Roth A, Häusser M (2001) Propagation of action potentials in dendrites depends on dendritic morphology. J Neurophysiol 85:926–937
Vladimirescu A, Zhang K, Newton AR, Pederson DO, Sangiovani-Vincentelli A (1981) SPICE. University of California, Berkeley, CA
Vos BP, Maex R, Volny-Luraghi A, De Schutter E (1999) Parallel fibers synchronize spontaneous activity in cerebellar Golgi cells. J Neurosci 19:RC6
Walter JT, Dizon M-J, Khodakhah K (2009) The functional equivalence of ascending and parallel fiber inputs in cerebellar computation. J Neurosci 29:8462–8473
Wang Y, Strahlendorf JC, Strahlendorf HK (1991) A transient voltage-dependent outward potassium current in mammalian cerbellar Purkinje cells. Brain Res 567:153–158
Watanabe S, Takagi H, Miyasho T, Inoue M, Kirino Y, Kudo Y, Miyakawa H (1998) Differential roles of two types of voltage-gated Ca2+ channels in the dendrites of rat cerebellar Purkinje neurons. Brain Res 791:43–55
Womack M, Khodakhah K (2002a) Active contribution of dendrites to the tonic and trimodal patterns of activity in cerebellar Purkinje neurons. J Neurosci 22:10603–10612
Womack MD, Khodakhah K (2002b) Characterization of large conductance Ca2+-activated K+ channels in cerebellar Purkinje neurons. Eur J Neurosci 16:1214–1222
Womack MD, Khodakhah K (2004) Dendritic control of spontaneous bursting in cerebellar Purkinje cells. J Neurosci 24:3511–3521
Womack MD, Chevez C, Khodakhah K (2004) Calcium-activated potassium channels are selectively coupled to P/Q-type calcium channels in cerebellar Purkinje neurons. J Neurosci 24:8818–8822
Yamazaki T, Nagao S (2012) A computational mechanism for unified gain and timing control in the cerebellum. PLoS One 7:e33319
Yamazaki T, Tanaka S (2007) A spiking network model for passage-of-time representation in the cerebellum. Eur J Neurosci 26:2279–2292
Yuen GL, Hockberger PE, Houk JC (1995) Bistability in cerebellar Purkinje cell dendrites modelled with high-threshold calcium and delayed-rectifier potassium channels. Biol Cybern 73:375–388
Zucker RS (1969) Field potentials generated by dendritic spikes and synaptic potentials. Science 165:409–413

Chapter 6
Calcium: The Answer to Life, the Universe, and Everything

Kim Blackwell

Abstract Calcium plays a critical role in numerous physiological processes, both inside and out of the nervous system, and thus is widely studied by both experimental and theoretical neuroscientists. While the role of calcium in the nervous system has been studied by experimentalists for many decades, the last 20 years has seen considerable growth in the use of computational modeling as a tool to unravel the complex cellular mechanisms requiring calcium. For example, computational modeling has enhanced our understanding of processes such as release of neurotransmitter and excitation–contraction coupling in myocytes. Long-term synaptic plasticity and the control of neuronal activity patterns are two additional functions of calcium that are of particular interest to computational neuroscientists. This chapter presents a brief history of computational modeling studies that investigate either the relationship between calcium and long-term synaptic plasticity, or the relationship between calcium and neuronal firing patterns. The focus is on the subset of models that made advancements either in the form of the model or by addressing a novel scientific question.

Introduction

Calcium plays a critical role in numerous physiological processes, both inside and out of the nervous system, and thus is widely studied by experimental and theoretical neuroscientists. The earliest studies demonstrated that calcium was the trigger for neurotransmitter release (Katz and Miledi 1967); however, the complete biophysical and biochemical mechanisms still are not delineated. In the past 20 years,

K. Blackwell (✉)
Molecular Neuroscience Department, Krasnow Institute for Advanced Study, 4400 University Drive, MS 2A1, Fairfax, VA 22030, USA
e-mail: kblackw1@gmu.edu

J.M. Bower (ed.), *20 Years of Computational Neuroscience*, Springer Series in Computational Neuroscience 9, DOI 10.1007/978-1-4614-1424-7_6,

even with advanced imaging techniques it has not been possible to visualize the calcium nanodomains directly surrounding the synaptic vesicles. Thus modeling of calcium in the axon terminal has addressed issues such as proximity of channels to vesicles, and mechanisms relating calcium concentration in the terminal to short-term plasticity (Schneggenburger and Neher 2005). Another critical function of calcium is excitation–contraction coupling in myocytes (Williams et al. 2010). Depolarization of myocytes triggers release of calcium through calcium-dependent calcium-permeable channels located on an intracellular organelle called the sarcoplasmic reticulum. The calcium then binds to proteins which modify the interaction of actin with myosin. In neurons and excitable endocrine cells the smooth endoplasmic reticulum (SER) is a store of calcium that is analogous to the sarcoplasmic reticulum. Release of calcium through two types of calcium-permeable channels, ryanodine receptor channels (RyR) and inositol triphosphate receptor channels (IP_3R), causes calcium spikes and waves. Modeling has been valuable for exploring interactions between calcium release mechanisms and calcium dynamics, such as demonstrating that gap junctions coordinate the rate of calcium oscillations in hepatocytes (Dupont et al. 2007).

Two additional functions of calcium of particular interest to computational neuroscientists are long-term synaptic plasticity and the control of neuronal activity patterns. Long-term potentiation (LTP) and long-term depression (LTD) are two types of synaptic plasticity both of which require elevations in intracellular calcium concentration for their induction (Teyler et al. 1994). Over the past 20 years modeling studies have addressed the relationship among calcium concentration, temporal pattern of stimulation, and direction of plasticity. In particular modeling studies have investigated whether the frequency of stimulation controls the level of calcium in hippocampal and neocortical pyramidal neurons, and why stimulation of two different sets of inputs is required for a large calcium elevation in cerebellar Purkinje neurons. A second experimental observation explored with computational models is that diverse neuronal firing patterns, such as spike frequency adaptation and burst firing, cannot be produced by the standard Hodgkin–Huxley sodium and potassium channels alone. Many other voltage-dependent ion channels are needed, including both calcium-permeable channels and calcium-dependent potassium channels. As these firing patterns may be important for information processing, modeling studies have evaluated the role of calcium in producing these patterns.

Accurate modeling of calcium dynamics involves numerous mechanisms for controlling intracellular concentration (De Schutter and Smolen 1998), and multiple feedback loops at multiple time scales are important (Fig. 6.1). Calcium flows into the cell through various types of voltage-dependent calcium channels (VDCC), and ligand-gated ion channels such as the NMDA type of glutamate receptor channels. Much of this calcium binds to various buffer proteins, such as calmodulin or calbindin, or is pumped out of the cytoplasm into the extracellular space by the plasma membrane calcium ATPase (PMCA) or the sodium calcium exchanger (NCX). Calcium also is released from the SER via IP_3R and RyR channels, and is pumped back into the SER by the smooth endoplasmic reticulum calcium ATPase (SERCA) pump. In addition, diffusion of either calcium or calcium buffers redistributes the

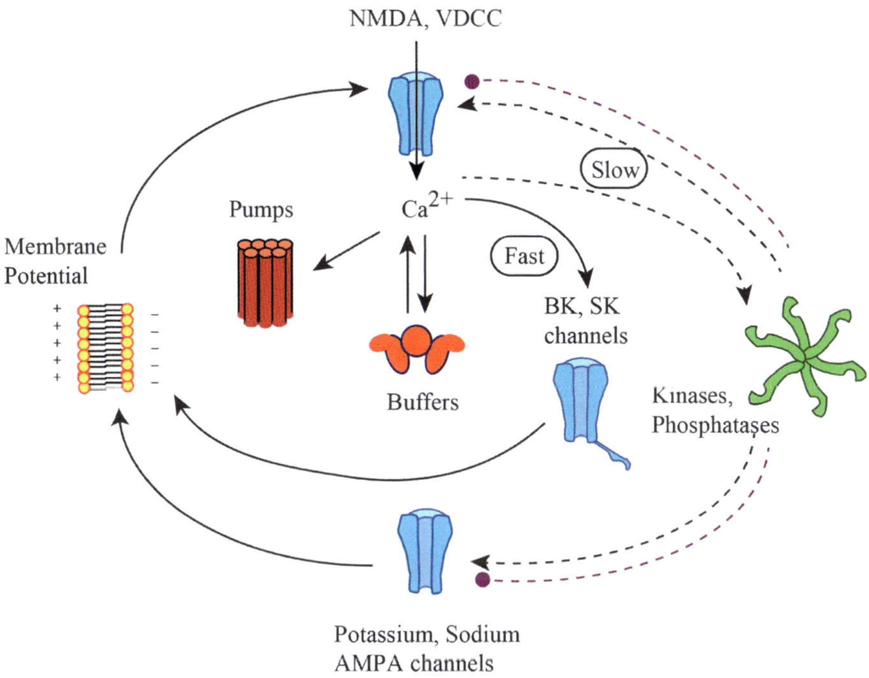

Fig. 6.1 Calcium enters the cell through NMDA receptor channels or voltage-dependent calcium channels. Most of the calcium binds to various buffer proteins or is pumped out of the cell. The remaining free calcium may activate calcium-dependent potassium channels, which oppose membrane depolarization (fast feedback loop). Alternatively, the calcium may activate various kinases and phosphatases, which modify ionic and synaptic channels, forming slow positive or negative feedback loops

calcium from high concentration regions to low concentration regions. The free calcium participates in a fast negative feedback loop which involves binding to calcium-dependent potassium channels. By hyperpolarizing the membrane, these potassium channels decrease influx of calcium through VDCC. A slow feedback loop involves activation of calcium-dependent phosphatases and kinases, which phosphorylate various ion channels, such as the AMPA and NMDA receptor channels, or sodium and potassium channels. By changing the kinetics or voltage dependence of channel opening, these calcium-mediated phosphorylation events modulate membrane excitability and subsequent calcium influx.

This review presents a brief history of computational modeling studies that investigate either the relationship between calcium and long-term synaptic plasticity, or the relationship between calcium and neuronal firing patterns. The focus is on the subset of models that made advancements either in the form of the model or by addressing a novel scientific question. Technical advancements include increasing the accuracy of the electrical model of the neuron, the types of calcium regulatory mechanisms included, and whether simulations were performed deterministically or stochastically. Note that progress has been highly nonlinear, with increased details in one aspect of the model being accompanied by simplifications in other aspects of the model. A further caveat is that technical advancements mentioned in the context of synaptic plasticity or firing patterns sometimes follow identical modeling approaches employed for investigating neurotransmitter release or calcium waves.

Models Investigating Long-Term Synaptic Plasticity

A critical question addressed with computational models of calcium dynamics is whether the elevation in intracellular calcium is sensitive to the characteristics of synaptic activation, as postulated by theories on synaptic plasticity. In the hippocampus, high frequency stimulation is thought to produce LTP because it produces a large calcium elevation, whereas low frequency stimulation produces a small calcium elevation, leading to LTD (Teyler et al. 1994; Bear and Malenka 1994). In the cerebellum, induction of LTD requires pairing of two different types of synaptic inputs (Ito 2001).

Early Models

The earliest models focused on calcium within a single dendritic spine to investigate the relationship between synaptic activation pattern and calcium influx through the NMDA receptor. The spine was subdivided into a small number of compartments, representing the postsynaptic density (PSD), spine head and spine neck and attached to a single dendritic compartment. Diffusion was modeled in one dimension, from the synaptic channels located at the PSD toward the dendrite, and calcium was regulated further by one or two buffer proteins and one or two membrane pumps (Fig. 6.2, Table 6.1). Gamble and Koch (1987) evaluated frequency-dependent elevation of calcium within the dendritic spine, and demonstrated that not only peak calcium but also calcium-bound-calmodulin was significantly greater with high frequency (100 Hz) stimulation as compared to low frequency (1 Hz) stimulation. Zador et al. (1990) coupled their model of calcium dynamics to a simplified 28-compartment electrical model of hippocampal CA1 neuron activity to investigate why both presynaptic activity and postsynaptic depolarization were required for LTP. They demonstrated that the postsynaptic depolarization was required to relieve the magnesium block of NMDA receptors and allow calcium to flow into the cell. Even more interesting, simulations showed that calcium influx was sensitive to the temporal interval between presynaptic glutamate release and postsynaptic depolarization, long before the experimental demonstration of spike timing-dependent plasticity (Bi and Wang 2002). Holmes and Levy (1990) were the first to evaluate the frequency dependence of LTP using a complete multi-compartment electrical model.

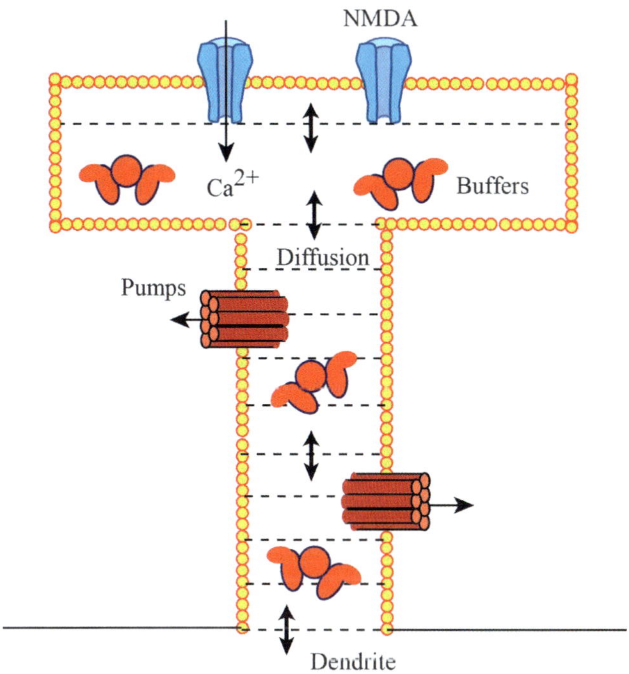

Fig. 6.2 Early models of calcium dynamics focused on calcium within single spine, and included one or two pumps, one or two buffers, axial diffusion influx through NMDA or VDCC

They demonstrated that the frequency dependence of calcium elevation crucially depended on the number of activated synapses, similar to the experimentally observed property of cooperativity (Wigstrom and Gustafsson 1986).

Two additional models published in the 1990s coupled calcium dynamics to electrical activity to explore additional characteristics of LTP. Schiegg et al. (1995) explored the experimental observation that calcium must be elevated for >2 s for successful LTP induction (Malenka et al. 1992). Specifically, they demonstrated that the calcium dynamics model of Zador et al. (1990) produces a fast calcium decay which is not compatible with the prolonged calcium elevation required for LTP. Then they included additional equations describing calcium release from intracellular stores, and showed that this calcium source maintained the calcium elevation for a sufficiently long time. Another model (Holmes and Levy 1997) investigated the experimental observation that weak and strong inputs must be close to each other on a hippocampal granule cell dendritic tree to produce associative LTP of the weak input pathway. This model represented a significant advance because the electrical model included voltage-dependent sodium, potassium, and calcium channels. This model demonstrated that the proximity requirement for the weak and strong inputs is due to shunting inhibition by GABA activation, which produces strong voltage attenuation.

Table 6.1 Characteristics of models used to study plasticity

Author, date	Pumps	Buffers	Diffusion	Calcium release	Electrical
Gamble and Koch (1987)	1	Calcineurin, calmodulin	1D	–	3 comp, AMPA, VDCC, KDr
Zador et al. (1990)	2	Calmodulin	1D	–	28 comp, NMDA, AMPA
Holmes and Levy (1990)	1	Calmodulin	1D	–	Many comp, NMDA, AMPA
Sala and Hernandez-Cruz (1990)	1	1 Diffusible, 1 fixed	1D	–	–
Nowycky and Pinter (1993)	1	1 Diffusible, 1 fixed	1D	–	–
Schiegg et al. (1995)	2	Calmodulin	1D	First order kinetics	9 comp, NMDA, AMPA
Holmes and Levy (1997)	1	Calmodulin, fast buffer	1D	–	Many comp, NMDA, AMPA, NaF, KDr, VDCC, KCa
Markram et al. (1998)	1	4	1D	–	–
Volfovsky et al. (1999)	PMCA, SERCA	Calcineurin, calmodulin, CG-1	2D	Constant	–
Kotaleski et al. (2002)	SERCA	RBA	1D	Li & Rinzel	–
Hernjak et al. (2005)	PMCA, SERCA	Parvalbumin, calbindin, CG-1	1D, 2D	Li & Rinzel	–
Blackwell (2004)	PMCA, NCX, SERCA	1	2D	De Young & Keizer, RyR	13 comp, VDCC, and KCa
Naoki et al. (2005)	PMCA, NCX, SERCA	Calmodulin, 4 others	1D	–	–
Kubota et al. (2007)	0	Calmodulin	3D, stochastic	–	–
Keller et al. (2008)	PMCA, NCX	Calmodulin, OGB-1 or calbindin	3D, stochastic	–	Many comp, NMDA, AMPA, NaF, KDr, VDCC
Schmidt and Eilers (2009)	1	Calbindin, parvalbumin, calmodulin, OGB-1	1D	–	–

Buffers are identified by name when provided in the publication; with the exception of Blackwell, all pumps are Michaelis–Menten formulation. Calcium release used equation from Li and Rinzel, De Young and Keizer, or a simpler equation unique to that publication.
RBA rapid buffer approximation

Thus, blocking $GABA_A$ inputs experimentally, or raising the membrane resistance of the model, eliminated this requirement, because the depolarization produced by the strong inputs then could spread throughout the dendritic tree.

Role of Calcium Buffers

The next set of models investigated how characteristics of calcium buffers, such as the rate of binding, modify calcium dynamics. Though these models were not coupled to neuron electrical activity, in some cases direct comparison to calcium imaging was used to constrain or validate the calcium model. The calcium dynamics were controlled by one-dimensional (1D) diffusion within a spherical neuron, a membrane pump, and several calcium buffers, which were characterized by three basic properties: diffusion rate, affinity, and the speed of binding to calcium. Sala and Hernandez-Cruz (1990) showed that the initial calcium decay was governed by diffusion, whereas a later decay phase depended on extrusion via membrane pumps. In the submembrane shell, calcium increased rapidly to a steady state plateau, whose amplitude was governed by the balance between calcium influx and binding to buffer. Consequently, an increase in buffer binding speed produced a decrease in the submembrane calcium concentration. Nowycky and Pinter (1993) further investigated the role of buffer properties, and showed that during slow changes in calcium (such as those that occur in the interior of the cell), the fraction of calcium bound to a buffer depended on its affinity, but in response to rapid or brief changes in calcium concentration (as typically occurs in the submembrane region), the binding rate was more important. Markram et al. (1998) reevaluated the role of buffer speed using patterns of calcium influx produced by single or trains of action potentials. They demonstrated that with strong calcium extrusion, the calcium due to a single action potential (AP) mostly bound to the fastest buffers. In contrast, during repetitive trains of action potentials, the slower calcium buffers ultimately bound a proportion of the calcium which depended on affinity and buffer quantity more than buffer speed. These studies all demonstrated that calcium dynamics could not be described adequately by a single time constant of decay; rather the buffers produced diverse calcium dynamics depending on the spatiotemporal input pattern.

Role of Diffusion and Morphology

The morphology of dendrites and spines, whose thin necks are diffusional barriers, influences calcium dynamics. The role of the spine neck on calcium dynamics was investigated (Volfovsky et al. 1999) using a model with 2D diffusion of calcium in a spine attached to a dendrite. Simulations studied how spine neck length and the presence of a calcium indicator dye influenced calcium concentration in response to caffeine, which causes calcium release from intracellular stores. Experiments and model simulations together demonstrated that the gradient from spine to dendrite was larger when the spine neck was longer. Simulations further showed that both peak calcium and the spine to dendrite gradient were larger in the absence of the calcium indicator dye. This role of spine neck as diffusional barrier also was observed in cerebellar Purkinje neurons (Schmidt and Eilers 2009) in response to burst stimulation of parallel fiber (PF) synapses. Nonetheless, the weaker diffusional barrier of a short stubby spine permitted sufficient diffusion of calcium-bound buffers to elevate calcium slightly in the dendrite. If neighboring spines were activated together, then the calcium accumulation in the dendrite could be sufficient to activate downstream signaling pathways. In summary, the strength of diffusional barriers critically depended on the quantities of immobile and diffusible calcium buffers.

Mechanisms Underlying the Requirement for Conjunctive Stimulation

Whereas the early models focus on hippocampal LTP, another set of models investigates synaptic plasticity in cerebellar Purkinje cells. Activity in Purkinje neurons is correlated with classical conditioning behavior (Gould and Steinmetz 1996), in which learning occurs in response to repeated pairings of a tone (conditioned stimulus or CS) with an air puff to the eye (unconditioned stimulus or US). In the cerebellum, information about the CS (tone) is conveyed by parallel fibers (PFs) while climbing fibers (CFs) carry the US information (Thompson and Steinmetz 2009). In vitro plasticity in Purkinje neurons resembles classical conditioning in the requirement for paired climbing fiber and parallel fiber inputs (Schreurs et al. 1996). The main question addressed by models of calcium dynamics in Purkinje neurons is "Why is paired stimulation required to trigger synaptic plasticity?"

One hypothesis tested with several computational models is that paired stimulation produces a supralinear calcium increase, which in turn activates critical downstream kinases. To test this hypothesis, the model of Kotaleski et al. (2002) included mechanisms for calcium elevation by both PF and CF. PF activation of metabotropic glutamate receptors produced IP_3 which leads to calcium release from intracellular stores (Li and Rinzel 1994). Activation of VDCC by climbing fiber stimulation was modeled as calcium influx into the dendrite, which then diffused into the spine. Simulations showed that the calcium elevation was significantly larger in response to conjunctive stimulation, but only if CF and PF had the appropriate temporal interval. Both the calcium elevation and activation of protein kinase C were sensitive to the temporal interval between PF and CF stimulation, similar to the sensitivity of classical conditioning behavior to temporal interval between CS and US (Fig. 6.3). This result was replicated in another model (Doi et al. 2005).

The issue of temporal sensitivity was reassessed together with spatial specificity of the calcium elevation in a 50 μm dendrite with attached spines (Hernjak et al. 2005). Simulations demonstrated that cooperativity of PF and CF for calcium

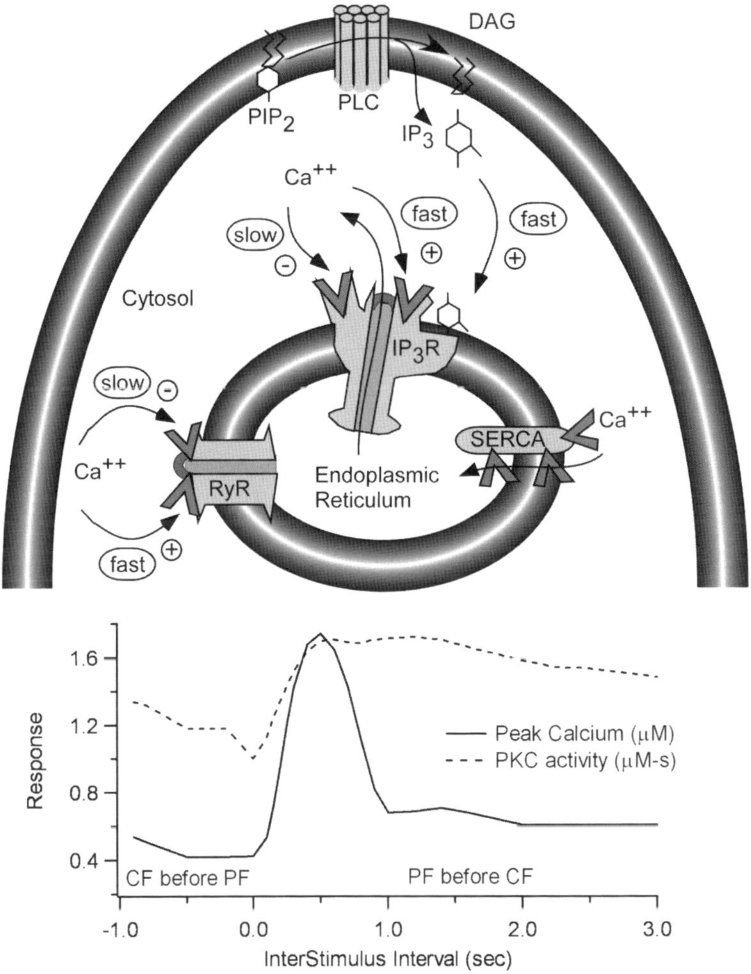

Fig. 6.3 In the cerebellar Purkinje cell, the calcium elevation is greatest for PF–CF intervals between 0.1 and 1.0 s. This calcium contributes to the temporal sensitivity of protein kinase C activation

depends on spine neck radius. If the spine neck radius was small, acting as a barrier for calcium diffusion, then conjunctive stimulation yielded a supralinear calcium increase, but a wide spine neck did not act as a barrier, yielding a small calcium elevation. The major advancement in this model was the morphology, with multiple spines attached to a long dendrite. Simulations using this morphology permitted investigation of spatial specificity, and demonstrated that calcium release did not occur in nearby spines because the diffusional barrier of the spine neck prevented sufficient elevation in IP_3 concentration in nearby spines.

Only one model tightly integrates realistic neuronal electrical activity with equations describing calcium release from intracellular stores to address the mechanisms underlying the requirement for conjunctive stimulation. Pairing light (CS) and vestibular stimulation (US) produces both classical conditioning behavior and plasticity of intrinsic excitability of the photoreceptors of the sea slug *Hermissenda crassicornis* (Alkon et al. 1982). A model of calcium dynamics (Blackwell 2004) addresses the hypothesis that vestibular stimulation causes a calcium elevation in the terminal branches, which propagates as a wave to the soma where it combines supralinearly with the light-induced calcium elevation. This model demonstrates that the two sources of calcium do *not* combine supralinearly; rather, they destructively interfere (Fig. 6.4). The critical difference between this result, showing a lack of cooperativity, and the cooperativity exhibited in the Purkinje cell models is due to the spatial distance between CS and US signals. In the Purkinje cell models, CF-induced depolarization causes a calcium elevation everywhere in the dendrite, and thus this calcium is close to PF-induced IP_3 production. In the *Hermissenda* photoreceptor, the CS and US signals are separated by more than 100 μm, too far to interact via diffusion alone. This inability to interact via wave propagation rules out one of the several alternative mechanisms underlying CS–US cooperativity.

Nanodomains of Calcium

The advancement in calcium imaging techniques has produced a revolution in our concepts of calcium domains which has been accompanied by a major transformation in the models of calcium dynamics. In particular, recent models focus on the generation of nanodomains of calcium within the spine, and readdress questions of synaptic plasticity. One outstanding conundrum is how the neuron can discriminate the calcium elevation required for LTD, which is induced with low frequency stimulation, from the calcium elevation required for LTP, which is induced with high frequency stimulation. Some experiments suggest that LTP and LTD have different sources of calcium, e.g., NMDA versus VDCC, and also different molecular targets. The difference in calcium source and molecular targets suggests that some calcium binding proteins may discriminate temporal patterns of calcium, or that molecular targets are colocalized with the appropriate calcium source and sense nanodomains of calcium. Due to the small size of spines, the numbers of molecules are small, and interactions between calcium and its binding proteins occur stochastically; thus, several of these recent models use stochastic simulation techniques.

The first model to address nanodomains related to synaptic plasticity (Naoki et al. 2005) investigated whether either calcium or calcium-bound-calmodulin could decode both frequency and amplitude of calcium influx. Model simulations showed that the PSD calcium concentration was sensitive to amplitude of calcium influx, but not frequency; whereas whole spine calcium concentration was sensitive to frequency of stimulation, but not amplitude of calcium influx. On the other hand, the quantity of calcium-bound-calmodulin was sensitive to both frequency

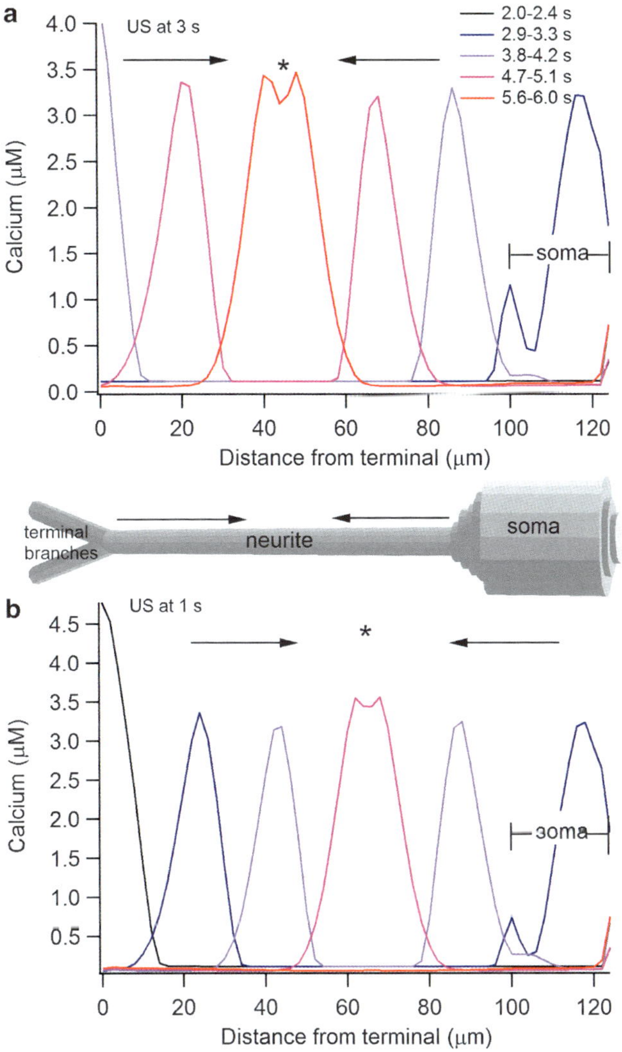

Fig. 6.4 Regardless of the intersimulus interval, light-induced calcium and turbulence-induced calcium propagate toward each other along the neurite (shown by *arrows*), and meet in the middle (*asterisk*). Neither intersimulus interval contributes to the calcium elevation in the soma. (**a**) Forward pairing: when the US occurs 1 s after the CS, the calcium waves meet 45 μm from the terminal branches. (**b**) Backward pairing: when the US occurs 1 s before the CS, the calcium waves meet 65 μm from the terminal branches. No calcium elevation is observed at 5.6–6 s due to destructive interference of the two waves at 4.7–5.1 s

and amplitude, which may be important for activation of downstream kinases and phosphatases. More recently, newly developed simulation software has facilitated investigation of nanodomains through stochastic simulation of calcium reaction–diffusion systems. Kubota et al. (2007) investigate the role of neurogranin, which binds to and acts as a buffer for calmodulin, in controlling calcium dynamics in a spine. Neurogranin has a lower affinity for the calcium-bound form of calmodulin than for the calcium-free form; conversely, neurogranin binding to calcium–calmodulin accelerates calcium dissociation from the C-terminal lobe of calmodulin. Model simulations show that by decreasing calcium binding to calmodulin, neurogranin delays the transition of calmodulin to the fully saturated state. Keller et al. (2008) couple a stochastic simulation of spine calcium dynamics to neuron electrical activity to investigate the experimental observation that the elevation in calcium depends on the timing of presynaptic glutamate release relative to an action potential (Bi and Wang 2002). Though the overall calcium elevation is greater when the glutamate occurs prior to the action potential, due to the voltage dependence of the NMDA receptor, this difference in calcium elevation does not appear at the base of the spine, but is emphasized at the PSD, suggesting that proteins located here would sense a huge difference in calcium concentration due to temporal interval.

Models Investigating Neuronal Firing Patterns

Another crucial issue addressed with computational models of calcium dynamics is the production of diverse firing patterns such as spike frequency adaptation, bursting, or pacing by interaction between various ionic channels. Experiments have shown that these firing patterns are generated by various calcium-dependent potassium channels and diverse VDCC in different cell types (Faber and Sah 2003; Bond et al. 2005). Computational models integrating calcium dynamics with electrical activity have been developed to identify critical characteristics of the channels and their interactions (Table 6.2). Numerous studies which simplify calcium regulatory mechanisms as a single time constant of decay are not addressed here.

Oscillations

The first set of models investigates the interaction between calcium concentration and ionic channels underlying the generation of burst firing and slow membrane potential oscillations (Canavier et al. 1991; Chay 1996a, b; Amini et al. 1999). Most of these studies represent the neuron as a single spherical compartment with multiple plasma membrane channels, and regulate the intracellular calcium using several pumps and a single calcium buffer. Action potentials are generated by the fast sodium and delayed rectifier potassium currents, but slower dynamics are rendered by diverse ionic channels.

The model of the Aplysia R15 (Canavier et al. 1991) includes a fast calcium-permeable channel and a slow calcium-permeable calcium-inactivated channel, both of which are activated by depolarization. Simulations demonstrate that a burst of action potentials produces an increase in calcium, which then inactivates the

Table 6.2 Characteristics of models used to study firing patterns

Author, date	Pumps	Buffers	Diffusion	Electrical comp; ion channels
Yamada et al. (1989)	1	1	1D	1 comp; NaF, KDr, VDCC, KCa
Canavier et al. (1991)	PMCA, NCX	1	–	1 comp; NaF, KDr, VDCC, SI, NS
Chay (1996a, b)	SERCA, PMCA	1	–	1 comp; NaF, KDr, VDCC, SOC, release
Amini et al. (1999)	PMCA, NCX	1	–	1 comp; NaF, KDr, KA, H, CaL, CaN, CaT, CaP/R, SK
Wilson and Callaway (2000)	PMCA	2 (RBA)	1D	1, 5, or many comp; VDCC, KDr, SK
Engel et al. (1999)	1	1	1D	4 comp; NaF, KDr, KA, VDCC, SK, BK
Shao et al. (1999)	1	1 (RBA)	1D	6 comp; NaF, KDr, KA, KD, KM, H, CaL, CaN, CaT, SK, BK
Gu et al. (2005, 2007)	1	1 (RBA)	1D	5 comp; NaF, NaP, KDr, KA, KD, KM, H, CaL, CaN, SK, BK
Goldberg et al. (2009)	1	3	1D	1 comp; SK

Buffers and pumps are identified by name when provided in the publication. Voltage-dependent calcium channels are identified either with a specific subtype or as VDCC if type is unspecified. *SOC* store-operated current, *SI* slow calcium-permeable calcium-inactivated current, *NS* nonspecific cation current, *H* hyperpolarization-activated current, *KA*, *KD* transient potassium currents, *KM* M type potassium current, *RBA* rapid buffer approximation

slow calcium-permeable channel to terminate the burst. During the subsequent repolarization the calcium declines; consequently, the fluctuations in calcium have the same period as the slow variation in potential on which the bursts were superimposed.

A slightly different mechanism to control bursting in Aplysia and Helix neurons was proposed (Chay 1996a, b). Similar to the previous model, this model contains both fast and slow calcium-permeable channels, but the latter is voltage-independent and activated by depletion of calcium in the SER, analogous to the store-operated current that has been found in other neurons. The model also contains a calcium release channel on the ER. During the burst, calcium in the SER increases (due to the SERCA pump) until it reaches a level which inactivates the slow calcium-permeable channel. While the cell repolarizes, SER calcium decreases due to a slow rate of release, until the slow calcium-permeable channel reactivates.

The phenomenon of bursting also was investigated in dopamine neurons of the midbrain (Amini et al. 1999). Elevation of calcium during the depolarizing phase of the oscillations activates the SK channels, which then hyperpolarize the membrane, turning off calcium influx and allowing a decrease in calcium concentration. In addition to SK channels, voltage-dependent activation of potassium channels also contributes to the oscillations but at a slower pace. Thus, inactivating the SK channels, both experimentally and in the model, produces much lower frequency oscillations.

All of these studies of oscillatory mechanisms deserve commendation because they further investigate how the dynamics depend on parameter values, and then simplify the models to use theoretical analysis to identify which mechanisms are essential for oscillations. In summary, the crucial elements producing slow oscillations are a calcium-permeable current that produces both a slow depolarization and slow accumulation in calcium, and either calcium inactivation of the depolarizing current or calcium activation of a hyperpolarizing current.

Independent of the exact ionic channels producing oscillations, the rate of change in calcium concentration depends on the balance between influx, which is proportional to surface area, and efflux, which depends both on surface area via the membrane pumps and volume via buffering. Consequently, the rate of calcium oscillations should be different in a large diameter dendrite compared to a small diameter dendrite. This model prediction was not supported by calcium imaging of dopamine neurons, which reveals a single rate of oscillations, with smaller amplitude oscillations in the larger diameter dendrites. This observation was investigated using a multi-compartment model (Wilson and Callaway 2000) of a single tapered dendrite. Simulations revealed that electrical coupling of the different diameter dendritic compartments synchronized both the electrical oscillations and the calcium oscillations, and compartment diameter instead had an effect on the amplitude of oscillations. This elegant intertwining of experimental measurements and model simulations revealed a parsimonious explanation of the experimental observation.

Spike Frequency Adaptation

Calcium-dependent potassium currents also are implicated in the phenomenon of spike frequency adaptation and control the shape of the action potential. In particular, the shape of the after hyperpolarization (AHP) was modified by calcium influx (Yamada et al. 1989) in a model of bullfrog sympathetic ganglion neuron which included one buffer, one pump, and radial diffusion. Though this model contained a single calcium-dependent potassium channel, another set of models investigated whether BK or SK types of calcium-dependent potassium channels were more important. One model (Engel et al. 1999) demonstrated that the slow SK channel, but not the faster BK channel, produced the observed spike frequency adaptation. Another model investigated the observation that blocking BK channels, but not SK channels, caused a broadening of the spikes, and decreased firing rate at large current injections (Shao et al. 1999; Gu et al. 2005, 2007); thus, BK channels contribute significantly to spike repolarization.

Channel Colocalization

The studies that demonstrated a role for BK channels colocalize the VDCC with BK channels and place both apart from SK channels, reflecting experimental studies suggesting colocalization (Berkefeld et al. 2006). One such experiment (Goldberg and Wilson 2005) reports that calcium influx through N type VDCC triggered by a single AP produces a medium AHP through activation of SK channels, whereas a calcium influx through L type VDCC due to a long, subthreshold depolarization produces a slow AHP through activation of a different calcium-dependent potassium channel. The compelling question is whether this effect requires that N type VDCC and SK channels be colocalized. The alternative is that different electrical activity produces different calcium dynamics, which lead to activation of different calcium-dependent potassium channels.

Goldberg et al. (2009) developed a model to test whether different temporal patterns of electrical activity could activate different calcium-dependent potassium channels *without* channel colocalization. The model has three types of calcium binding proteins: a fast, high affinity, mobile protein representing calmodulin; a fast, low affinity protein; and a slow, high affinity, mobile protein. SK channels in the model have the same binding properties as calmodulin. Simulations show that a brief, large amplitude calcium influx, as occurs with a single action potential, binds mostly to the fast proteins, including SK channels. Calcium does redistribute to slower binding proteins, but with low concentration because total calcium influx is small. In contrast, a low amplitude, long duration calcium influx, as occurs during a long, subthreshold depolarization, is too low in amplitude to activate the SK channel. Instead, most of the calcium binds to the slow, high affinity binding protein. Thus, this elegant study demonstrates that channel colocalization is not required to explain preferential coupling between subtypes of calcium channels and calcium-dependent potassium channels.

Discussion/Conclusion

The future direction of the field of calcium dynamics can be seen from the past evolution of calcium dynamics models, not only those used for studying plasticity and firing patterns but also those developed for calcium release and neurotransmitter release. In model simulations as in experiments, the trend is to investigate functions of calcium within ever smaller regions of space. Interestingly, model advancements both follow and lead advancements in experiments. As experimental technique improves, the data provides better constraints for the model, allowing more precise implementation of realistic mechanisms. The models also lead the experiments by evaluating even smaller spatial and temporal domains, or by evaluating response to in vivo-like spatiotemporal input patterns.

Whether the investigation involves plasticity or firing patterns, the trend is toward more realistic mechanisms. Why do many advancements comprise increasing complexity of the calcium dynamics? Two valid disadvantages of complex models are the difficulty in estimating parameters and the difficulty in fully understanding the mechanisms that produce the essential effect. However, the highly nonlinear nature of calcium mechanisms also makes it extremely difficult to derive simplified models that capture the myriad interactions. For example, models that use a single time constant of decay for calcium do not exhibit calcium dynamics in the spine head that are sensitive to the diameter of the spine neck. Diffusion is required to explain this observation. Similarly, the speed of binding interacts with the dynamics of the calcium influx to control whether calcium binds to one type of potassium channel versus another type. Models with one or two decay time constants or a single calcium buffer cannot explore alternative mechanisms for why certain calcium-permeable channels seem to be coupled to certain calcium-dependent potassium channels.

One of the most important questions still not adequately answered is the relationship between calcium concentration and direction of synaptic plasticity. One experimental investigation has demonstrated that the level of calcium elevation does not sufficiently predict the direction of plasticity (Nevian and Sakmann 2006). Modeling approaches are evaluating activation of other molecules that are critical for induction of synaptic plasticity (Kotaleski and Blackwell 2010) (Chap. 9, this volume). Alternatively, calcium concentration might predict the direction of plasticity if it were measured or simulated in the correct local nanodomains. Similarly, the spatio-temporal pattern of calcium influx may indeed control which downstream calcium binding proteins are activated. Future models of calcium dynamics in nanodomains with appropriately localized calcium channels and binding proteins are certain to readdress these critical questions of neuronal function.

Acknowledgements Support is gratefully acknowledged from ONR grant MURI N00014-10-1-0198 and the CRCNS program through NIAAA R01 AA180660, and NIAAA R01 AA016022. The author thanks Rebekah Evans for kindly reviewing and providing comments on the manuscript.

References

Alkon DL, Lederhendler II, Shoukimas JL (1982) Primary changes of membrane currents during retention of associative learning. Science 215:693–695

Amini B, Clark JW Jr, Canavier CC (1999) Calcium dynamics underlying pacemaker-like and burst firing oscillations in midbrain dopaminergic neurons: a computational study. J Neurophysiol 82:2249–2261

Bear MF, Malenka RC (1994) Synaptic plasticity: LTP and LTD. Curr Opin Neurobiol 4:389–399

Berkefeld H, Sailer CA, Bildl W, Rohde V, Thumfart JO, Eble S, Klugbauer N, Reisinger E, Bischofberger J, Oliver D, Knaus HG, Schulte U, Fakler B (2006) BKCa-Cav channel complexes mediate rapid and localized Ca2+-activated K+ signaling. Science 314:615–620

Bi GQ, Wang HX (2002) Temporal asymmetry in spike timing-dependent synaptic plasticity. Physiol Behav 77:551–555

Blackwell KT (2004) Paired turbulence and light do not produce a supralinear calcium increase in Hermissenda. J Comput Neurosci 17:81–99

Bond CT, Maylie J, Adelman JP (2005) SK channels in excitability, pacemaking and synaptic integration. Curr Opin Neurobiol 15:305–311

Canavier CC, Clark JW, Byrne JH (1991) Simulation of the bursting activity of neuron R15 in Aplysia: role of ionic currents, calcium balance, and modulatory transmitters. J Neurophysiol 66:2107–2124

Chay TR (1996a) Electrical bursting and luminal calcium oscillation in excitable cell models. Biol Cybern 75:419–431

Chay TR (1996b) Modeling slowly bursting neurons via calcium store and voltage-independent calcium current. Neural Comput 8:951–978

De Schutter E, Smolen P (1998) Calcium dynamics in large neuronal models. In: Koch C, Segev I (eds) Methods in Neuronal Modeling. MIT Press, Cambridge, Massachusetts, pp 211–250

Doi T, Kuroda S, Michikawa T, Kawato M, Doi T, Kuroda S, Michikawa T, Kawato M (2005) Inositol 1,4,5-trisphosphate-dependent Ca2+ threshold dynamics detect spike timing in cerebellar Purkinje cells. J Neurosci 25:950–961

Dupont G, Combettes L, Leybaert L (2007) Calcium dynamics: spatio-temporal organization from the subcellular to the organ level. Int Rev Cytol 261:193–245

Engel J, Schultens HA, Schild D (1999) Small conductance potassium channels cause an activity-dependent spike frequency adaptation and make the transfer function of neurons logarithmic. Biophys J 76:1310–1319

Faber ES, Sah P (2003) BK channel inactivation contributes to spike broadening during repetitive firing in the rat lateral amygdala. J Physiol 552(Pt 2):483–497

Gamble E, Koch C (1987) The dynamics of free calcium in dendritic spines in response to repetitive synaptic input. Science 236:1311–1315

Goldberg JA, Wilson CJ (2005) Control of spontaneous firing patterns by the selective coupling of calcium currents to calcium-activated potassium currents in striatal cholinergic interneurons. J Neurosci 25:10230–10238

Goldberg JA, Teagarden MA, Foehring RC, Wilson CJ (2009) Nonequilibrium calcium dynamics regulate the autonomous firing pattern of rat striatal cholinergic interneurons. J Neurosci 29:8396–8407

Gould TJ, Steinmetz JE (1996) Changes in rabbit cerebellar cortical and interpositus nucleus activity during acquisition, extinction, and backward classical eyelid conditioning. Neurobiol Learn Mem 65:17–34

Gu N, Vervaeke K, Hu H, Storm JF (2005) Kv7/KCNQ/M and HCN/h, but not KCa2/SK channels, contribute to the somatic medium after-hyperpolarization and excitability control in CA1 hippocampal pyramidal cells. J Physiol 566:689–715

Gu N, Vervaeke K, Storm JF (2007) BK potassium channels facilitate high-frequency firing and cause early spike frequency adaptation in rat CA1 hippocampal pyramidal cells. J Physiol 580:859–882

Hernjak N, Slepchenko BM, Fernald K, Fink CC, Fortin D, Moraru II, Watras J, Loew LM (2005) Modeling and analysis of calcium signaling events leading to long-term depression in cerebellar Purkinje cells. Biophys J 89:3790–3806

Holmes WR, Levy WB (1990) Insights into associative long-term potentiation from computational models of NMDA receptor-mediated calcium influx and intracellular calcium changes. J Neurophysiol 63:1148–1168

Holmes WR, Levy W (1997) Quantifying the role of inhibition in associative long-term potentiation in dentate granule cells with computational models. J Neurophysiol 78:103–115

Ito M (2001) Cerebellar long-term depression: characterization, signal transduction, and functional roles. Physiol Rev 81:1143–1195

Katz B, Miledi R (1967) A study of synaptic transmission in the absence of nerve impulses. J Physiol 192:407–436

Keller DX, Franks KM, Bartol TM Jr, Sejnowski TJ (2008) Calmodulin activation by calcium transients in the postsynaptic density of dendritic spines. PLoS One 3:e2045

Kotaleski JH, Blackwell KT (2010) Modelling the molecular mechanisms of synaptic plasticity using systems biology approaches. Nat Rev Neurosci 11:239–251

Kotaleski JH, Lester DS, Blackwell KT (2002) Subcellular interactions between parallel fibre and climbing fibre signals in Purkinje cells predict sensitivity of classical conditioning to interstimulus interval. Integr Physiol Behav Sci 37:265–292

Kubota Y, Putkey JA, Waxham MN (2007) Neurogranin controls the spatiotemporal pattern of postsynaptic Ca2+/CaM signaling. Biophys J 93:3848–3859

Li Y-X, Rinzel J (1994) Equations for InsP3 receptor-mediated [Ca^{2+}]; oscillations derived from a detailed kinetic model: a Hodgkin-Huxley like formalism. J Theor Biol 166:461–473

Malenka RC, Lancaster B, Zucker RS (1992) Temporal limits on the rise in postsynaptic calcium required for the induction of long-term potentiation. Neuron 9:121–128
Markram H, Roth A, Helmchen F (1998) Competitive calcium binding: implications for dendritic calcium signaling. J Comput Neurosci 5:331–348
Naoki H, Sakumura Y, Ishii S (2005) Local signaling with molecular diffusion as a decoder of Ca2+ signals in synaptic plasticity. Mol Syst Biol 1:2005.0027
Nevian T, Sakmann B (2006) Spine Ca2+ signaling in spike-timing-dependent plasticity. J Neurosci 26:11001–11013
Nowycky MC, Pinter MJ (1993) Time courses of calcium and calcium-bound buffers following calcium influx in a model cell. Biophys J 64(77–91):77–91
Sala F, Hernandez-Cruz A (1990) Calcium diffusion modeling in a spherical neuron Relevance of buffering properties. Biophys J 57:313–324
Schiegg A, Gerstner W, Ritz R, Leo van Hemmen J (1995) Intracellular Ca2+ stores can account for the time course of LTP Induction: a model of CA2+ dynamics in dendritic spines. Am Physiol Soc 74:1046–1055
Schmidt H, Eilers J (2009) Spine neck geometry determines spino-dendritic cross-talk in the presence of mobile endogenous calcium binding proteins. J Comput Neurosci 27:229–243
Schneggenburger R, Neher E (2005) Presynaptic calcium and control of vesicle fusion. Curr Opin Neurobiol 15:266–274
Schreurs BG, Oh MM, Alkon DL (1996) Pairing-specific long-term depression of Purkinje cell excitatory postsynaptic potentials results from a classical conditioning procedure in rabbit. J Neurophysiol 75(1051–1060):1051–1060
Shao LR, Halvorsrud R, Borg-Graham L, Storm JF (1999) The role of BK-type Ca2+-dependent K+ channels in spike broadening during repetitive firing in rat hippocampal pyramidal cells. J Physiol 521(Pt 1):135–146
Teyler TJ, Cavus I, Coussens C, DiScenna P, Grover L, Lee YP, Little Z (1994) Multideterminant role of calcium in hippocampal synaptic plasticity. Hippocampus 4:623–634
Thompson RF, Steinmetz JE (2009) The role of the cerebellum in classical conditioning of discrete behavioral responses. Neuroscience 162:732–755
Volfovsky N, Parnas H, Segal M, Korkotian E (1999) Geometry of dendritic spines affects calcium dynamics in hippocampal neurons: theory and experiments. J Neurophysiol 82:450–462
Wigstrom H, Gustafsson B (1986) Postsynaptic control of hippocampal long-term potentiation. J Physiol Paris 81:228–236
Williams GS, Smith GD, Sobie EA, Jafri MS (2010) Models of cardiac excitation-contraction coupling in ventricular myocytes. Math Biosci 226:1–15
Wilson CJ, Callaway JC (2000) Coupled oscillator model of the dopaminergic neuron of the substantia nigra. J Neurophysiol 83:3084–3100
Yamada WM, Koch C, Adams PR (1989) Multiple channels and calcium dynamics. In: Koch C, Segev I (eds) Methods in Neuronal Modeling (Chap 4). MIT Press, Cambridge, pp 97–133
Zador A, Koch C, Brown TH (1990) Biophysical model of a Hebbian synapse. Proc Natl Acad Sci USA 87:6718–6722

Chapter 7
The More We Look, the More Biological Variation We See: How Has and Should This Influence Modeling of Small Networks?

Ronald L. Calabrese

Abstract Models of the small neuronal networks from invertebrates, especially rhythmically active central pattern generators, have not only been useful experimental tools for circuit analyses but also been instrumental in revealing general principles of neuronal network function. This ability of small network models to illuminate basic mechanisms attests to their heuristic power. In the 20 years since the first CNS meeting, theoretical studies, now supported abundantly by experimental analyses in several different networks and species, have shown that functional network activity arises in animals and models even though parameters (e.g., the intrinsic membrane properties (maximal conductances) of the neurons and the strengths of the synaptic connections) show two to fivefold animal-to-animal variability.

Models of the small neuronal networks from invertebrates, especially rhythmically active central pattern generators (CPGs), have not only been useful experimental tools for circuit analyses but also been instrumental in revealing general principles of neuronal network function. This ability of small network models to illuminate basic mechanisms attests to their heuristic power.

In the 20 years since the first CNS meeting, theoretical studies, now supported abundantly by experimental analyses in several different networks and species, have shown that functional network activity arises in animals and models even though parameters (e.g., the intrinsic membrane properties (maximal conductances) of the neurons and the strengths of the synaptic connections) show two to fivefold animal-to-animal variability (Golowasch et al. 2002; Prinz et al. 2004; Bucher et al. 2005; Marder and Goaillard 2006; Marder et al. 2007; Prinz 2007; Schulz et al. 2007; Goaillard et al. 2009; Tobin et al. 2009; Doloc-Mihu and Calabrese 2011; Norris et al. 2011; Roffman et al. 2012).

R.L. Calabrese (✉)
Department of Biology, Emory University, 1510 Clifton Road, Atlanta, GA 30322, USA
e-mail: ronald.calabrese@emory.edu

J.M. Bower (ed.), *20 Years of Computational Neuroscience*, Springer Series in Computational Neuroscience 9, DOI 10.1007/978-1-4614-1424-7_7,

In our own work, presented at CNS 2010, we reviewed experiments, using the leech heartbeat CPG, which explore the consequences of animal-to-animal variability in synaptic strength for coordinated motor output (Fig. 7.1). Our experiments focused on a set of segmental heart motor neurons that all receive inhibitory synaptic input from the same four premotor interneurons (Norris et al. 2011) (Fig. 7.1A). These four premotor inputs fire in a phase progression and the motor neurons also fire in a phase progression because of differences in synaptic strength profiles of the four inputs among segments (Fig. 7.1B). Our experiments showed that relative synaptic strengths of the different premotor inputs to each motor neuron vary across animals yet functional output is maintained. Moreover, animal-to-animal variations in strength of particular inputs do not correlate strongly with output phase. We measured the precise temporal pattern of the premotor inputs, the segmental synaptic strength profiles of their connections onto motor neurons, and the temporal pattern (phase progression) of those motor neurons all in single animals and compiled a database of 12 individual animals (Fig. 7.1C1, 2). We analyzed input and output in this database and our results suggest that the number (four) of inputs to each motor neuron and the variability of the temporal pattern of input from the CPG across individuals weaken the influence of the strength of individual inputs so that correlations are not easily detected. Additionally, the temporal pattern of the output, albeit in all cases consistent with heart function, varies as much across individuals as that of the input. It seems then that each animal arrives at a unique solution for how the network produces functional output. This work has been supplemented by dynamic clamp analysis of pharmacologically isolated heart motor neurons using synaptic input patterns derived from the 12 individual of our database that further support these conclusions (Wright and Calabrese 2011a, b).

All the observations summarized above have contributed to the growing consensus that to understand a neuronal network through biophysical modeling, we must construct populations of models with multiple sets of parameter values corresponding to parameters from different individuals (Prinz 2010; Marder 2011; Marder and Taylor 2011). Thus the computational effort needed to produce a state-of-the-art biophysical model is vastly increased. The situation is clearly still fluid, and the reaction in the modeling community has ranged from a continued pursuance "ideal parameter sets" or sticking to averaged values for parameters to what Prinz (2010) calls ensemble modeling, where multiple functional instances are identified and examined. We have not come to this situation smoothly but by fits and starts, and the purpose of this chapter is to highlight two papers that were presented at CNS 1993 that seem now dated but indeed presage this understanding.

Looking Back

At CNS 1993 two papers were presented and book chapters written in "Computation in Neurons and Neural Systems" edited by Frank H. Eeckman were inspired by work on invertebrate CPGs (LoFaro et al. 1994; Skinner et al. 1994). These papers

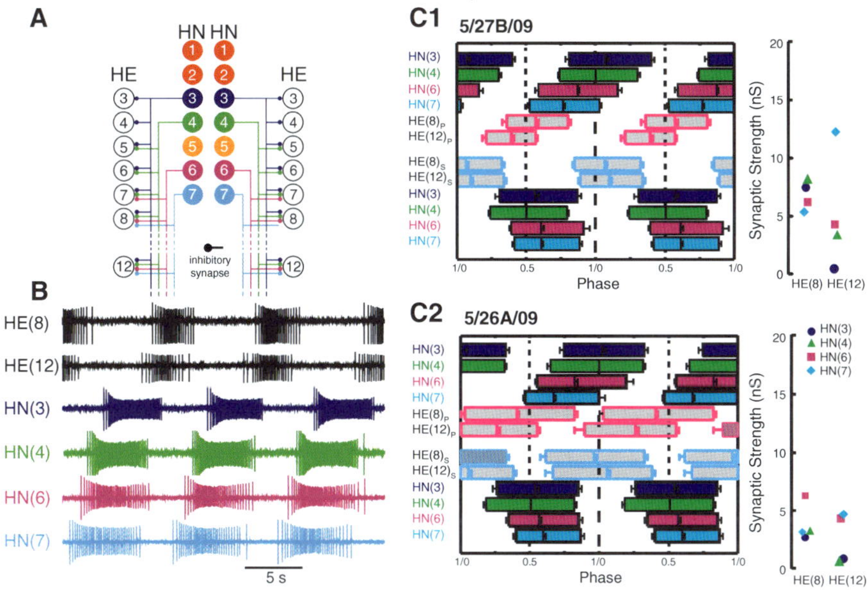

Fig. 7.1 (**A**) Bilateral circuit diagram from the heartbeat control system of medicinal leeches including all the identified heart (HN) interneurons of the core CPG showing the inhibitory connections from the heart interneurons of the leech heartbeat CPG onto heart (HE) motor neurons in the first 12 midbody segmental ganglia. The ipsilateral HN(3) and HN(4) front premotor interneurons and the ipsilateral HN(6) and HN(7) middle premotor interneurons provide input to heart motor neurons (HE(3)–HE(12)) (Norris et al. 2007a). The *large filled circles* are cell bodies and associated input processes. *Lines* indicate cell processes and *small filled circles* indicate inhibitory chemical synapses. Connections among the interneurons of the CPG are not indicated. Standard colors for the heart interneurons are used in the rest of the figure. (**B**) There are two coordination modes (peristaltic and synchronous) of the heart motor neurons and heart interneurons one on either body side that switch sides regularly (Norris et al. 2006, 2007b). Simultaneous extracellular recordings are shown of ipsilateral HN(3), HN(4), HN(6), and HN(7) premotor interneurons (inputs) (standard colors) and HE(8) and HE(12) motor neurons (outputs) (*black*) in peristaltic (p) coordination mode—similar recordings, not shown, were made in the synchronous (s) coordination mode. (**C1**, **2**) Complete analysis of input and output temporal patterns and synaptic strength profiles for two different animals from our sample of 12. Summary phase diagram (temporal patterns of inputs and outputs) of the premotor interneurons (standard colors) and the HE(8) and HE(12) motor neurons in both the peristaltic (boxes outlined in *pink*) and synchronous (boxes outlined in *light blue*) coordination modes for two different preparations. Phase diagrams were determined from recordings like in (**B**). The segmental synaptic strength profiles of the inputs were determined in the same preparations by voltage clamping each of the motor neurons (HE(8) and HE(12)) and performing spike-triggered averaging of IPSCs, and are shown to the *right* of each phase diagram. Standard colors are used. Animals are specified by the day on which they were recorded; letters accompany the designation of day, if more than one animal was recorded on that day. Note that both the temporal patterns (both input and output) and synaptic strength profiles vary between the two animals as in the rest of the sample of 12 animals. Adapted from Norris et al. (2011)

reflect the time in which they were written yet they point to the present day. They illustrate the limits of our technological ability to model small neuronal networks and the naiveté of our theoretical understanding of what a realistic neuronal network model was. They also illustrate how the then novel technique of dynamic current clamping would be brought to bear in future studies of small networks. Using these papers as a starting point, I will discuss how my own thinking and that of the field has evolved since then. In both these papers, parameter variation in reduced models of half-center oscillators (oscillatory networks with reciprocal inhibition between two neurons (or groups of neurons)) is shown to lead to interesting changes in network activity.

In the first case, a two-cell network was modeled with one cell an inherent burster and the other not, the presence of I_h is shown to be critical for the non-bursting neuron to assume an integer bursting ratio smaller than 1:1 as the level of injected current in the non-bursting neuron is adjusted. The theoretical analysis was motivated and augmented by electrophysiological experiments in the crustacean stomatogastric nervous system (STN) focusing on the well-characterized pyloric CPG. The LP neuron in the isolated STN is capable of plateau production: it is not an autonomous bursting neuron but it is engaged in reciprocal inhibitory connections to the bursting PD neurons. Normally these cells produce alternating bursts but with appropriate hyperpolarizing current injection into the LP neurons they assume a 6:1 (−3.6 nA) or 12:1 (−5.1 nA) burst ratio. The theoretical analysis modeled each neuron using the Morris–Lecar formalism (Morris and Lecar 1981) tuned so that the PD neuron was spontaneously oscillatory (a burster) whereas the LP neuron was silent in the absence of input from the PD but plateau forming. The LP neuron was additionally given the I_h current. A two-cell network was then constructed with reciprocal inhibitory synapses, thus forming a half-center oscillator, with one cell (PD) an inherent burster and the other (LP) not. The presence of I_h in the non-bursting LP neuron was shown to be critical for it to assume an integer bursting ratio smaller than 1:1 as the level of injected current in the non-bursting neuron was adjusted.

This study was naïve in that simplified neuron models were used and only one parameter was considered in determining how burst ratios less than 1:1 could be achieved—considering the desktop computational ability available at the time it is hardly surprising that simplified neuron models were used and other parameters were not also analyzed. The study was forward-looking in that it was clearly tied to an interesting experimentally observed phenomenon—frequency demultiplication—and implicated a specific ionic current as producing the phenomenon. The real implications of this tenuous step were seen in strong experimental, modeling, and hybrid analysis with dynamic clamp that followed and which led to fundamental insights into how fast and slow rhythms in neuronal networks can interact not only in the crustacean STN (Bartos and Nusbaum 1997; Bartos et al. 1999) but also in general (Marder et al. 1998).

In the second case, again a half-center oscillator was formed between two oscillatory Morris–Lecar model neurons (Morris and Lecar 1981)—and the mechanisms promoting the transitions during alternate "bursting" were explored. The most interesting aspect of the analysis was a determination of the effect of synaptic threshold on the period of the half-center oscillator's activity. The main finding was that there was a middle range where period was prolonged and relatively insensitive to synaptic threshold but fell of sharply on either side of this range. This theoretic analysis was given some experimental backing by forming a half-center oscillator between pharmacologically isolated leech heart interneurons using artificial inhibitory synapses implemented with dynamic clamp. The hybrid half-center oscillators again showed a period maximum at a middle range of synaptic threshold with period falling off on either side.

Like the previous study, this one was naïve in that simplified neuron models were used and only one parameter was considered in determining how burst period was controlled in a half-center oscillator. The study was very forward-looking in that it introduces the profound interaction of theory and experiment that is possible when hybrid systems are created with dynamic clamp. This analysis planted the seed for more sophisticated and systematic hybrid systems analysis of half-center oscillators that have defined the role of currents like I_h and low-threshold Ca current in producing half-center oscillations (Sorensen et al. 2004; Olypher et al. 2006). Other studies have used modeling and dynamic clamp and similar techniques to more fully explore the synaptic dynamics of mutually inhibitory neurons in controlling network period (Mamiya and Nadim 2004; Nadim et al. 2011). Yet more interesting and germane to the current interest in how neuronal and synaptic variability affects circuit performance are more recent hybrid system analyses of half-center oscillators that employ ensemble modeling and database techniques to systematically explore the parameter space of the half-center oscillator and make an attempt to make sense of animal-to-animal variability in neuronal properties that confront all experimentalists (Grashow et al. 2009, 2010; Brookings et al. 2012).

Concluding Thoughts

Models of neuronal networks essentially consist of differential equations that describe the dynamics of state variables, e.g., membrane potential (V_m) and the gating variables of voltage-gated conductances and the variables controlling activation of synaptic conductances. Embedded in these equations are a number of parameters, including maximal conductances, half-activation voltages and time constants of channel gates, and parameters controlling synaptic dynamics. Some of these parameters are considered free, or variable between instances, while the remaining parameters are fixed. For example, in the pioneering work of Prinz et al. (2003, 2004), only maximal conductances were considered free parameters. Indeed maximal conductances have been shown to be quite variable among animals (Bucher et al. 2005; Schulz et al. 2007; Goaillard et al. 2009; Tobin et al. 2009; Norris et al. 2011; Roffman et al. 2012). But it is clear that the other parameters mentioned will also show animal-to-animal variability though these have not been as widely studied

(Marder et al. 2007; Marder 2011). Even with powerful computing resources, it is not possible or desirable to consider all instances of a model. Making a model then involves deciding on a neuronal structure (single or multiple compartments), network connectivity, descriptive equations (often derivatives of the Hodgkin–Huxley formalism), which parameters are free, and the range over which each may vary. These decisions will all be driven by the data available and by the investigators' intuition for which parameters are likely to be significant in controlling neuronal activity. In short, the ability to consider multiple instances of a model does not free one from making a good model, and making a good model requires detailed knowledge of the system and judgment about what details can be ignored and parameters fixed.

There are many pertinent issues which models of small networks can still help clarify many interesting issues. Although some studies have suggested that variability in cellular intrinsic properties becomes less important when neurons are embedded in networks (Grashow et al. 2010; Brookings et al. 2012), others suggest that the interaction network topology and neuronal dynamics are critical (Gaiteri and Rubin 2011). Moreover, we know that networks are subject to frequent environmental perturbations and that neuromodulation plays an important role in pattern generation in many networks. Nevertheless, the question of how networks can produce functional output despite perturbations and modulatable parameters and yet not crash has barely been addressed especially at the experimental level (Grashow et al. 2009; Marder and Tang 2010; Tang et al. 2010). The ensemble modeling approach to address such questions is likely to expand as we move forward, despite the caveat expressed above, especially given the ever-increasing computational capabilities available. The analysis experimental and computational of small neuronal networks like invertebrate CPGs is likely to lead the way in this endeavor for several years to come.

References

Bartos M, Manor Y, Nadim F, Marder E, Nusbaum MP (1999) Coordination of fast and slow rhythmic neuronal circuits. J Neurosci 19:6650–6660

Bartos M, Nusbaum MP (1997) Intercircuit control of motor pattern modulation by presynaptic inhibition. J Neurosci 17:2247–2256

Brookings T, Grashow R, Marder E (2012) Statistics of neuronal identification with open- and closed-loop measures of intrinsic excitability. Front Neural Circuits 6:19

Bucher D, Prinz AA, Marder E (2005) Animal-to-animal variability in motor pattern production in adults and during growth. J Neurosci 25:1611–1619

Doloc-Mihu A, Calabrese R (2011) A database of computational models of a half-center oscillator for analyzing how neuronal parameters influence network activity. J Biol Phys 37:263–283

Gaiteri C, Rubin JE (2011) The interaction of intrinsic dynamics and network topology in determining network burst synchrony. Front Comput Neurosci 5:10

Goaillard JM, Taylor AL, Schulz DJ, Marder E (2009) Functional consequences of animal-to-animal variation in circuit parameters. Nat Neurosci 12:1424–1430

Golowasch J, Goldman MS, Abbott LF, Marder E (2002) Failure of averaging in the construction of a conductance-based neuron model. J Neurophysiol 87:1129–1131

Grashow R, Brookings T, Marder E (2010) Compensation for variable intrinsic neuronal excitability by circuit-synaptic interactions. J Neurosci 30:9145–9156

Grashow R, Brookings T, Marder E (2009) Reliable neuromodulation from circuits with variable underlying structure. Proc Natl Acad Sci USA 106:11742–11746
LoFaro T, Koppell N, Marder E, Hooper SL (1994) The Effect of ih currents on bursting patterns of pairs of coupled neurons. In: Eeckmann FH (ed) Computation in neurons and neural systems. Kluwer Academic Publishers. Boston, pp 15–20
Mamiya A, Nadim F (2004) Dynamic interaction of oscillatory neurons coupled with reciprocally inhibitory synapses acts to stabilize the rhythm period. J Neurosci 24:5140–5150
Marder E, Manor Y, Nadim F, Bartos M, Nusbaum MP (1998) Frequency control of a slow oscillatory network by a fast rhythmic input: pyloric to gastric mill interactions in the crab stomatogastric nervous system. Ann N Y Acad Sci 860:226–238
Marder E, Goaillard JM (2006) Variability, compensation and homeostasis in neuron and network function. Nat Rev Neurosci 7:563–574
Marder E, Tobin AE, Grashow R (2007) How tightly tuned are network parameters? Insight from computational and experimental studies in small rhythmic motor networks. Prog Brain Res 165:193–200
Marder E, Tang LS (2010) Coordinating different homeostatic processes. Neuron 66:161–163
Marder E (2011) Variability, compensation, and modulation in neurons and circuits. PNAS 108(suppl 3):15542–15548
Marder E, Taylor AL (2011) Multiple models to capture the variability in biological neurons and networks. Nat Neurosci 14:133–138
Morris C, Lecar H (1981) Voltage oscillations in the barnacle giant muscle fiber. Biophys J 35:193–213
Nadim F, Zhao S, Zhou L, Bose A (2011) Inhibitory feedback promotes stability in an oscillatory network. J Neural Eng 8:065001
Norris BJ, Weaver AL, Wenning A, Garciá PA, Calabrese RL (2007a) A central pattern generator producing alternative outputs: pattern, strength and dynamics of premotor synaptic input to leech heart motor neurons. J Neurophysiol 98(5):2992–3005
Norris BJ, Weaver AL, Wenning A, Garciá PA, Calabrese RL (2007b) A central pattern generator producing alternative outputs: phase relations of leech heart motor neurons with respect to premotor synaptic input. J Neurophysiol 98(5):2983–2991
Norris BJ, Weaver AL, Morris LG, Wenning A, Garcia PA, Calabrese RL (2006) A central pattern generator producing alternative outputs: temporal pattern of premotor activity. J Neurophysiol 96:309–326
Norris BJ, Wenning A, Wright TM, Calabrese RL (2011) Constancy and variability in the output of a central pattern generator. J Neurosci 31:4663–4674
Olypher A, Cymbalyuk G, Calabrese RL (2006) Hybrid systems analysis of the control of burst duration by low-voltage-activated calcium current in leech heart interneurons. J Neurophysiol 96:2857–2867
Prinz AA, Billimoria CP, Marder E (2003) Alternative to hand-tuning conductance-based models: construction and analysis of databases of model neurons. J Neurophysiol 90:3998–4015
Prinz AA, Bucher D, Marder E (2004) Similar network activity from disparate circuit parameters. Nat Neurosci 7:1345–1352
Prinz AA (2007) Computational exploration of neuron and neural network models in neurobiology. Methods Mol Biol 401:167–179
Prinz AA (2010) Computational approaches to neuronal network analysis. Philos Trans R Soc Lond B Biol Sci 365:2397–2405
Roffman RC, Norris BJ, Calabrese RL (2012) Animal-to-animal variability of connection strength in the leech heartbeat central pattern generator. J Neurophysiol 107:1681–1693
Schulz DJ, Goaillard JM, Marder EE (2007) Quantitative expression profiling of identified neurons reveals cell-specific constraints on highly variable levels of gene expression. PNAS 104:13187–13191
Skinner FK, Gramoll S, Calabrese RL, Kopell N, Marder E (1994) Frequency control in biological half-center oscillators. In: Eeckmann FH (ed) Computation in neurons and neural systems. Kluwer Academic Publishers. Boston, pp 223–228

Sorensen M, DeWeerth S, Cymbalyuk G, Calabrese RL (2004) Using a hybrid neural system to reveal regulation of neuronal network activity by an intrinsic current. J Neurosci 24:5427–5438
Tang LS, Goeritz ML, Caplan JS, Taylor AL, Fisek M, Marder E (2010) Precise temperature compensation of phase in a rhythmic motor pattern. PLoS Biol 8
Tobin AE, Cruz-Bermudez ND, Marder E, Schulz DJ (2009) Correlations in ion channel mRNA in rhythmically active neurons. PLoS One 4:e6742
Wright TM Jr, Calabrese RL (2011a) Contribution of motoneuron intrinsic properties to fictive motor pattern generation. J Neurophysiol 106:538–553
Wright TM Jr, Calabrese RL (2011b) Patterns of presynaptic activity and synaptic strength interact to produce motor output. J Neurosci 31:17555–17571

Chapter 8
20 Years of "Noise": Contributions of Computational Neuroscience to the Exploration of the Effect of Background Activity on Central Neurons

Alain Destexhe

Abstract The central nervous system is subject to many different forms of noise, which have fascinated researchers since the beginning of electrophysiological recordings. In cerebral cortex, the largest amplitude noise source is the "synaptic noise," which is dominant in intracellular recordings in vivo. The consequences of this background activity are a classic theme of modeling studies. In the last 20 years, this field tremendously progressed as the synaptic noise was measured for the first time using quantitative methods. These measurements have allowed computational models not only to be more realistic and closer to the biological data but also to investigate the consequences of synaptic noise in more quantitative terms, measurable in experiments. As a consequence, the "high-conductance state" conferred by this intense activity in vivo could also be replicated in neurons maintained in vitro using dynamic-clamp techniques. In addition, mathematical approaches of stochastic systems provided new methods to analyze synaptic noise and obtain critical information such as the optimal conductance patterns leading to spike discharges. It is only through such a combination of different disciplines, such as experiments, computational models, and theory, that we will be able to understand how noise participates to neural computations.

Introduction

The central nervous system is subject to many different forms of noise, which have fascinated researchers since the beginning of electrophysiological recordings. In cerebral cortex, the largest amplitude noise source is the "synaptic noise,"

A. Destexhe (✉)
Unité de Neuroscience Information et Complexité (UNIC),
CNRS, Gif-sur-Yvette 91198, France
e-mail: destexhe@unic.cnrs-gif.fr

J.M. Bower (ed.), *20 Years of Computational Neuroscience*, Springer Series in Computational Neuroscience 9, DOI 10.1007/978-1-4614-1424-7_8,

which is dominant in intracellular recordings in vivo. Indeed, one of the most striking characteristics of awake and attentive states is the highly complex nature of cortical activity. Global measurements, such as the electroencephalogram (EEG) or local field potentials (LFPs), display low amplitude and very irregular activity, the so-called desynchronized EEG (Steriade 2003). This activity has very low spatiotemporal coherence between multiple sites in cortex, which contrasts with the widespread synchronization in slow-wave sleep (Destexhe et al. 1999). Local measurements, such as extracellular (unit activity) or intracellular recordings of single neurons, also demonstrate very irregular spike discharge and high levels of fluctuations similar to noise (Steriade et al. 2001), as shown in Fig. 8.1. Multiple unit activity (Fig. 8.1A) shows that the firing is irregular and of low correlation between different cells, while intracellular recordings (Fig. 8.1B) reveal that the membrane potential (V_m) is dominated by intense fluctuations ("noise").

How neurons integrate synaptic inputs in such noisy conditions is a problem which was identified in early work on motoneurons (Barrett and Crill 1974; Barrett 1975), which was followed by studies in Aplysia (Bryant and Segundo 1976) and cerebral cortex (Holmes and Woody 1989). This early work motivated further studies using compartmental models in cortex (Bernander et al. 1991) and cerebellum (Rapp et al. 1992; De Schutter and Bower 1994). These studies pointed out that the integrative properties of neurons can be drastically different in such noisy states. However, at the time, no precise experimental measurements were available to characterize the noise sources in neurons.

How neurons integrate their inputs in such states and, more generally, how entire populations of neurons represent and process information in such noisy states are still highly debated. In this chapter, we will describe recent measurements and associated progress to characterize the nature and the impact of this noisy activity. We will show that a series of major progress have been made in the last 20 years, and that computational neuroscience has played a particularly important role in this exploration.

Characterization of Synaptic Noise In Vivo

A first major advance was that this amount of "noise" was characterized and measured for the first time using quantitative methods. Figure 8.2 illustrates such measurements (Paré et al. 1998; Destexhe and Paré 1999). This first quantitative characterization was done using the "up-states" of ketaminexylazine anesthesia, which display very similar network activity as the awake brain (they were later measured in awake animals; Rudolph et al. 2007). The experiments were designed such that the same cell could be recorded before and after total suppression of network activity. A powerful blocker of network activity (tetrodotoxin, TTX) was micro-perfused during the intracellular recordings, enabling characterization of the membrane state before and after TTX infusion (Fig. 8.2, top scheme). The comparison between these two states included measuring the membrane potential (Fig. 8.2A), input resistance (Fig. 8.2B), and voltage distributions (Fig. 8.2C). These experiments

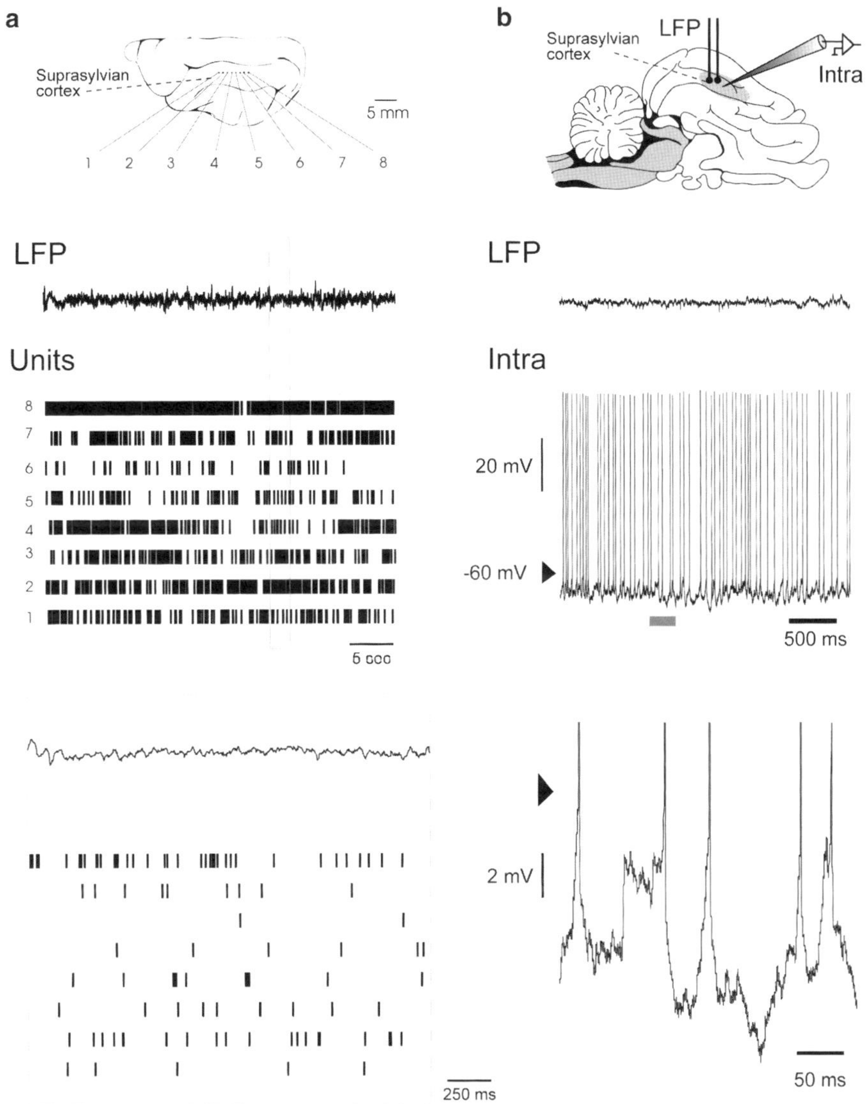

Fig. 8.1 Highly complex and "noisy" cortical activity during wakefulness. (**a**) Irregular firing activity of eight multiunits shown at the same time as the local field potential (LFP) recorded in electrode 1 (scheme on *top*). During wakefulness, the LFP is of low amplitude and irregular activity ("desynchonized") and unit activity is sustained and irregular (see magnification below; 20 times higher temporal resolution). (**b**) Intracellular activity in the same brain region during wakefulness. Spiking activity was sustained and irregular, while the membrane potential displayed intense fluctuations around a relatively depolarized state (around −65 mV in this cell; see magnification below). (**a**) Modified from Destexhe et al. 1999; (**b**) modified from Steriade et al. 2001

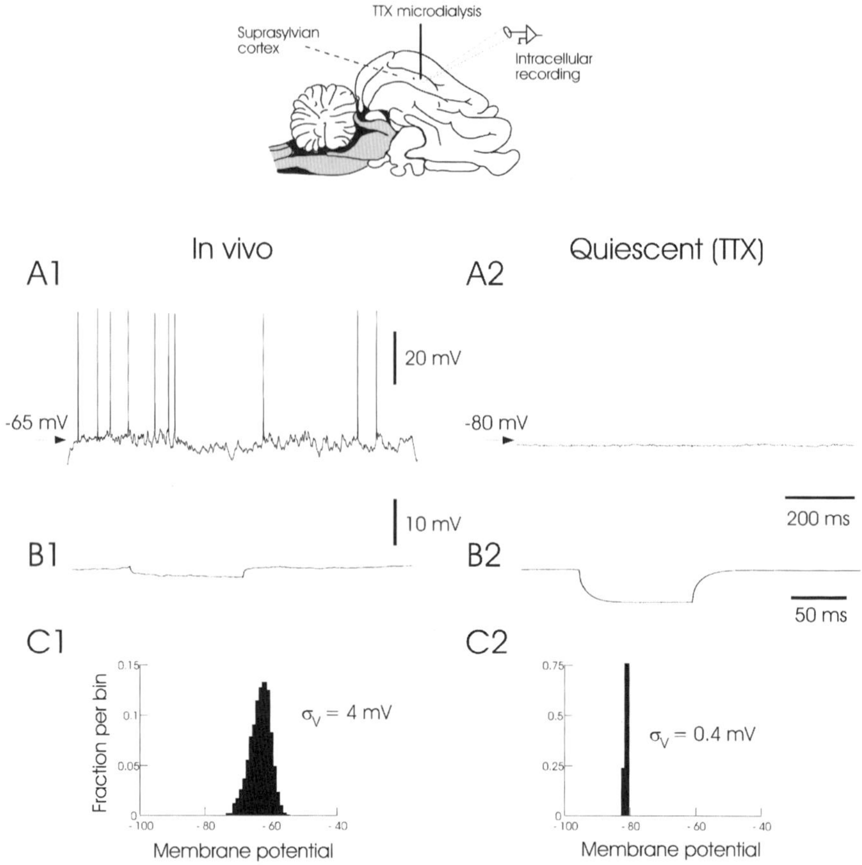

Fig. 8.2 Characterization of synaptic noise by suppression of network activity using micro-perfusion of tetrodotoxin (TTX). *Top*: experimental setup; a micro-perfusion pipette was used to infuse TTX into the cortex in vivo, at the same time of the intracellular recording. *Left*: characterization of network states in vivo. *Right*: same measurements after dialysis of TTX. The different measurements are the membrane potential (**A**), the averaged response to hyperpolarizing pulses (**B**), and the voltage distribution (**C**). (**A–C**) Modified from Destexhe and Paré 1999

revealed that about 80 % of the membrane conductance is attributable to synaptic activity (Paré et al. 1998; Destexhe and Paré 1999), demonstrating that neurons in vivo operate in a "high-conductance state."

Detailed Biophysical Models of Synaptic Noise

Investigating the consequences of noisy background activity is a classic theme which started by studies in motoneurons (Barrett and Crill 1974; Barrett 1975) and followed by model studies of neurons in cerebral cortex (Holmes and Woody 1989;

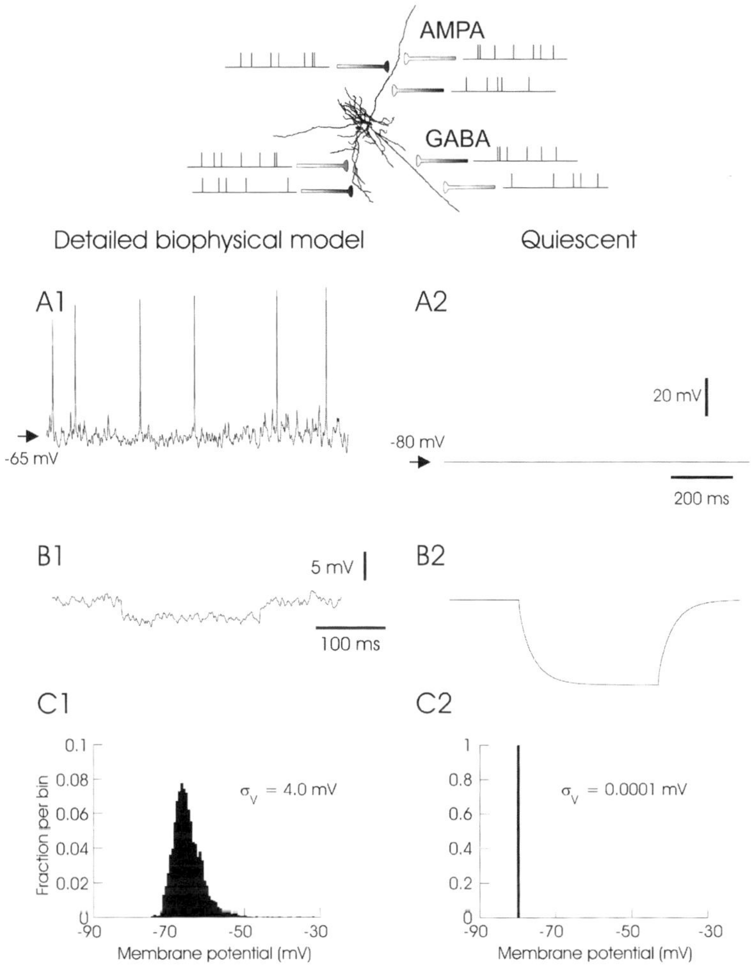

Fig. 8.3 Detailed biophysical models of synaptic background activity in cortical pyramidal neurons. *Top*: scheme of the model, based on a reconstructed cell morphology from cat parietal cortex. The model can reproduce the main features of in vivo measurements ((**A**)–(**C**) arranged similarly as Fig. 8.2). Figure modified from Destexhe et al. 2001

Bernander et al. 1991) and cerebellum (Rapp et al. 1992; De Schutter and Bower 1994). The measurements of synaptic background activity outlined above (Paré et al. 1998) have allowed computational models not only to be more realistic and closer to the biological data but also to investigate the consequences of synaptic noise in more quantitative terms. Figure 8.3 summarizes a first approach consisting of biophysically detailed models based on morphologically accurate reconstructions of cortical pyramidal neurons, combined with realistic patterns of synaptic input and intrinsic voltage-dependent conductances (see details and parameters in Destexhe and Paré 1999). These models could be tuned to reproduce all experimental measurements (Fig. 8.3A–C).

Such detailed biophysical models have been used to investigate the consequences of synaptic background activity in cortical neurons, starting with the first investigation of this kind by Bernander et al. (1991). This study revealed that the presence of background activity, although at the time nonconstrained by experimental measurements, was able to change several features of the integrative properties of the cell, such as coincidence detection.

Using models constrained from experiments, such as that of Fig. 8.3, enabled the derivation of several interesting properties, which we enumerate here.

1. *Enhanced responsiveness*. The presence of background activity was found to markedly change the cell's excitability, and produce a detectable response to inputs that are normally subthreshold (Hô and Destexhe 2000). This prediction was verified in dynamic-clamp experiments (see section "Synaptic Noise in Dynamic-Clamp").
2. *Location-independence*. The effectiveness of synaptic inputs becomes much less dependent on their position in dendrites, as found in cerebellar (De Schutter and Bower 1994) and cortical neurons (Rudolph and Destexhe 2003b), although based on very different mechanisms.
3. *Different integrative mode*. As initially predicted by Bernander et al. (1991), this important property was indeed confirmed with models constrained by experimental measurements (Rudolph and Destexhe 2003b).
4. *Enhanced temporal processing*. As a direct consequence of the "high-conductance state" of the neurons under background activity, the faster membrane time constant allows the neuron to perform finer discrimination, which is essential for coincidence detection (Softky 1994; Rudolph and Destexhe 2003b; Destexhe et al. 2003) or detecting brief changes of correlation (Rudolph and Destexhe 2001). The latter prediction was also verified experimentally (Fellous et al. 2003).
5. *Modulation of intrinsic properties*. It was found that in the presence of synaptic background activity, the responsiveness of bursting neurons is strongly affected (Wolfart et al. 2005). This aspect will be considered in more detail below.

These properties have been summarized and detailed in different review papers and books (Destexhe et al. 2003; Destexhe 2007; Haider and McCormick 2009; Destexhe and Rudolph 2012) which should be consulted for more information.

Simplified Models of Synaptic Noise

A second major step was to obtain simplified representations that capture the main properties of the synaptic "noise." This advance is important, because simple models have enabled real-time applications such as the dynamic-clamp (see section "Synaptic Noise in Dynamic-Clamp"). Simple models also have enabled a number of mathematical treatments, some of which resulted in methods to analyze experiments, as outlined in sections "Stochastic Systems Analysis of Synaptic Noise" and

"Estimating the Optimal Conductance Patterns Leading to Spikes in 'Noisy' States." These approaches relied on a simplified model of synaptic noise, called the "point-conductance model" (Destexhe et al. 2001), which can be written as:

$$C\frac{dv}{dt} = -g_{\mathrm{L}}(V - E_{\mathrm{L}}) - g_{\mathrm{e}}(V - E_{\mathrm{e}}) - g_{\mathrm{i}}(V - E_{\mathrm{i}}) + I_{\mathrm{ext}} \tag{8.1}$$

$$\frac{dg_{\mathrm{e}}(t)}{dt} = -\frac{1}{\tau_{\mathrm{e}}}\left[g_{\mathrm{e}}(t) - g_{\mathrm{e0}}\right] + \sqrt{\frac{2\sigma_{\mathrm{e}}^2}{\tau_{\mathrm{e}}}}\xi_{\mathrm{e}}(t) \tag{8.2}$$

$$\frac{dg_{\mathrm{i}}(t)}{dt} = -\frac{1}{\tau_{\mathrm{i}}}\left[g_{\mathrm{e}}(t) - g_{\mathrm{i0}}\right] + \sqrt{\frac{2\sigma_{\mathrm{i}}^2}{\tau_{\mathrm{i}}}}\xi_{\mathrm{i}}(t) \tag{8.3}$$

where C denotes the membrane capacitance, I_{ext} a stimulation current, g_{L} the leak conductance, and E_{L} the leak reversal potential. $g_{\mathrm{e}}(t)$ and $g_{\mathrm{i}}(t)$ are stochastic excitatory and inhibitory conductances with respective reversal potentials E_{e} and E_{i}. The excitatory synaptic conductance is described by Ornstein–Uhlenbeck (OU) stochastic processes (8.2), where g_{e0} and σ_{e}^2 are, respectively, the mean value and variance of the excitatory conductance, τ_{e} is the excitatory time constant, and $\xi_{\mathrm{e}}(t)$ is a Gaussian white noise source with zero mean and unit standard deviation. The inhibitory conductance $g_{\mathrm{i}}(t)$ is described by an equivalent equation (8.3) with parameters g_{i0}, σ_{i}^2, τ_{i}, and noise source $\xi_{\mathrm{i}}(t)$. Note that all conductances are here expressed in absolute units (in nS) but a formulation in terms of conductance densities is also possible.

In many previous models, synaptic activity was modeled by a source of current noise in the neuron (Tuckwell 1988), and thus the membrane potential is equivalent to a stochastic process. In contrast, in the point-conductance model, the conductances are the stochastic processes, and the V_{m} fluctuations result from the combined action of two of such fluctuating conductances. This model is thus capable of reproducing all features of the high-conductance state found in cortical neurons in vivo, such as large-amplitude fluctuations, low input resistance, and depolarized V_{m} (Fig. 8.4). In addition, it also captures the correct power spectral structure of the synaptic conductances (see Destexhe et al. 2001).

Synaptic Noise in Dynamic-Clamp

An elegant technique to investigate the effect of synaptic noise on neurons is to use the dynamic-clamp technique (Robinson and Kawai 1993; Sharp et al. 1993; for a recent review, see Destexhe and Bal 2009). This technique can be used to artificially reproduce stochastic synaptic activity by injecting the corresponding computer-generated conductance in a living neuron (Destexhe et al. 2001; Chance et al. 2002;

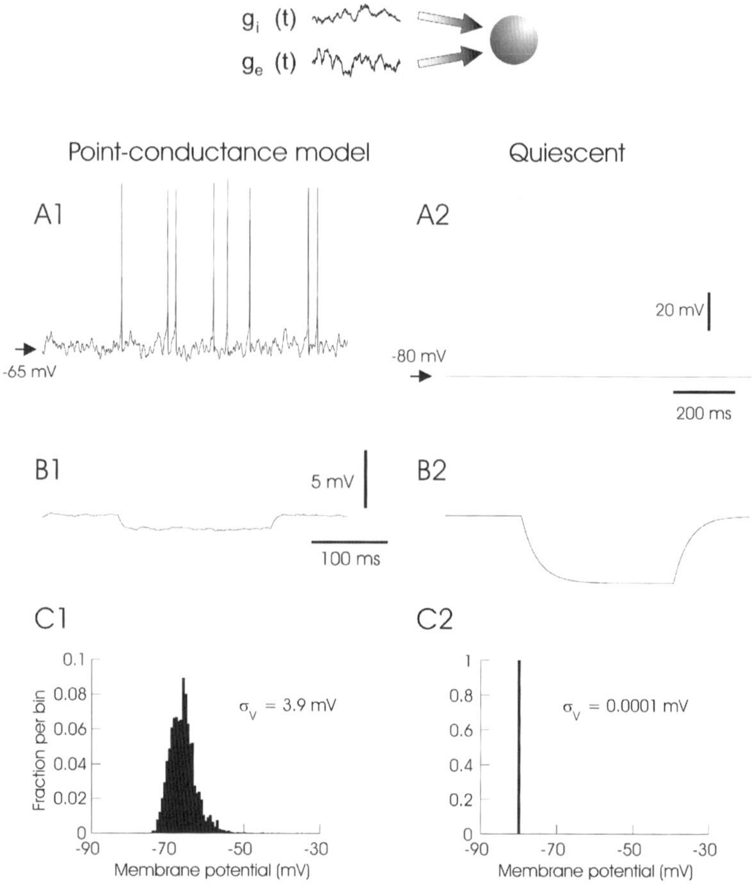

Fig. 8.4 Point-conductance model of synaptic background activity in cortical neurons. *Top*: scheme of the point-conductance model, where two stochastically varying conductances determine the V_m fluctuations through their (multiplicative) interaction. This simplified model reproduces the main features of in vivo measurements (same arrangement of (**A**)–(**C**) as in Fig. 8.2). Figure modified from Destexhe et al. 2001

Fellous et al. 2003; Mitchell and Silver 2003; Prescott and De Koninck 2003; Shu et al. 2003). This approach was first applied to cortical neurons, and revealed an important effect of the stochastic synaptic activity on neuronal responsiveness (Destexhe et al. 2001; Chance et al. 2002; Mitchell and Silver 2003; Prescott and De Koninck 2003; Shu et al. 2003; Higgs et al. 2006), similar to computational model predictions (Hô and Destexhe 2000). Some of these properties are reminiscent of the "stochastic resonance" phenomenon, which is an optimal signal-to-noise ratio in nonlinear systems subject to noise, and which was long studied by physicists (Wiesenfeld and Moss 1995; Gammaitoni et al. 1998).

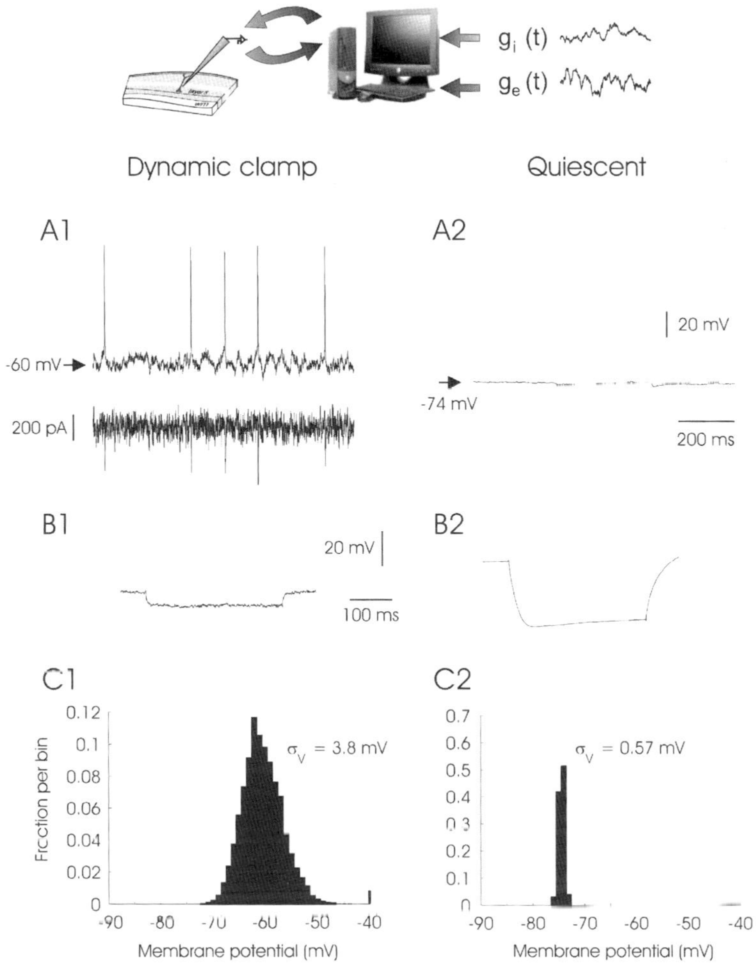

Fig. 8.5 Dynamic-clamp recreation of high-conductance states in neurons in vitro. *Top*: scheme of the dynamic-clamp, the point-conductance model is simulated and the excitatory and inhibitory conductances are injected in a living neuron using dynamic-clamp. This technique enables obtaining states very similar to in vivo measurements (similar arrangement of panels as Fig. 8.2). Figure modified from Destexhe et al. 2001

Figure 8.5 shows the "high-conductance state" conferred by intense synaptic activity, as replicated in neurons maintained in vitro using the dynamic-clamp technique. As for models, this technique enables the experimentalist to reproduce (and modulate at will) a background activity with similar properties as found in vivo.

Perhaps the most unexpected property of synaptic noise was found when investigating the effect of noise on thalamic neurons (Wolfart et al. 2005). These neurons are classically known to display two distinct firing modes, a single-spike (tonic) mode and a burst mode at more hyperpolarized levels (Llinas and Jahnsen 1982).

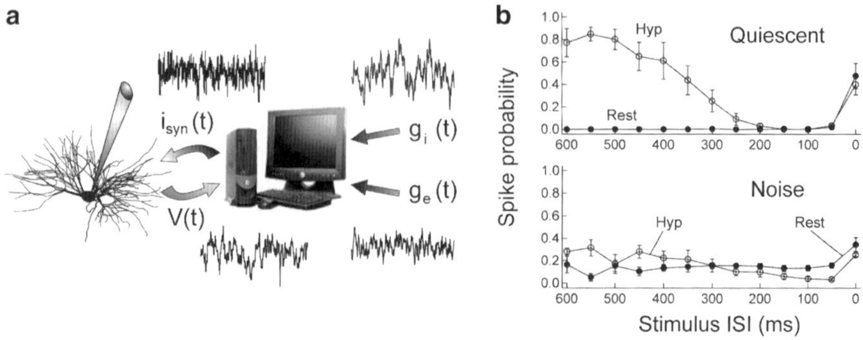

Fig. 8.6 Dynamic-clamp investigation of the transfer function of thalamic neurons in vitro. (**a**) Scheme of the dynamic-clamp experiment, in which stochastic conductances are injected in the neuron. (**b**) Effect of synaptic noise in thalamic neurons. The conductance noise interacts with burst generation to generate transfer response curves that are roughly independent on the V_m. (**b**) Modified from Wolfart et al. 2005

However, thalamic neurons are also known to receive large amounts of synaptic noise through their numerous direct synaptic connections from descending corticothalamic fibers, and this activity accounts for about half of the input resistance of thalamic neurons (Contreras et al. 1996). Based on these measurements, the effect of synaptic noise was simulated using dynamic-clamp on thalamic neurons in slices, and remarkably it was found that under such in vivo-like conditions, the duality of firing modes disappears because single spikes and bursts now appear at all V_m levels (Wolfart et al. 2005). But more interestingly, if one calculates the full transfer function of the neuron, the amount of spikes transmitted to cortex becomes independent of the V_m level (Fig. 8.6). This property is due to the fact that for hyperpolarized V_m, the low-threshold Ca^{2+} current generates more bursts, and thus "compensates" for hyperpolarization. This remarkable property shows that both the intrinsic properties and synaptic noise are necessary to understand the transfer function of central neurons in vivo.

Stochastic Systems Analysis of Synaptic Noise

Another consequence of the simplicity of the point-conductance model is that it enables mathematical approaches. In particular, if one could obtain an analytic expression of the steady-state voltage distribution (such that shown in Fig. 8.2C1), fitting such an expression to experimental data could yield estimates of conductances and other parameters of background activity. This idea was formulated for the first time less than 10 years ago (Rudolph and Destexhe 2003a) and subsequently gave rise to a method called the "VmD method" (Rudolph et al. 2004), which we outline here.

The method to obtain an analytical expression for the voltage distribution is to consider the point-conductance model ((8.1), (8.2), and (8.3)) and evaluate the probability density of finding the system at a value V at time t, denoted $\rho(V, t)$. The time evolution of this probability density is given by a Fokker–Planck equation (Risken 1984), and at steady-state, the probability density gives the voltage distribution $\rho(V)$. So obtaining an analytic estimate of this voltage distribution requires finding the steady-state solution of the Fokker–Planck equation for the system ((8.1), (8.2), and (8.3)). However, this system is nonlinear due to the presence of conductances and their multiplicative effect on the membrane potential, so the corresponding Fokker–Planck equation is not solvable, and one has to rely on approximations. This problem was studied by several groups who proposed different approximations to this problem (Rudolph and Destexhe 2003a, 2005; Richardson 2004; Lindner and Longtin 2006; for a comparative study, see Rudolph and Destexhe 2006).

One of these expressions is invertible (Rudolph and Destexhe 2003a, 2005), which enables one to directly estimate the parameters (g_{e0}, g_{i0}, σ_e, σ_i) from experimentally calculated V_m distributions. This constitutes the basis of the VmD method (Rudolph et al. 2004).

One main assumption behind this method is that the conductance variations are Gaussian-distributed, and thus this distribution can be described by the mean (g_{e0}, g_{i0}) and the standard deviations (σ_e, σ_i) for each conductance. We use the following expression for V_m fluctuations

$$\rho(V) \sim \exp\left[-\frac{(V-\bar{V})^2}{2\sigma_V^2}\right]$$

where $\bar{V}$ is the average V_m and σ_V its standard deviation. This expression provides an excellent approximation of the V_m distributions obtained from models and experiments (Rudolph et al. 2004), because the V_m distributions obtained experimentally show little asymmetry (for up-states and activated states; for specific examples, see Rudolph et al. 2004, 2005, 2007).

This Gaussian distribution can be inverted, which leads to expressions of the synaptic noise parameters as a function of the V_m measurements, $\bar{V}$ and σ_V. To extract the four parameters, means (g_{e0}, g_{i0}) and standard deviations (σ_e, σ_i), from the V_m requires to measure two V_m distributions obtained at two different constant levels of injected current. In this case, the Gaussian fit of the two distributions gives two mean V_m values, $\bar{V}_1$ and $\bar{V}_2$, and two standard deviation values, σ_{V_1} and σ_{V_2}. The system can be solved for four unknowns, leading to expressions of g_{e0}, g_{i0}, σ_e, σ_i from the values of $\bar{V}_1$, $\bar{V}_2$, σ_{V_1}, and σ_{V_2} (for details, see Rudolph et al. 2004).

This method was tested using controlled conductance injection in neurons using the dynamic-clamp technique, as shown in Fig. 8.7. In this experiment, cortical neurons were recorded in slices displaying spontaneous "up-states" of activity. These up-states were analyzed by computing their V_m distribution, which was then used to evaluate the synaptic conductance parameters according to the VmD method. This estimate of conductances was then used to generate synthetic conductance noise traces, which were injected in the same neuron during silent states.

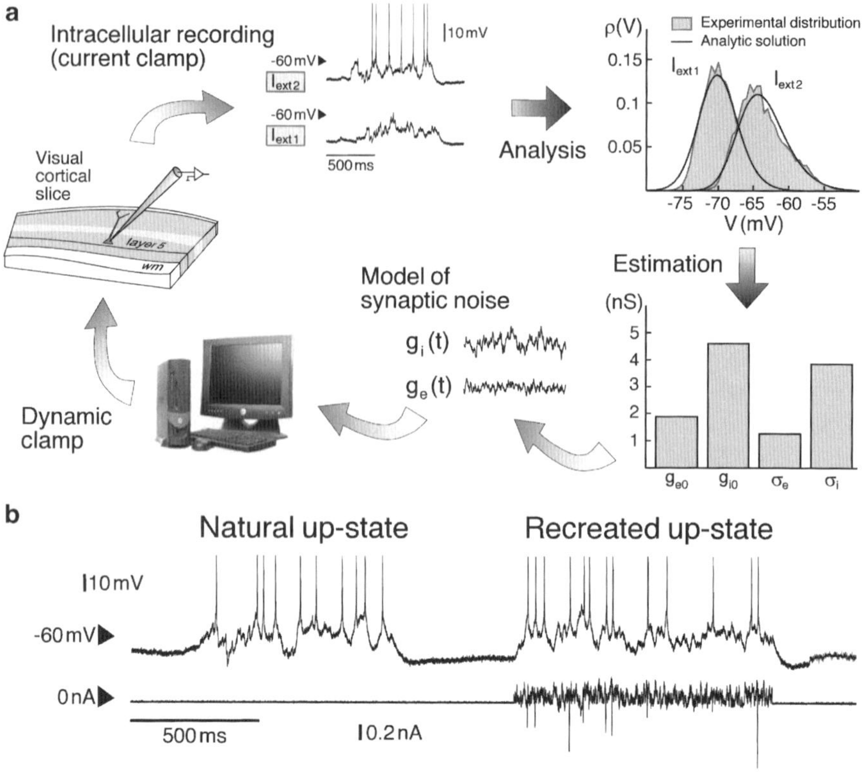

Fig. 8.7 VmD method and test using dynamic-clamp experiments. (**a**) VmD conductance estimation and test of the estimates. *Top left*: spontaneous active network states ("up-states") were recorded intracellularly in ferret visual cortex slices at two different injected current levels (I_{ext1}, I_{ext2}). *Top right*: the V_{m} distributions (*gray*) were computed from experimental data and used to estimate synaptic conductances using the VmD method (analytic expression of V_{m} distribution shown by *solid lines*). *Bottom right*: histogram of the mean and standard deviation of excitatory and inhibitory conductances obtained from the fitting procedure (*gray*). *Bottom left*: a dynamic-clamp protocol was used to inject stochastic conductances consistent with these estimates, therefore recreating artificial up-states in the same neuron. (**b**) Example of natural and recreated up-states in the same cell as in (**a**). This procedure recreated V_{m} activity similar to the active state. Figure modified from Rudolph et al. 2004

The match between the original V_{m} distribution with the one obtained synthetically demonstrated that the VmD method provides good conductance estimates.

The main advantage of the VmD method is that it provides a full characterization of the stochastic conductances. Like other "classic" methods of conductance estimation (reviewed in Monier et al. 2008), the VmD method provides estimates of the total (mean) level of excitatory and inhibitory conductances (g_{e0}, g_{i0}). In addition, it also provides estimates of the *conductance fluctuations*, through the standard deviation of conductances (σ_{e}, σ_{i}). This information is not readily obtained by other methods but is important because it provides estimates of the respective contributions of excitation and inhibition to the V_m fluctuations, and thus offers a quantitative characterization of the "synaptic noise."

Another advantage of the VmD method is that it does not require to record in voltage-clamp mode, which considerably simplifies the experimental protocols, as everything can be estimated from recordings of the V_m activity (current-clamp). However, action potentials must be removed, because the associated Na^+ and K^+ conductances can significantly bias the VmD estimates, so the V_m distributions must be estimated exclusively by accumulating periods of subthreshold activity in-between spikes. Using such a procedure, the VmD method was applied to intracellular recordings in vivo during anesthetized states (Rudolph et al. 2005) and in awake cats (Rudolph et al. 2007). The latter provided the first quantitative conductance estimates in awakc animals.

Estimating the Optimal Conductance Patterns Leading to Spikes in "Noisy" States

The estimation of conductance fluctuations by the VmD method had an important consequence: it opened the route to experimentally characterize the influencc of fluctuations on action potential generation. This was the object of a recent method to estimate the spike-triggered average (STA) conductance patterns from V_m recordings (Pospischil et al. 2007). This "STA method" is also based on the point-conductance model, and requires the prior knowledge of the parameters of mean excitatory and inhibitory conductances (g_{e0}, g_{i0}) and their variances (σ_e, σ_i), which can be provided by the VmD method. Using this knowledge, one can use a maximum likelihood estimator to compute the STA conductance patterns. Similar to the VmD method, the STA method was also tested using dynamic-clamp experiments and was shown to provide accurate estimates (Pospischil et al. 2007; Piwkowska et al. 2008).

Figure 8.8 illustrates STA estimates in a computational model reproducing two extreme conditions found experimentally. First, states where both excitatory and inhibitory conductances are of relatively low and comparable amplitude ("Equal conductance," left panels in Fig. 8.8), similar to some measurements (Shu et al. 2003; Haider et al. 2006). Second, cases where the inhibitory conductance can be up to several-fold larger than the excitatory conductance ("Inhibition-dominated," right panels in Fig. 8.8), which was observed in other measurements in anesthetized (Borg-Graham et al. 1998; Hirsch et al. 1998; Destexhe et al. 2003; Rudolph et al. 2005) or awake preparations (Rudolph et al. 2007). These two extreme cases produce similar mean V_m and V_m fluctuations, but they predict different patterns of conductance STA, as shown in Fig. 8.8B. In the "Equal conductance" condition, the total conductance increases before the spike, and this increase is necessarily due to excitation. In "Inhibition-dominated" neurons, the opposite pattern is seen: there is

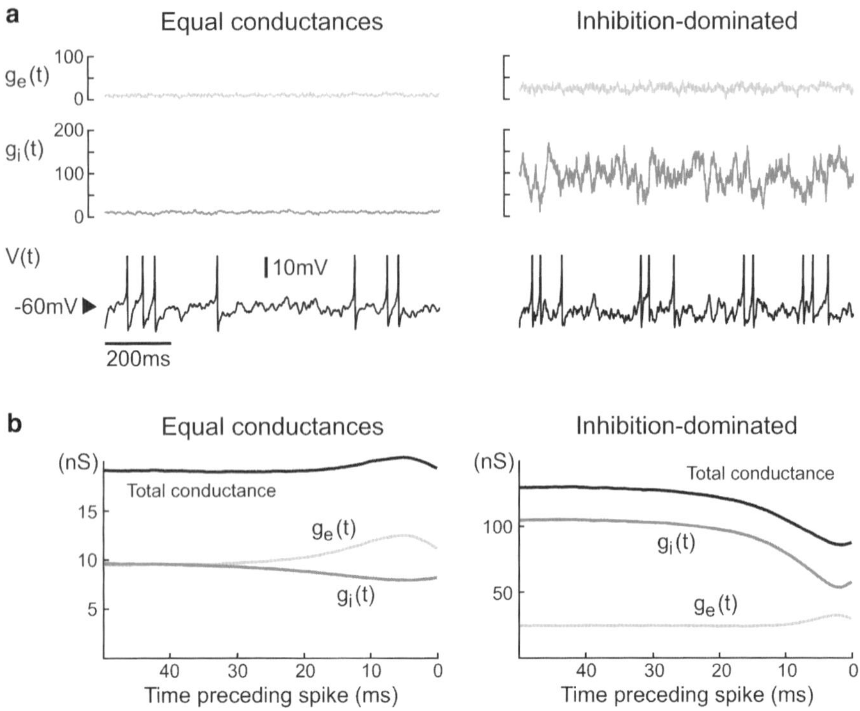

Fig. 8.8 Two patterns of conductances associated to generating spikes in model neurons. Two different "states" are displayed, both leading to comparable V_m fluctuations. *Left*: "Equal conductance" pattern, where g_e and g_i are of comparable amplitude and statistics. *Right*: "Inhibition-dominated" pattern, where g_{e0} is stronger than with equal conductances, but g_{i0} needs to be several-fold larger to maintain the V_m at a similar level. (**a**) g_e, g_i, and V_m activity. (**b**) Spike-triggered conductance patterns associated to each state. Figure modified from Rudolph et al. 2007

a decrease of total conductance prior to the spike, and this decrease necessarily comes from the decrease of inhibition before the spike.

To determine which conductance pattern is seen in cortical neurons in vivo, we applied the STA method to intracellular recordings in awake cats (Rudolph et al. 2007). From intracellular recordings of electrophysiologically identified RS cells, we evaluated the STA of excitatory and inhibitory conductances, as well as the total conductance preceding the spike for neurons recorded in awake (Fig. 8.9A, top) or naturally sleeping (Fig. 8.9A, bottom) cats (see details in Rudolph et al. 2007). In most cells tested (7 out of 10 cells in awake, 6 out of 6 cells in slow-wave sleep, and 2 out of 2 cells in REM sleep), the total conductance drops before the spike, in yielded STAs qualitatively equivalent to that of the model when inhibition is dominant (Fig. 8.8B, right panels).

Note that this pattern is opposite to what is expected from feed-forward inputs. A feed-forward drive would predict an increase of excitation closely associated to an

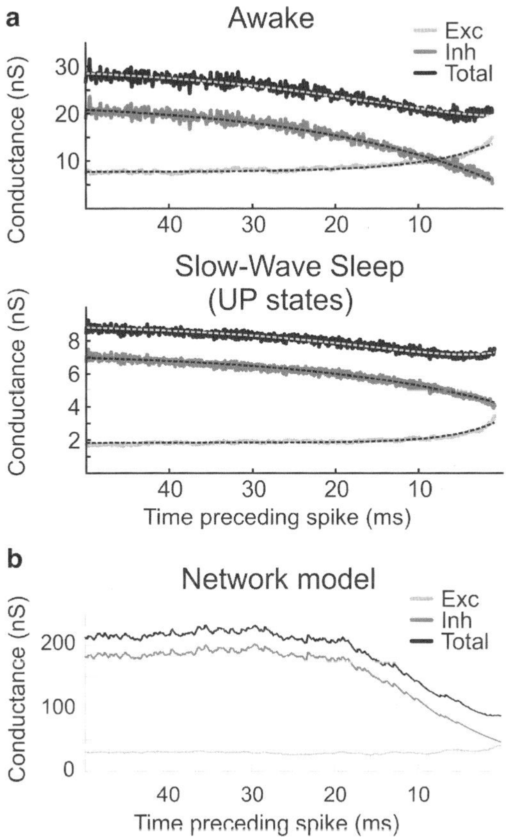

Fig. 8.9 Evidence for "Inhibition-dominated" states in wake and sleep states, as well as in network models. (**a**) Spike-triggered average (STA) of the excitatory, inhibitory, and total conductances obtained from intracellular data of regular-spiking neurons in an awake (*top*) and sleeping (slow-wave sleep up-states, *bottom*) cat. The estimated conductance time courses showed in both cases a drop of the total conductance caused by a marked drop of inhibitory conductance within about 20 ms before the spike. (**b**) STA of conductances in a representative neuron in a network model displaying self-sustained asynchronous irregular states. A 10,000-cell network of integrate-and-fire neurons with conductance-based synaptic interactions was used (see details in El Boustani et al. 2007). (**a**) Modified from Rudolph et al. 2007; (**b**) modified from El Boustani et al. 2007

increase of inhibition, as seen in many instances of evoked responses during sensory processing (Borg-Graham et al. 1998; Monier et al. 2003; Wehr and Zador 2003; Wilent and Contreras 2005). There is no way to account for a concerted g_e increase and g_i drop without invoking recurrent activity, except if the inputs evoked a strong disinhibition, but this was so far not observed in conductance measurements. Indeed, this pattern with inhibition drop was found in self-generated irregular states in networks of integrate-and-fire neurons (Fig. 8.9B; see details in El Boustani et al. 2007). This constitutes direct evidence that most spikes in neocortex in vivo are caused by recurrent (internal) activity, and not by evoked (external) inputs.

Discussion

In this chapter, we have overviewed several recent developments of the exploration of the integrative properties of central neurons in the presence of "noise." This theme has been popular in modeling studies, starting from seminal work (Barrett and Crill 1974; Barrett 1975; Bryant and Segundo 1976; Holmes and Woody 1989), which was followed by compartmental model studies (Bernander et al. 1991; Rapp et al. 1992; De Schutter and Bower 1994). In the last 2 decades, significant progress was made in several aspects of this problem.

The first aspect which we overviewed here is that background activity was measured quantitatively for the first time in "activated" network states in vivo (Paré et al. 1998). Based on these quantitative measurements, constrained models could be built (Destexhe and Paré 1999) to investigate integrative properties in realistic in vivo-like activity states. Consequences on dendritic integration, such as coincidence detection and enhanced temporal processing, as predicted (Bernander et al. 1991; Softky 1994), were confirmed (Rudolph and Destexhe 2003b). New consequences were also found, such as enhanced responsiveness (Hô and Destexhe 2000) and location-independent synaptic efficacy (Rudolph and Destexhe 2003b). The first of these predictions was confirmed by dynamic-clamp experiments (Destexhe et al. 2001; Chance et al. 2002; Fellous et al. 2003; Mitchell and Silver 2003; Prescott and De Koninck 2003; Shu et al. 2003; Higgs et al. 2006).

We reviewed another aspect that tremendously progressed, namely the formulation of simplified models that replicate the in vivo measurements, as well as important properties such as the typical Lorentzian spectral structure of background activity. This point-conductance model (Destexhe et al. 2001) had many practical consequences, such as to enable dynamic-clamp. Indeed, many of the aforementioned dynamic-clamp studies used the point-conductance model to recreate in vivo-like activity states in neurons maintained in vitro. In addition to confirm model predictions, dynamic-clamp experiments also took these concepts further and investigated important properties such as gain modulation (Chance et al. 2002; Fellous et al. 2003; Mitchell and Silver 2003; Prescott and De Koninck 2003). An inverse form of gain modulation can also be observed (Fellous et al. 2003) and may be explained by potassium conductances (Higgs et al. 2006). It was also found that the intrinsic properties of neurons combine with synaptic noise to yield unique responsiveness properties (Wolfart et al. 2005).

It must be noted that although the point-conductance model was the first model of fluctuating synaptic conductances injected in living neurons using dynamic-clamp, other models are also possible. For example, models based on the convolution of Poisson processes with exponential synaptic waveforms ("shot noise") have also been used (e.g., see Chance et al. 2002; Prescott and De Koninck 2003). However, it can be shown that these models are in fact equivalent, as the point-conductance model can be obtained as a limit case of a shot-noise process with exponential conductances (Destexhe and Rudolph 2004).

An important consequence, specific to the point-conductance model, is that its mathematical simplicity enabled formulation of a number of variants of the Fokker–Planck equation for the membrane potential probability density (Rudolph and Destexhe 2003a, 2005; Richardson 2004; Lindner and Longtin 2006), which led to a method to estimate synaptic conductances from V_m recordings (Rudolph et al. 2004). This "VmD method" decomposed the V_m fluctuations into excitatory and inhibitory contributions, estimating their mean and variance. This method was successfully tested in dynamic-clamp experiments (Rudolph et al. 2004) as well as in voltage-clamp (Greenhill and Jones 2007; see also Ho et al. 2009). The most interesting aspect of the VmD method is that it provides estimates of the variance of conductances or, equivalently, conductance fluctuations. This type of estimate was made for cortical neurons during artificially activated brain states (Rudolph et al. 2005) or in awake animals (Rudolph et al. 2007). The latter provided the first quantitative characterization of synaptic conductances and their fluctuations in aroused animals.

Finally, this approach was extended to estimate dynamic properties related to action potential initiation. If the information about synaptic conductances and their fluctuations is available (for example following VmD estimates), then one can use maximum likelihood methods to evaluate the spike-triggered conductance patterns. This information is very important to determine which optimal conductance variations determine the "output" of the neuron, which is a fundamental aspect of integrative properties. We found that in awake and naturally sleeping animals, the majority of spikes are statistically related to disinhibition, which plays a permissive role. This type of conductance dynamics is opposite to the conductance patterns evoked by external input, but can be replicated by models displaying self-generated activity. This suggests that most spikes in awake animals are due to internal network activity. This argues for a dominant role of the network state in vivo and that inhibition is a key player. Both aspects should be investigated by future studies.

Thus, the last 20 years have seen a tremendous theoretical and experimental characterization of the synaptic "noise," and its consequences on neurons and networks. Computational models have played—and still continue to play—a pivotal role in this exploration.

Acknowledgments The experimental data shown here were obtained in collaboration with Thierry Bal, Diego Contreras, Jean-Marc Fellous, Denis Paré, Zuzanna Piwkowska, Mircea Steriade and Igor Timofeev. The models and analyses were done in collaboration with Sami El Boustani, Martin Pospischil, Michelle Rudolph and Terrence Sejnowski. Research supported by the CNRS, ANR (HR-CORTEX project), HFSP and the European Community (FACETS project FP6-15879; BrainScales project FP7-269921).

References

Barrett JN (1975) Motoneuron dendrites: role in synaptic integration. Fed Proc 34:1398–1407

Barrett JN, Crill WE (1974) Influence of dendritic location and membrane properties on the effectiveness of synapses on cat motoneurones. J Physiol 293:325–345

Bernander O, Douglas RJ, Martin KA, Koch C (1991) Synaptic background activity influences spatiotemporal integration in single pyramidal cells. Proc Natl Acad Sci USA 88:11569–11573

Borg-Graham LJ, Monier C, Frégnac Y (1998) Visual input evokes transient and strong shunting inhibition in visual cortical neurons. Nature 393:369–373
Bryant HL, Segundo JP (1976) Spike initiation by transmembrane current: a white-noise analysis. J Physiol 260:279–314
Chance FS, Abbott LF, Reyes AD (2002) Gain modulation from background synaptic input. Neuron 35:773–782
Contreras D, Timofeev I, Steriade M (1996) Mechanisms of long lasting hyperpolarizations underlying slow sleep oscillations in cat corticothalamic networks. J Physiol 494:251–264
De Schutter E, Bower JM (1994) Simulated responses of cerebellar Purkinje cells are independent of the dendritic location of granule cell synaptic inputs. Proc Natl Acad Sci USA 91:4736–4740
Destexhe A (2007) High-conductance state. Scholarpedia 2:1341. http://www.scholarpedia.org/article/High-Conductance State
Destexhe A, Bal T (eds) (2009) The dynamic-clamp: from principles to applications. Springer, New York
Destexhe A, Paré D (1999) Impact of network activity on the integrative properties of neocortical pyramidal neurons in vivo. J Neurophysiol 81:1531–1547
Destexhe A, Rudolph M (2004) Extracting information from the power spectrum of synaptic noise. J Comput Neurosci 17:327–345
Destexhe A, Rudolph M (2012) Neuronal noise. Springer, New York
Destexhe A, Contreras D, Steriade M (1999) Spatiotemporal analysis of local field potentials and unit discharges in cat cerebral cortex during natural wake and sleep states. J Neurosci 19:4595–4608
Destexhe A, Rudolph M, Fellous J-M, Sejnowski TJ (2001) Fluctuating synaptic conductances recreate in vivo-like activity in neocortical neurons. Neuroscience 107:13–24
Destexhe A, Mand R, Paré D (2003) The high-conductance state of neocortical neurons *in vivo*. Nat Rev Neurosci 4:739–751
El Boustani S, Pospischil M, Rudolph-Lilith Mand Destexhe A (2007) Activated cortical states: experiments, analyses and models. J Physiol Paris 101:99–109
Fellous JM, Rudolph M, Destexhe A, Sejnowski TJ (2003) Synaptic background noise controls the input/output characteristics of single cells in an in vitro model of in vivo activity. Neuroscience 122:811–829
Gammaitoni L, Hanggi P, Jung P, Marchesoni F (1998) Stochastic resonance. Rev Mod Phys 70:223–287
Greenhill SD, Jones RS (2007) Simultaneous estimation of global background synaptic inhibition and excitation from membrane potential fluctuations in layer III neurons of the rat entorhinal cortex in vitro. Neuroscience 147:884–892
Haider B, McCormick DA (2009) Rapid neocortical dynamics: cellular and network mechanisms. Neuron 62:171–189
Haider B, Duque A, Hasenstaub AR, McCormick DA (2006) Neocortical network activity in vivo is generated through a dynamic balance of excitation and inhibition. J Neurosci 26: 4535–4545
Higgs MH, Slee SJ, Spain WJ (2006) Diversity of gain modulation by noise in neocortical neurons: regulation by the slow after-hyperpolarization conductance. J Neurosci 26:8787–8799
Hirsch JA, Alonso JM, Reid RC, Martinez LM (1998) Synaptic integration in striate cortical simple cells. J Neurosci 18:9517–9528
Hô N, Destexhe A (2000) Synaptic background activity enhances the responsiveness of neocortical pyramidal neurons. J Neurophysiol 84:1488–1496
Ho EC, Zhang L, Skinner FK (2009) Inhibition dominates in shaping spontaneous CA3 hippocampal network activities in vitro. Hippocampus 19:152–165
Holmes WR, Woody CD (1989) Effects of uniform and non-uniform synaptic "activation-distributions" on the cable properties of modeled cortical pyramidal neurons. Brain Res 505:12–22
Lindner B, Longtin A (2006) Comment on "Characterization of subthreshold voltage fluctuations in neuronal membranes", by M. Rudolph and A. Destexhe. Neural Comput 18:1896–1931

Llinas RR, Jahnsen H (1982) Electrophysiology of thalamic neurones in vitro. Nature 297:406–408

Mitchell SJ, Silver RA (2003) Shunting inhibition modulates neuronal gain during synaptic excitation. Neuron 38:433–445

Monier C, Chavane F, Baudot P, Graham LJ, Frégnac Y (2003) Orientation and direction selectivity of synaptic inputs in visual cortical neurons: a diversity of combinations produces spike tuning. Neuron 37:663–680

Monier C, Fournier J, Frégnac Y (2008) In vitro and in vivo measures of evoked excitatory and inhibitory conductance dynamics in sensory cortices. J Neurosci Methods 169:323–365

Paré D, Shink E, Gaudreau H, Destexhe A, Lang EJ (1998) Impact of spontaneous synaptic activity on the resting properties of cat neocortical neurons in vivo. J Neurophysiol 79: 1450–1460

Piwkowska Z, Pospischil M, Brette R, Sliwa J, Rudolph-Lilith M, Bal T, Destexhe A (2008) Characterizing synaptic conductance fluctuations in cortical neurons and their influence on spike generation. J Neurosci Methods 169:302–322

Pospischil M, Piwkowska Z, Rudolph M, Bal T, Destexhe A (2007) Calculating event-triggered average synaptic conductances from the membrane potential. J Neurophysiol 97:2544–2552

Prescott SA, De Koninck Y (2003) Gain control of firing rate by shunting inhibition: roles of synaptic noise and dendritic saturation. Proc Natl Acad Sci USA 100:2076–2081

Rapp M, Yarom Y, Segev I (1992) The impact of parallel fiber background activity on the cable properties of cerebellar Purkinje cells. Neural Comput 4:518–533

Richardson MJ (2004) Effects of synaptic conductance on the voltage distribution and firing rate of spiking neurons. Phys Rev E 69:051918

Risken H (1984) The Fokker Planck equation: methods of solution and application. Springer, Berlin

Robinson HP, Kawai N (1993) Injection of digitally synthesized synaptic conductance transients to measure the integrative properties of neurons. J Neurosci Methods 49:157–165

Rudolph M, Destexhe A (2001) Correlation detection and resonance in neural systems with distributed noise sources. Phys Rev Lett 86:3662–3665

Rudolph M, Destexhe A (2003a) Characterization of subthreshold voltage fluctuations in neuronal membranes. Neural Comput 15:2577–2618

Rudolph M, Destexhe A (2003b) A fast-conducting, stochastic integrative mode for neocortical dendrites *in vivo*. J Neurosci 23:2466–2476

Rudolph M, Destexhe A (2005) An extended analytic expression for the membrane potential distribution of conductance-based synaptic noise. Neural Comput 17:2301–2315

Rudolph M, Destexhe A (2006) On the use of analytic expressions for the voltage distribution to analyze intracellular recordings. Neural Comput 18:917–922

Rudolph M, Piwkowska Z, Badoual M, Bal T, Destexhe A (2004) A method to estimate synaptic conductances from membrane potential fluctuations. J Neurophysiol 91:2884–2896

Rudolph M, Pelletier J-G, Paré D, Destexhe A (2005) Characterization of synaptic conductances and integrative properties during electrically-induced EEG-activated states in neocortical neurons in vivo. J Neurophysiol 94:2805–2821

Rudolph M, Pospischil M, Timofeev I, Destexhe A (2007) Inhibition determines membrane potential dynamics and controls action potential generation in awake and sleeping cat cortex. J Neurosci 27:5280–5290

Sharp AA, O'Neil MB, Abbott LF, Marder E (1993) The dynamic clamp: artificial conductances in biological neurons. Trends Neurosci 16:389–394

Shu Y, Hasenstaub A, Badoual M, Bal T, McCormick DA (2003) Barrages of synaptic activity control the gain and sensitivity of cortical neurons. J Neurosci 23:10388–10401

Softky W (1994) Sub-millisecond coincidence detection in active dendritic trees. Neuroscience 58:13–41

Steriade M (2003) Neuronal substrates of sleep and epilepsy. Cambridge University Press, Cambridge, UK

Steriade M, Timofeev I, Grenier F (2001) Natural waking and sleep states: a view from inside neocortical neurons. J Neurophysiol 85:1969–1985
Tuckwell HC (1988) Introduction to theoretical neurobiology. Cambridge University Press, Cambridge, UK
Wehr M, Zador AM (2003) Balanced inhibition underlies tuning and sharpens spike timing in auditory cortex. Nature 426:442–446
Wiesenfeld K, Moss F (1995) Stochastic resonance and the benefits of noise: from ice ages to crayfish and SQUIDs. Nature 373:33–36
Wilent W, Contreras D (2005) Dynamics of excitation and inhibition underlying stimulus selectivity in rat somatosensory cortex. Nat Neurosci 8:1364–1370
Wolfart J, Debay D, Le Masson G, Destexhe A, Bal T (2005) Synaptic background activity controls spike transfer from thalamus to cortex. Nat Neurosci 8:1760–1767

Chapter 9
Still Looking for the Memories: Molecules and Synaptic Plasticity

Upinder S. Bhalla

Abstract Computational neuroscientists have been playing around with plastic synapses for several decades. Interestingly, mechanistically detailed models of synaptic plasticity started around the same time as the CNS meetings. This was when the associative properties of the N-methyl-D-aspartate (NMDA) receptor were demonstrated, first setting out the molecular and mechanistic underpinnings of synaptic plasticity. Some 20 years ago there was little reason to expect that the underlying biology would turn out to be as outrageously complicated as we now find it. Associativity seemed to be established by the NMDA receptor especially through the work of Collingridge, and there were already a couple of candidate mechanisms for how to maintain synaptic weights: the CaMKII autocatalytic process found by several people and first modeled by Lisman, and the PKA story from Kandel. These leads led into a maze. Even 10 years ago, there were over a 100 known molecules implicated in synaptic plasticity. The first major molecular models of synaptic plasticity had some dozen signaling pathways—a far cry from what was known. The field as a whole is still playing catch-up. Nevertheless, most of the key properties of plasticity have had a good share of models, at various levels of detail. I suggest that there has been a recent shift in perspective, from enumerating molecules to looking at functional roles that may involve different, often overlapping sets of molecules. It is the identification and integration of these diverse functions of the synapse that is the key conceptual direction of the field. This has combined with technical and data-driven advances in managing and modeling multiscale phenomena spanning single-molecule reaction–diffusion, through chemistry, electrical and structural effects, and the network. As many of us felt, 20 years ago, we are again at a fascinating time where the experiments, the databases, and the computational tools are just coming together to address these questions.

U.S. Bhalla (✉)
National Centre for Biological Sciences, Tata Institute of Fundamental Research, Bellary Road, Bangalore 560065, India
e-mail: bhalla@ncbs.res.in

J.M. Bower (ed.), *20 Years of Computational Neuroscience*, Springer Series in Computational Neuroscience 9, DOI 10.1007/978-1-4614-1424-7_9,

Introduction

Memory, as we think of it today, bridges many concepts. We are all familiar with the immediate personal working of memory, its sometimes difficult recall, and the sensory completeness inherent in "reliving" an experience. We are also used to memory expressed as information storage and retrieval in computing devices. The gap between these two concepts is large. It is a long way indeed from information storage to human memory, and a considerable section of computational neuroscience has sought to span this divide. Many of the early inspirations in the field derive from this effort. These include "learning" networks like multilayer perceptrons with back-propagation (Minsky and Papert 1969; Rosenblatt 1962; Rumelhart et al. 1986) and Hopfield networks for associative memory (Hopfield 1982). The role of the synapse as the likely locus of these mechanisms has been apparent since before Hebb (1949) articulated it in the form we still use: "When an axon of cell A is near enough to excite a cell B and repeatedly or persistently takes part in firing it, some growth process or metabolic change takes place in one or both cells such that A's efficiency, as one of the cells firing B, is increased."

What is this "growth process or metabolic change?" This leads to another venerable thread of memory research, dating from the earliest days of computational neuroscience, to understand the most basic mechanisms for information storage in the brain. What, in other words, are the biological equivalents of transistors and gates for memory? Here I will trace the evolution of our understanding of these mechanisms, as articulated in computational models. I will do so in the form of three snapshots, of each a decade apart, reflecting the early, middle, and current periods of the computational neuroscience (CNS) meetings.

Snapshot 1: 1990

Around 1990, synaptic plasticity was already well established and was accepted as the cellular correlate of Hebb's rule (Bliss and Collingridge 1993). In the typical experiment to measure synaptic plasticity, one records from a thin slice of tissue from the hippocampus of a rat, cut so as to preserve cell bodies from input and output regions, and also the fibers and synapses between them (Fig. 9.1). Widely spaced test stimuli are given on the fiber bundles to "tickle" the synapses, and the baseline response in the target cells is measured. Then a series of strong stimuli are delivered, following which the test stimuli are resumed. The response is now about twice the original baseline, and this elevated response continues for hours, as long as the slice remains healthy. Elaborations of this experiment, specially using patch recording for greater precision, established the key tenets of Hebb's rule: associativity, correlations between inputs and outputs, and input specificity. This body of effects is widely referred to as long-term potentiation, LTP. While it took a long while, another key theoretical prediction had also been established by 1990: the existence

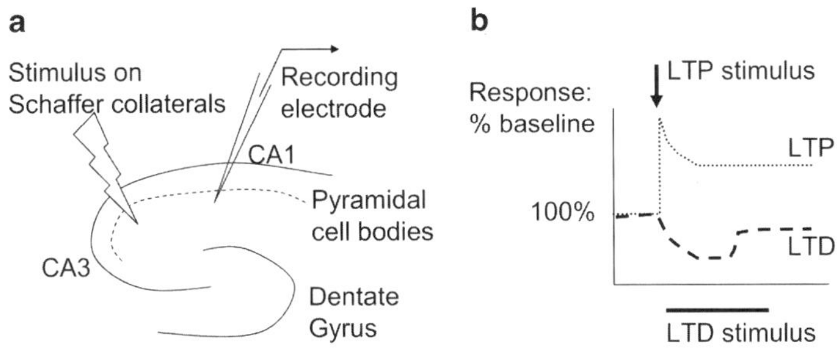

Fig. 9.1 Hippocampal long-term potentiation (LTP)/LTD experiments. (**a**) Schematic of hippocampal brain slice experiment. A stimulating electrode is placed on the Schaffer collaterals toward the CA3. A recording electrode is placed in the dendritic region (stratum oriens) of the CA1. (**b**) Recordings. An initial baseline is established by delivering test pulses to the stimulating electrode, and recording field or intracellular responses. Following this the plasticity stimuli are given, for example, a tetanic burst for LTP or a 1 Hz series of pulses for 900 s for LTD. After these stimuli, the test stimuli are resumed and the response is monitored to observe increase or decrease with reference to baseline

of the complementary process for reversal of the enhanced response (Bienenstock et al. 1982; Ito 1989). LTD was first characterized in cerebellum, but hippocampal forms soon followed (Dudek and Bear 1992; Mulkey and Malenka 1992; Stanton and Sejnowski 1989). Instead of brief strong input, the most commonly elicited form of hippocampal LTD requires the delivery of sustained low-frequency (1 Hz) stimuli (Fig. 9.1). Many other brain regions, other stimulus patterns, and almost all nervous systems were found to exhibit some variant of these forms of synaptic plasticity. At least at face value, the synapse offered all that could be asked by Hebb's rule. So how does it do it?

In 1984, the first key aspect of Hebb's rule was reduced to a single molecule. It was shown that the excitatory neurotransmitter glutamate caused the opening of a specific receptor protein that exhibited associativity (Nowak et al. 1984). This receptor, called the N-methyl-D-aspartate (NMDA) receptor, did so through a particularly simple mechanism (Fig. 9.2). The channel opened when neurotransmitter was present: this was the presynaptic activity from Hebb's rule. However, it also required postsynaptic activity to depolarize the synapse enough to release a block caused by Mg^{2+} plugging the channel. Thus the NMDA receptor acted as a molecular association device. It would only open when both pre- and postsynaptic activity were present. Numerous papers explored the effectiveness of this associativity (e.g., Zador et al. 1990).

The second key aspect of synaptic plasticity, its persistence, also was reduced to a possible set of molecular event terms in the late 1980s. Many researchers had been intrigued by the properties of the calcium–calmodulin-dependent type II kinase (CaMKII) (Malinow et al. 1989). Among other things, this molecule is present in outrageously high concentrations at the synapse, and it has the peculiar ability to activate itself. John Lisman was one of the first to point out using computational studies that this could readily form the basis for an autocatalytic feedback switch

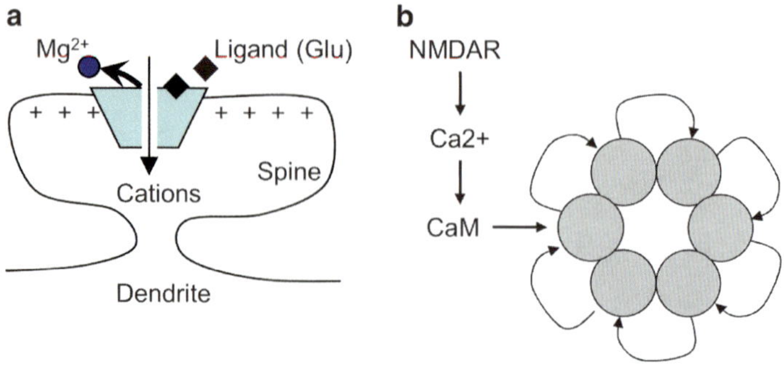

Fig. 9.2 The two key mechanisms proposed for long-term synaptic plasticity. (**a**) Associativity due to the N-methyl-D-aspartate (NMDA) receptor. Presynaptic input resulting in glutamate release must be associated with postsynaptic depolarization of the dendrite (indicated by *plus* symbols). The postsynaptic depolarization releases the block of the open channel due to magnesium ions, and this allows cations to flow through the receptor into the dendritic spine. (**b**) Activation and autocatalysis of CaM kinase II. Calcium enters through the NMDA receptor and binds to CaM, which binds to and activates individual subunits of the CaMKII enzyme (*circles*). The CaMKII holoenzyme has 12 subunits stacked in rings of 6, each of which can activate its neighbor

(Lisman 1989). Better still, the CaMKII switch could be triggered by the calcium influx through the NMDA channel. While the applicability of the putative CaMKII switch remains a matter for contention, the concept of molecular switches in the synapse has proven popular and most current molecular models for long-term plasticity incorporate some kind of a bistable switch. In chemical terms, a bistable switch is a system of reactions which has two stable steady states. Note that these are not equilibrium states, as energy is required to maintain them. Since each state is steady, the system will tend to restore itself to that state even when perturbed by metabolic changes, chemical signals, or even noise. It takes a large stimulus to push the system over from one state to another (Fig. 9.2).

How does one achieve bistability through chemical reactions? Many theoretical studies have addressed this question over the years (Ferrell and Xiong 2001; Ramakrishnan and Bhalla 2008), but rather fewer are backed by experimental correlates. The core attribute of a bistable chemical system is the presence of feedback interactions with a net positive sign around the feedback loop. This is not sufficient. The slope of the respective dose–response curves for mutual activation of the components of the loop must be steep enough that they intersect 3 times. These intersection points define fixed points of the system. The outer two fixed points are the stable states, and the intermediate fixed point behaves like a transition point or threshold. If you push the system above threshold, it will settle to the upper fixed point. If you now push it down past threshold again, it will settle to the lower one.

Why a bistable switch and not just some kind of slow-decaying molecular capacitor? An example of the latter would be a one-time event which inserted a lot of receptors into the postsynaptic density, and then left them there. If the receptors were to stay there stably, this one-time insertion event itself would be sufficient to establish a change in synaptic weight. The slow-decay model is attractive both for its simplicity and the fact that it can store analog values for synaptic weight. Unfortunately, it won't work for long-term memory. From experiments and simple calculations, it was clear long before the 1990s that there were at least three rather formidable problems for any kind of long-term molecular storage of synaptic state information.

1. All molecules in the brain turn over, typically with a time constant of 1 or 2 days or less (Ehlers 2003). This means that one cannot store information through accumulation of receptor or regulatory molecules, like charge in a capacitor: it will decay rather quickly.
2. Molecules in the brain diffuse and undergo transport. So even if the molecules didn't turn over, they might move away.
3. Synapses are so small that the molecular reactions within them (involving receptors, for example) are highly stochastic. A simple calculation shows that a typical 0.5 μm spine head would have around five free calcium ions. More recent measurements show that the number of receptor proteins in a typical synapse is in the low hundreds—again, a likely range for stochasticity.

With these constraints in mind, the advantage of bistable systems over other forms of molecular memory maintenance becomes more apparent. Specifically, molecular turnover has little impact. Turnover times of days or even hours are a rather small perturbation on the kinds of molecular switches that have been proposed for the synapse. The other two issues of diffusion and stochasticity turned out to be more difficult to analyze computationally, and indeed these established fertile grounds for computational research that still continue.

In summary, the beginning years of computational neuroscience also coincided with a molecular simplicity of memory mechanisms that seemed irresistible. The key questions of synaptic plasticity seemed to be resolved: associativity and long-term information storage had promising molecular correlates, and the field seemed ready to move on to using these building blocks to make circuits that could bridge the rest of the great divide between synapses and systems memory.

Snapshot 2: 2000

The decade of the 1990s was a triumphal period for molecular biology. Genomes began to be sequenced, and this flood of new data, new techniques, and new molecules transformed neuroscience. Many aspects of computational neuroscience were hardly affected by this, but anything to do with synaptic plasticity underwent huge changes. Specifically, the nascent field of computational studies of mechanisms of synaptic plasticity had to struggle to keep itself afloat in the flood. Computational neuroscientists often wish they had more data. Here was a situation where there was way too much data, and not quite of the kind that they were hoping for. It would have been nice to have gotten better data about reaction rates, mechanisms, and concentrations. Instead, what the field did get was more molecules, more pathways, and far more complexity.

Two particularly telling illustrations of the change are provided by a review and a paper that came out in the year 1999. The review posed the question: "Can molecules explain long-term potentiation?" (Sanes and Lichtman 1999). Of course this question is one that the computational neuroscientists in the area were scrambling to answer in the affirmative. The review traced the very same developments in the field that we have touched upon here. The authors listed over 100 molecules that were then known to be important for LTP (Table 9.1). They remarked, somewhat ominously, that there were indications that the number might be far greater. This has turned out to be true, in the light of findings in the decade since this review.

The same influx of molecular and pathway data that had overwhelmed the study of signaling also provided an impetus to the new (or at least newly renamed) field of systems biology. The key attribute of this somewhat broadly defined area was its use of quantitative techniques to analyze complex biological problems. This, of course, is what computational neuroscientists had been doing all along. The second illustration I pick is about a project at this intersection of systems biology and computational neuroscience. It was a computational analysis of the emergent properties of the then best understood signaling pathways involved in synaptic plasticity (Bhalla and Iyengar 1999). The analysis used straightforward ordinary differential equation modeling to represent the time-evolution of signaling responses following different kinds of synaptic input. Ironically, though the study utilized relatively recent molecular insights for the topology of signaling networks, the "hard data" for the calculations (rate constants and concentrations) were available only from classic, tedious, test-tube biochemistry. This reiterated the complaint of data, data everywhere, but none of the numbers that we really needed.

Even though it covered only a subset of the species in the Sanes and Lichtman review, our signaling model incorporated over 200 molecules, many of them phosphorylation or binding states of primary signaling molecules (Fig. 9.3). Not coincidentally, this early synaptic model incorporated not one but two bistable switches. One of these was, of course, the classic CaMKII autophosphorylation feedback switch. The other was a prediction of a feedback cycle involving two of the major kinases, protein kinase C (PKC) and microtubule-associated protein kinase (MAPK). Remarkably, the prediction of bistability in this cycle was subsequently supported by experiments (Bhalla et al. 2002; Tanaka and Augustine 2008). The key point of the signaling simulations was that putting together a lot of signaling pathways didn't just add up their properties, there were unexpected and emergent behaviors of the system as a whole.

The common subtext to both these studies was the demise of the simple, optimistic view of synaptic plasticity as a process that could be understood by a few clear-cut mechanisms. Synaptic plasticity no longer offered even the hope of being a closed system: both studies discussed here emphasized their own incompleteness. The open-ended and somewhat fractious character of the field of experimental synaptic plasticity research reinforced this pessimistic outlook.

Beyond the broad sweep of these two illustrative studies, there was a steady stream of studies that began to nibble away at the pile of molecules in specific domains. Some of these studies took specific phenomena, such as calcium

Table 9.1 Molecules involved in LTP (reproduced from Sanes and Lichtman 1999)

Glutamate receptors	*Calcium/calmodulin binding proteins*	*Kinases*
GluR1	Calmodulin	Inositol 1,4,5-triphosphate-3-kinase
GluR2	RC3/neurogranin	MAPK
mGluR1	GAP43/B50/neuromodulin	src
mGluR4	S100	fyn
mGluR5		ERK
mGluR7		Protein kinase A C1 β subunit
NMDA NR1		Protein kinase A R1 β subunit
NMDA NR2A		Protein kinase C-gamma
NMDA NR2D		Protein kinase G
		Protein kinase M-zeta
		CaM kinase I
		CaM kinase IV
		CaM kinase II
		Ecto protein kinase
Other neurotransmitters, neuromodulators, and their receptors	*Ion channels*	*Proteases and their inhibitors*
Norepinephrine and β-adrenergic receptors	L-type calcium channels	Calpain
Adenosine and adenosine 2A receptors	Olfactory cyclic nucleotide-gated channels	Calpastatin
Dopamine and D1 dopamine receptors		Protease nexin 1
μ-opioid receptors		Tissue plasminogen activator
Δ 1 opioid receptors		E6-AP ubiquitin ligase
Acetylcholine and muscarinic receptors		Plasmin
GABA and GABA-B receptors		
Anandamine and CB1 cannabinoid receptors		
Orphanin NQ and nociceptin receptors		
Serotonin and 5HT3-receptors		
sn-2 arachidonylglyceral (2-AG)		
Endothelin-1		
γ-hydroxybutynic acid (GHB) and GHB receptors		

(continued)

Table 9.1 (continued)

Intercellular messengers, their synthetic enzymes, and their receptors	*Transcription factors*	*Other enzymes*
CO	Retinoic acid receptor β	Phospholipase A2
NO	CREB	Phospholipase C β
EGF	Krox 20	Phospholipase C γ
Basic FGF	Krox 24	ADP ribosyl transferase
Superoxide		Calcineurin
Neuregulin		Acetylcholinesterase
erB4		Protein phosphatase I
NGF		Guanylate cyclase
BDNF		Adenylate cyclase
TrkB		
nNOS		
eNOS		
Arachidonic acid		
Platelet activating factor		
Interleukin 1 β		
H_2S		
β-Activin		
Carbohydrates	*Adhesion molecules*	*Miscellaneous*
Polysialic acid	ephA5	Spectrin/fodrin
Ganglioside GQ1B	Ephrin A5	GFAP
Ganglioside GM1	NCAM	Stathmin RB3/XB3
	E-cadherin	EBI-1 G protein-coupled receptor
	N-cadherin	
	thy-1	Mas G protein-coupled receptor
	Telencephalin	Vest
	L1/NgCAM	
	HB-GAM/pleitrophin	
	Integrins	
	Integrin-associated protein	
	Tenascin C	

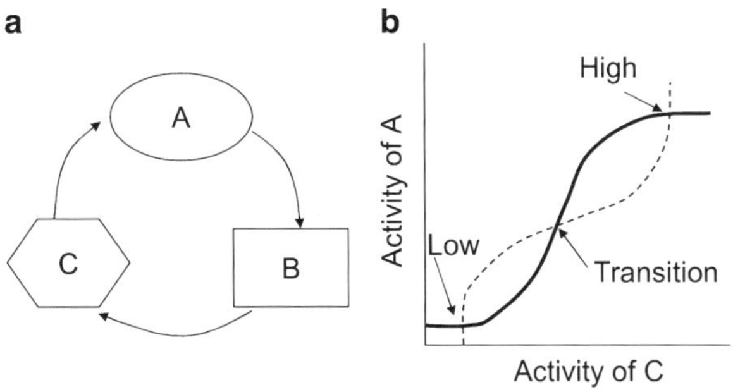

Fig. 9.3 Positive feedback and bistability. (**a**) Illustrative positive feedback loop involving three molecular species, A, B, and C, each of which activates the next. (**b**) Criterion for bistability. The solid curve is a dose–response curve indicating the strength of response of A to different values of its input C. The dashed line is the converse line, indicating the activation of C for a given activity of A. Intersections between these two curves are fixed points of the system. The low and high fixed points are stable states, and the intermediate fixed point is an unstable transition point. Any two molecules in the feedback loop could be plotted to get the stable points

dynamics, and incorporated more molecular, physiological, and spatial details (Gold and Bear 1994; Jaeger et al. 1997; Zador et al. 1990). In addition to considerable work in the area of subcellular calcium dynamics, there were numerous studies on synaptic release, including detailed stochastic and spatial simulations (e.g., Anglister et al. 1994; Bennett et al. 1995). There was an interesting confluence of experimental and modeling studies on homeostasis, and its implications for synaptic normalization and memory retention (Liu et al. 1998; Turrigiano 1999). As discussed below, these studies began a shift in thinking toward key functional roles that had to be assembled to understand synapses.

Snapshot 3: 2010

In the second decade of this history and of the computational neuroscience meeting, the model building efforts of many groups finally began to make headway. The flood of new molecules and interactions continued unabated, but the fields of systems biology and computational neuroscience had begun to organize their troops. These included a well-defined numerical framework for tackling and modeling complex signaling networks, the recruitment of control theory techniques to this analysis, and many computational tools. There were also the beginnings of standards for data handling and sharing (Crook et al. 2007b; Hucka et al. 2003), and a new wave of experimental inputs, notably imaging, that began to provide comparable quantitative data to back up the qualitative, topological information from molecular work. Most importantly—and this was a specific contribution of the computational viewpoint—it became possible to recognize functional motifs in signaling pathways. This meant that instead of puzzling out a continually growing tapestry of molecules and interactions, one could view the system in terms of larger, logically coherent operations (Alon 2007; Bhalla 2003; Tyson et al. 2003). This was in parallel with the experimental realization that the process of synaptic plasticity had much more to it than the bald reiteration of Hebb's rule. Here I will trace some of this evolution, much of which was reflected in work presented at the CNS meetings.

Bistability

Bistable biochemical switches were one of the key early ideas about the mechanisms for long-lasting synaptic change. Molecular and genomic techniques brought in an influx of molecular candidates and signaling pathway networks, and many of the key synaptic players turned out to be plausible candidates for bistable switches. The most obvious way to flag a network for possible bistability is the presence of positive feedback loops (Fig. 9.4). New positive feedback motifs suggested in this period included switches involving PKA (Lindskog et al. 2006; Nakano et al. 2010; Song et al. 2006), protein synthesis (Aslam et al. 2009; Jain and Bhalla 2009), and gene switches (Sanyal et al. 2002) . Models involving such feedback systems have been proposed to account for long-term plasticity in systems ranging from Aplysia to Drosophila and mammalian hippocampus. Research has also continued to consolidate the ideas around the venerable CaMKII autophosphorylation loop (Graupner and Brunel 2007; Lisman and Zhabotinsky 2001; Miller et al. 2005; Pepke et al. 2010). The MAPK–PKC feedback loop received both theoretical and experimental support (Bhalla et al. 2002), and its possible role was extended to LTD in Purkinje neurons (Kuroda et al. 2001; Tanaka and Augustine 2008). Not all bistable systems have overt positive feedback. For example, multisite phosphorylation cascades can exhibit multistability (Markevich et al. 2004; Smolen et al. 2008). The process of receptor trafficking, which has very direct correlations with synaptic strength, was found to be a potential bistable through a computational study (Hayer and Bhalla 2005). This has yet to be experimentally validated.

These illustrations of nonobvious bistables suggested a reverse computational approach to finding possible signaling systems that had bistability. Instead of looking for feedback motifs in known chemical networks and then testing for bistability, we tested for bistability in a systematic sweep through all possible permutations of chemical topologies up to a certain size (Ramakrishnan and Bhalla 2008). The resultant bistables were then examined to look for reaction topologies that might be bistable. This study suggested that as many as 10 % of all possible reaction topologies might give rise to bistability.

Non-bistable Memories

Notwithstanding the arguments for bistability as the basis for long-lasting plastic changes, there have been several proposals for non-bistable memories. Many studies

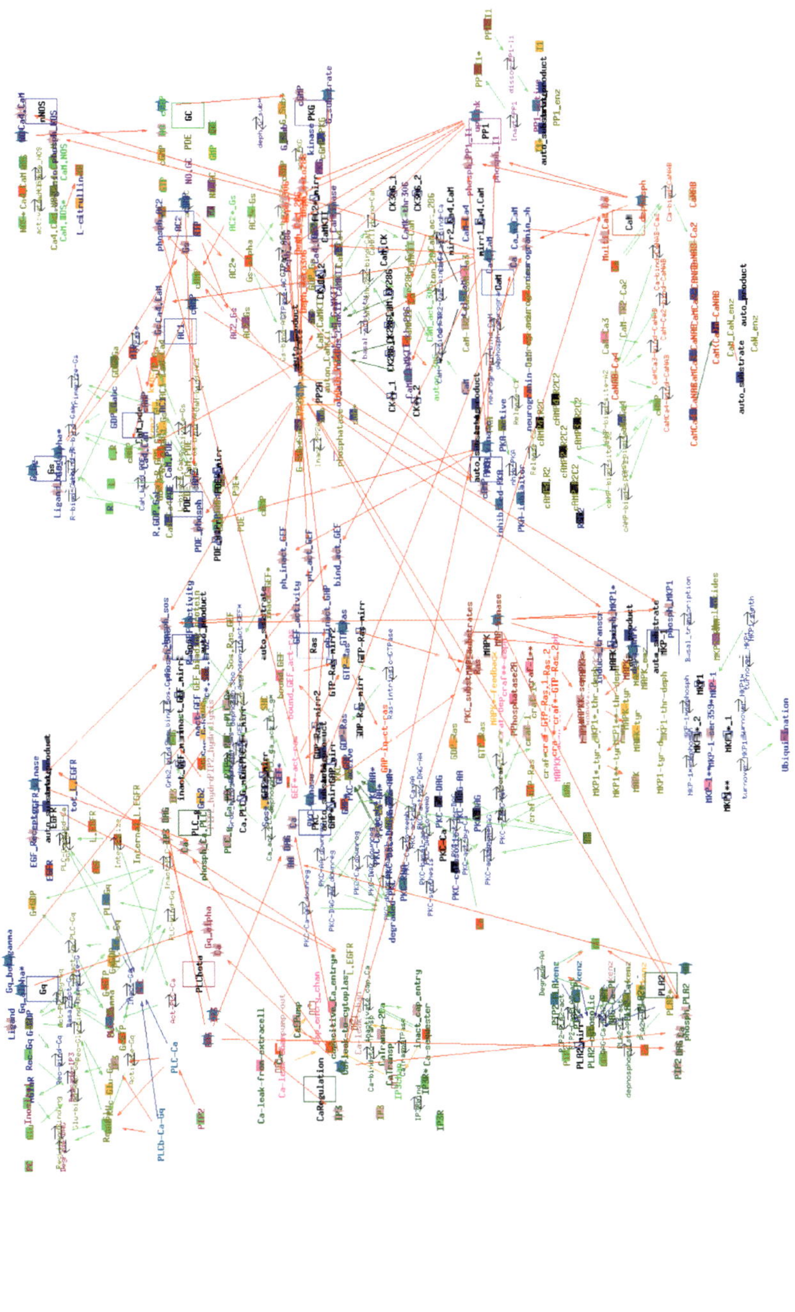

Fig. 9.4 Schematic of chemical interactions in a 1999 model of signaling in the synapse (Bhalla and Iyengar 1999). This represents only a small fraction of known molecular players at the synapse

analyze processes for triggering synaptic change and assume persistence till the next volley of plasticity-inducing input (D'Alcantara et al. 2003). This formulation has a counterpart in a large body of work on network-level plasticity, where the synapse is envisioned as having a continually changing synaptic strength. One proposed way to achieve long-term synaptic stability (>1 year) without bistability, in the presence of stochasticity and turnover, is through receptor clustering (Shouval 2005). The key idea is that receptor replacement in a vacated position in the cluster is much faster than for isolated receptors.

Stochasticity and Memory

Stochasticity has already been pointed out as a particularly challenging aspect of synaptic signaling that any theory of chemical memory must address. The classic way of doing such calculations was devised by Gillespie (1977). This lets one take a chemical reaction system and treat rate terms as probabilities for reaction transitions. Using this, one can uniformly and exactly sample possible trajectories of the evolution of the chemical system in time. With enough such trajectories, one can find the probability distribution of the time-evolution of the system. Given the obvious relevance of stochasticity to synaptic plasticity, several studies looked at what would happen to chemical switches in synaptic volumes. The first order expectation of a stochastic biochemical switch is that it should spontaneously flip state at a rate that depends on the level of noise, or inversely on the reaction volume if concentrations and rates are held fixed. Theoretical calculations showed that, in principle, the right combination of rates and concentrations should allow for state stability for a century or more, even with synaptic volumes (Bialek 2001). It turned out to be much harder to account for such stability using existing models. The MAPK feedback loop, for example, turned out to have a spontaneous switching time of a few hours or less, if one just made the original model stochastic (Bhalla 2004). Eventually a specific set of parameters and mechanisms for the CaMKII switch were found that were able to hold its state for at least a century, while retaining the ability to switch in response to a calcium stimulus of around a second (Miller et al. 2005). The receptor trafficking switch discussed above also has century-long state retention with its physiologically derived parameters (Hayer and Bhalla 2005). Interlinked positive feedback loops have also been suggested as a mechanism for producing long-term stability as well as fast switching (Smolen et al. 2009).

Input Pattern Selectivity

Synaptic plasticity is, both from a theoretical and experimental viewpoint, a highly selective phenomenon. Theoretically, the system has to distinguish between "learning" signals and regular ongoing activity which should pass through the synapse

Table 9.2 Synaptic plasticity stimuli and their effects

Stimulus	Description	Effect
Massed tetanic stimuli	1–3 strong bursts of pulses (e.g., 100 Hz) of 1 s each, separated by 10–20 s	S-LTP. Plasticity of the order of 50–100 %, lasting an hour or so. Does not trigger protein synthesis
Spaced tetanic stimuli	3–4 strong bursts of pulses (100 Hz of 1 s each) separated by 5–10 min	L-LTP. Plasticity around 100 %, lasting indefinitely, depends on protein synthesis
Theta burst stimulation		L-LTP, as above
Pronged low-frequency stimulation	900 s of 1 Hz stimulation	LTD. Reduces synaptic strength by 30–50 %, lasts indefinitely, depends on protein synthesis

without affecting its weights. One way of doing this is through the addition of external modulators such as acetylcholine (Barkai and Hasselmo 1994; Hasselmo et al. 1992). However, it is also interesting to consider how the synapse can itself select between different patterns of input to decide whether to change, in what direction, and by how much. Partly inspired by this theoretical argument, a large variety of input stimuli have been tried out in experiments, and four major patterns have become standards in the field. Interestingly, these patterns seem to elicit quite different plasticity mechanisms, with different time-courses and signaling pathways. I will discuss spike-timing-dependent plasticity (STDP) below. The four major patterns are in Table 9.2.

The challenge for computational neuroscientists here was to find out if and how signaling networks could select between these different input patterns.

The most pronounced distinction in this list was between LTD and the various LTP stimuli. It was already apparent from experiments and theory that the LTP stimuli tended to be brief but strong, whereas the LTD stimulus was lower but prolonged. Converted to calcium influx, this observation bore a striking resemblance to the classic Bienenstock–Cooper–Munroe (BCM) curve (Bienenstock et al. 1982). A suggestion by John Lisman and others (Lisman 1989) was that kinases such as CaMKII were triggered by high-threshold calcium sensors, whereas phosphatases such as calcineurin required lower calcium levels for activation.

Signaling pathways turn out also to be good at more subtle discriminations of input pattern. A series of theoretical (Bhalla 2002a, b) and experimental (Ajay and Bhalla 2004) studies showed that the MAPK pathway discriminated between massed and spaced stimulus patterns. The mGluR pathway (Steuber and Willshaw 2004) and the PKA pathway are also capable of similar pattern selectivity (Kim et al. 2010; Lee et al. 2009).

STDP brought a much finer level of timing precision to input patterning than had previously been expected (Bi and Poo 1998; Magee and Johnston 1997; Markram et al. 1997). Instead of pattern discrimination of the order of many seconds or minutes, STDP depended on millisecond differences between pre- and postsynaptic spike timings. How did the cell do it? This question has engaged many experimental and computational groups (Appleby and Elliott 2005; Badoual et al. 2006; Froemke et al. 2005; Graupner and Brunel 2007; Zou and Destexhe 2007). The answer turned out to be suggestive of the directions for the field: both electrical and chemical signaling had to be considered to achieve bidirectional plasticity with millisecond time-resolution. Higher order spike combinations such as triplets add another layer to the synaptic pattern decoding (Badoual et al. 2006; Pfister and Gerstner 2006; Zou and Destexhe 2007). This branch of research remains lively, with numerous network implications as well as growing activity at the interface between chemical and electrical signaling.

Homeostasis

Cellular homeostasis is another major theme of synaptic plasticity research that highlights the intersection between chemical and electrical scales. Plasticity in individual synapses changes excitability in a very small part of the cell. However, the cell has to retain an overall balance of excitability otherwise it (and the network) risks going into an epileptic state. Theoretical constructs, such as synaptic renormalization, have achieved this for network models in an abstracted way. An elegant series of experimental and computational studies has examined how this homeostatic signaling may be achieved within a cell (Liu et al. 1998; Marder and Goaillard 2006; Olypher and Prinz 2010; Turrigiano 2008).

In summary, the computational neuroscience of synapses and molecules began the decade with an overabundance of qualitative data—molecules and interactions—which quite swamped the earlier simplicity of ideas about synaptic plasticity. Over the course of the decade, a few key functional motifs kept coming up from the profusion of interactions. Bistability, long-term stability, pattern selection, and homeostasis were some of these core motifs. These functions often shared molecular players, and often had alternate pathways for their mechanisms. This functional perspective was a powerful way to step back from the molecular complexity and better understand how different aspects of synaptic plasticity might work.

Directions

Extrapolating from work in the past few years, the trajectory of computational studies of synaptic plasticity seems to be to push toward greater completeness, of two kinds. On the one hand there is a continuing process of modeling and understanding further pathways crucial for plasticity, as the experimental literature identifies and characterizes them. In this vein there have been recent studies on dendritic protein synthesis control during synaptic plasticity (Aslam et al. 2009; Clopath et al. 2008; Jain and Bhalla 2009), and the role of PKM zeta (Ajay and Bhalla 2004; Shema et al. 2007). The older pathways too still have life in them, and models there are being further refined as better data come in (e.g., Byrne et al. 2009; Kim et al. 2010).

The other direction for completeness is toward more inclusive detail. Better data and better simulation tools have made it possible to now model synaptic events at literally single-molecule resolution (Andrews et al. 2010; Coggan et al. 2005; Oliveira et al. 2010). Models of this kind take enormously detailed three-dimensional structural data for synaptic geometry and populate the space with the best current information about receptor and other molecular localization. Due to the extreme technical requirements, detailed chemistry in such models is currently traded off for better spatial resolution, with different studies favoring one or the other. Progress is clearly being made on both fronts.

Detail is also improving at the intersection between different kinds of cellular events. Multiscale models of this kind consider electrical and chemical events simultaneously, so as to more completely understand processes such as plasticity which are inherently multilevel. The close coupling between electrical events and calcium concentration change, calcium concentration change and signaling, and signaling and modification of ion channels is one that has mostly been avoided. The fortuitous separation of time-scales between these events has let us get away with approximations of quasi-steady states in many models. Simulation hardware and software are now at the point where one can tackle these multiscale problems through brute force, and this opens out new forms of multiscale feedback and coupling. Among these are highly detailed single-molecule views of coupled reaction–electrodiffusion systems (Santamaria et al. 2006; Wils and De Schutter 2009), and coarser grained models incorporating reaction–diffusion signaling with compartmental electrical models (Ajay and Bhalla 2007). Further along this line, we are likely to see how signaling interacts with mechanical events, and that to structure (Crook et al. 2007a), leading to yet more ways to alter the synapse.

In closing, we are still looking for the memories. Computational neuroscience has had the privilege of being part of a transformation of understanding about synaptic plasticity: from the apparently simple to the overwhelmingly complex, and from there to a better appreciation of conceptual and scientific frameworks that can deal with the complexity. This journey has been shared with many other parts of biology, and there has been a fertile cross-talk between systems modelers of different persuasions. The memories themselves are still a little further away, perhaps we can gather them in when we have had another decade or two to digest how information flows back and forth between molecules, ions, synapses, cells, and networks.

References

Ajay SM, Bhalla US (2004) A role for ERKII in synaptic pattern selectivity on the time-scale of minutes. Eur J Neurosci 20(10):2671–2680. doi:10.1111/j.1460-9568.2004.03725.x

Ajay SM, Bhalla US (2007) A propagating ERKII switch forms zones of elevated dendritic activation correlated with plasticity. HFSP J 1(1):49–66. doi:10.2976/1.2721383

Alon U (2007) Network motifs: theory and experimental approaches. Nat Rev Genet 8(6):450–461. doi:10.1038/nrg2102
Andrews SS, Addy NJ, Brent R, Arkin AP (2010) Detailed simulations of cell biology with Smoldyn 2.1. PLoS Comput Biol 6(3):e1000705. doi:10.1371/journal.pcbi.1000705
Anglister L, Stiles JR, Salpeter MM (1994) Acetylcholinesterase density and turnover number at frog neuromuscular junctions, with modeling of their role in synaptic function. Neuron 12(4):783–794
Appleby PA, Elliott T (2005) Synaptic and temporal ensemble interpretation of spike-timing-dependent plasticity. Neural Comput 17(11):2316–2336. doi:10.1162/0899766054796879
Aslam N, Kubota Y, Wells D, Shouval HZ (2009) Translational switch for long-term maintenance of synaptic plasticity. Mol Syst Biol 5:284. doi:10.1038/msb.2009.38
Badoual M, Zou Q, Davison AP, Rudolph M, Bal T, Frégnac Y, Destexhe A (2006) Biophysical and phenomenological models of multiple spike interactions in spike-timing dependent plasticity. Int J Neural Syst 16(2):79–97
Barkai E, Hasselmo ME (1994) Modulation of the input/output function of rat piriform cortex pyramidal cells. J Neurophysiol 72(2):644–658
Bennett MR, Gibson WG, Robinson J (1995) Probabilistic secretion of quanta: spontaneous release at active zones of varicosities, boutons, and endplates. Biophys J 69(1):42–56. doi:10.1016/S0006-3495(95)79873-3
Bhalla US (2002a) Biochemical signaling networks decode temporal patterns of synaptic input. J Comput Neurosci 13(1):49–62
Bhalla US (2002b) Mechanisms for temporal tuning and filtering by postsynaptic signaling pathways. Biophys J 83(2):740–752. doi:10.1016/S0006-3495(02)75205-3
Bhalla US (2003) Understanding complex signaling networks through models and metaphors. Prog Biophys Mol Biol 81(1):45–65
Bhalla US (2004) Signaling in small subcellular volumes. II. Stochastic and diffusion effects on synaptic network properties. Biophys J 87(2):745–753. doi:10.1529/biophysj.104.040501
Bhalla US, Iyengar R (1999) Emergent properties of networks of biological signaling pathways. Science 283(5400):381–387
Bhalla US, Ram PT, Iyengar R (2002) MAP kinase phosphatase as a locus of flexibility in a mitogen-activated protein kinase signaling network. Science 297(5583):1018–1023
Bi GQ, Poo MM (1998) Synaptic modifications in cultured hippocampal neurons: dependence on spike timing, synaptic strength, and postsynaptic cell type. J Neurosci 18(24):10464–10472
Bialek W (2001) Stability and noise in biochemical switches. Adv Neural Inf Process Syst 13:103–109
Bienenstock EL, Cooper LN, Munro PW (1982) Theory for the development of neuron selectivity: orientation specificity and binocular interaction in visual cortex. J Neurosci 2(1):32–48
Bliss TV, Collingridge GL (1993) A synaptic model of memory: long-term potentiation in the hippocampus. Nature 361(6407):31–39. doi:10.1038/361031a0
Byrne MJ, Putkey JA, Waxham MN, Kubota Y (2009) Dissecting cooperative calmodulin binding to CaM kinase II: a detailed stochastic model. J Comput Neurosci 27(3):621–638. doi:10.1007/s10827-009-0173-3
Clopath C, Ziegler L, Vasilaki E, Büsing L, Gerstner W (2008) Tag-trigger-consolidation: a model of early and late long-term-potentiation and depression. PLoS Comput Biol 4(12):e1000248. doi:10.1371/journal.pcbi.1000248
Coggan JS, Bartol TM, Esquenazi E, Stiles JR, Lamont S, Martone ME, Berg DK et al (2005) Evidence for ectopic neurotransmission at a neuronal synapse. Science 309(5733):446–451. doi:10.1126/science.1108239
Crook SM, Dur-E-Ahmad M, Baer SM (2007a) A model of activity-dependent changes in dendritic spine density and spine structure. Math Biosci Eng 4(4):617–631
Crook S, Gleeson P, Howell F, Svitak J, Silver RA (2007b) MorphML: level 1 of the NeuroML standards for neuronal morphology data and model specification. Neuroinformatics 5(2):96–104
D'Alcantara P, Schiffmann SN, Swillens S (2003) Bidirectional synaptic plasticity as a consequence of interdependent Ca2+-controlled phosphorylation and dephosphorylation pathways. Eur J Neurosci 17(12):2521–2528

Dudek SM, Bear MF (1992) Homosynaptic long-term depression in area CA1 of hippocampus and effects of N-methyl-D-aspartate receptor blockade. Proc Natl Acad Sci USA 89(10):4363–4367

Ehlers MD (2003) Activity level controls postsynaptic composition and signaling via the ubiquitin-proteasome system. Nat Neurosci 6(3):231–242. doi:10.1038/nn1013

Ferrell JE, Xiong W (2001) Bistability in cell signaling: how to make continuous processes discontinuous, and reversible processes irreversible. Chaos 11(1):227–236

Froemke RC, Poo M, Dan Y (2005) Spike-timing-dependent synaptic plasticity depends on dendritic location. Nature 434(7030):221–225

Gillespie DT (1977) Exact stochastic simulation of coupled chemical reactions. J Phys Chem 81(25):2340–2361. doi:10.1021/j100540a008

Gold JI, Bear MF (1994) A model of dendritic spine Ca2+ concentration exploring possible bases for a sliding synaptic modification threshold. Proc Natl Acad Sci USA 91(9):3941–3945

Graupner M, Brunel N (2007) STDP in a bistable synapse model based on CaMKII and associated signaling pathways. PLoS Comput Biol 3(11):e221. doi:10.1371/journal.pcbi.0030221

Hasselmo ME, Anderson BP, Bower JM (1992) Cholinergic modulation of cortical associative memory function. J Neurophysiol 67(5):1230–1246

Hayer A, Bhalla US (2005) Molecular switches at the synapse emerge from receptor and kinase traffic. PLoS Comput Biol 1(2):137–154

Hebb D (1949) The organization of behavior. Wiley, New York

Hopfield JJ (1982) Neural networks and physical systems with emergent collective computational abilities. Proc Natl Acad Sci USA 79(8):2554–2558

Hucka M, Finney A, Sauro HM, Bolouri H, Doyle JC, Kitano H, Arkin AP et al (2003) The systems biology markup language (SBML): a medium for representation and exchange of biochemical network models. Bioinformatics 19(4):524–531

Ito M (1989) Long-term depression. Annu Rev Neurosci 12:85–102. doi:10.1146/annurev.ne.12.030189.000505

Jaeger D, De Schutter E, Bower JM (1997) The role of synaptic and voltage-gated currents in the control of Purkinje cell spiking: a modeling study. J Neurosci 17(1):91–106

Jain P, Bhalla US (2009) Signaling logic of activity-triggered dendritic protein synthesis: an mTOR gate but not a feedback switch. PLoS Comput Biol 5(2):e1000287. doi:10.1371/journal.pcbi.1000287

Kim M, Huang T, Abel T, Blackwell KT (2010) Temporal sensitivity of protein kinase a activation in late-phase long term potentiation. PLoS Comput Biol 6(2):e1000691. doi:10.1371/journal.pcbi.1000691

Kuroda S, Schweighofer N, Kawato M (2001) Exploration of signal transduction pathways in cerebellar long-term depression by kinetic simulation. J Neurosci 21(15):5693–5702

Lee CJ, Anton M, Poon C, McRae GJ (2009) A kinetic model unifying presynaptic short-term facilitation and depression. J Comput Neurosci 26(3):459–473. doi:10.1007/s10827-008-0122-6

Lindskog M, Kim M, Wikström MA, Blackwell KT, Kotaleski JH (2006) Transient calcium and dopamine increase PKA activity and DARPP-32 phosphorylation. PLoS Comput Biol 2(9):e119. doi:10.1371/journal.pcbi.0020119

Lisman J (1989) A mechanism for the Hebb and the anti-Hebb processes underlying learning and memory. Proc Natl Acad Sci USA 86(23):9574–9578

Lisman JE, Zhabotinsky AM (2001) A model of synaptic memory: a CaMKII/PP1 switch that potentiates transmission by organizing an AMPA receptor anchoring assembly. Neuron 31(2): 191–201

Liu Z, Golowasch J, Marder E, Abbott LF (1998) A model neuron with activity-dependent conductances regulated by multiple calcium sensors. J Neurosci 18(7):2309–2320

Magee JC, Johnston D (1997) A synaptically controlled, associative signal for Hebbian plasticity in hippocampal neurons. Science 275(5297):209–213

Malinow R, Schulman H, Tsien RW (1989) Inhibition of postsynaptic PKC or CaMKII blocks induction but not expression of LTP. Science 245(4920):862–866

Marder E, Goaillard J (2006) Variability, compensation and homeostasis in neuron and network function. Nat Rev Neurosci 7(7):563–574. doi:10.1038/nrn1949

Markevich NI, Hoek JB, Kholodenko BN (2004) Signaling switches and bistability arising from multisite phosphorylation in protein kinase cascades. J Cell Biol 164(3):353–359

Markram H, Lübke J, Frotscher M, Sakmann B (1997) Regulation of synaptic efficacy by coincidence of postsynaptic APs and EPSPs. Science 275(5297):213–215

Miller P, Zhabotinsky AM, Lisman JE, Wang X (2005) The stability of a stochastic CaMKII switch: dependence on the number of enzyme molecules and protein turnover. PLoS Biol 3(4):e107

Minsky M, Papert S (1969) Perceptrons: an introduction to computational geometry. MIT Press, Cambridge, MA

Mulkey RM, Malenka RC (1992) Mechanisms underlying induction of homosynaptic long-term depression in area CA1 of the hippocampus. Neuron 9(5):967–975

Nakano T, Doi T, Yoshimoto J, Doya K (2010) A kinetic model of dopamine- and calcium-dependent striatal synaptic plasticity. PLoS Comput Biol 6(2):e1000670. doi:10.1371/journal.pcbi.1000670

Nowak L, Bregestovski P, Ascher P, Herbet A, Prochiantz A (1984) Magnesium gates glutamate-activated channels in mouse central neurones. Nature 307(5950):462–465

Oliveira RF, Terrin A, Di Benedetto G, Cannon RC, Koh W, Kim M, Zaccolo M et al (2010) The role of type 4 phosphodiesterases in generating microdomains of cAMP: large scale stochastic simulations. PLoS One 5(7):e11725. doi:10.1371/journal.pone.0011725

Olypher AV, Prinz AA (2010) Geometry and dynamics of activity-dependent homeostatic regulation in neurons. J Comput Neurosci 28(3):361–374. doi:10.1007/s10827-010-0213-z

Pepke S, Kinzer-Ursem T, Mihalas S, Kennedy MB (2010) A dynamic model of interactions of Ca2+, calmodulin, and catalytic subunits of Ca2+/calmodulin-dependent protein kinase II. PLoS Comput Biol 6(2):e1000675. doi:10.1371/journal.pcbi.1000675

Pfister J, Gerstner W (2006) Triplets of spikes in a model of spike timing-dependent plasticity. J Neurosci 26(38):9673–9682

Ramakrishnan N, Bhalla US (2008) Memory switches in chemical reaction space. PLoS Comput Biol 4(7):e1000122. doi:10.1371/journal.pcbi.1000122

Rosenblatt F (1962) Principles of perceptrons. Spartan, Washington, DC

Rumelhart DE, Hinton GE, Williams RJ (1986) Learning representations by back-propagating errors. Nature 323:533–536

Sanes JR, Lichtman JW (1999) Can molecules explain long-term potentiation? Nat Neurosci 2(7):597–604. doi:10.1038/10154

Santamaria F, Wils S, De Schutter E, Augustine GJ (2006) Anomalous diffusion in Purkinje cell dendrites caused by spines. Neuron 52(4):635–648. doi:10.1016/j.neuron.2006.10.025

Sanyal S, Sandstrom DJ, Hoeffer CA, Ramaswami M (2002) AP-1 functions upstream of CREB to control synaptic plasticity in Drosophila. Nature 416(6883):870–874. doi:10.1038/416870a

Shema R, Sacktor TC, Dudai Y (2007) Rapid erasure of long-term memory associations in the cortex by an inhibitor of PKM zeta. Science 317(5840):951–953. doi:10.1126/science.1144334

Shouval HZ (2005) Clusters of interacting receptors can stabilize synaptic efficacies. Proc Natl Acad Sci USA 102(40):14440–14445. doi:10.1073/pnas.0506934102

Smolen P, Baxter DA, Byrne JH (2008) Bistable MAP kinase activity: a plausible mechanism contributing to maintenance of late long-term potentiation. Am J Physiol Cell Physiol 294(2):C503–C515. doi:10.1152/ajpcell.00447.2007

Smolen P, Baxter DA, Byrne JH (2009) Interlinked dual-time feedback loops can enhance robustness to stochasticity and persistence of memory. Phys Rev E Stat Nonlin Soft Matter Phys 79(3 Pt 1):031902

Song H, Smolen P, Av-Ron E, Baxter DA, Byrne JH (2006) Bifurcation and singularity analysis of a molecular network for the induction of long-term memory. Biophys J 90(7):2309–2325. doi:10.1529/biophysj.105.074500

Stanton PK, Sejnowski TJ (1989) Associative long-term depression in the hippocampus induced by hebbian covariance. Nature 339(6221):215–218. doi:10.1038/339215a0

Steuber V, Willshaw D (2004) A biophysical model of synaptic delay learning and temporal pattern recognition in a cerebellar Purkinje cell. J Comput Neurosci 17(2):149–164. doi:10.1023/B:JCNS.0000037678.26155.b5

Tanaka K, Augustine GJ (2008) A positive feedback signal transduction loop determines timing of cerebellar long-term depression. Neuron 59(4):608–620. doi:10.1016/j.neuron.2008.06.026
Turrigiano GG (1999) Homeostatic plasticity in neuronal networks: the more things change, the more they stay the same. Trends Neurosci 22(5):221–227
Turrigiano GG (2008) The self-tuning neuron: synaptic scaling of excitatory synapses. Cell 135(3):422–435. doi:10.1016/j.cell.2008.10.008
Tyson JJ, Chen KC, Novak B (2003) Sniffers, buzzers, toggles and blinkers: dynamics of regulatory and signaling pathways in the cell. Curr Opin Cell Biol 15(2):221–231
Wils S, De Schutter E (2009) STEPS: modeling and simulating complex reaction-diffusion systems with python. Front Neuroinform 3:15. doi:10.3389/neuro.11.015.2009
Zador A, Koch C, Brown TH (1990) Biophysical model of a Hebbian synapse. Proc Natl Acad Sci USA 87(17):6718–6722
Zou Q, Destexhe A (2007) Kinetic models of spike-timing dependent plasticity and their functional consequences in detecting correlations. Biol Cybern 97(1):81–97

Chapter 10
20 Years of the Dynamics of Memory: The Long and Winding Road Linking Cellular Mechanisms to Behavior

Michael E. Hasselmo

Abstract The first Computational Neuroscience meetings in the 1990s fostered an increasing focus on biologically realistic modeling of neurons to understand the function of neural circuits. This chapter reviews some of the developments over the past 20 years, relating papers presented at the early meetings to subsequent developments. The review addresses developments in research on associative memory function, hippocampal memory function, the functional role of theta rhythm oscillations, and the discovery and modeling of grid cells.

Impact of the Computational Neuroscience Meeting

I remember the feeling of excitement associated with the first Computational Neuroscience meetings in the early 1990s. I had a sense of a field coalescing from many different disciplines, building on research that had started decades earlier. I anticipated great accomplishments to take place over the subsequent 20 years from those first meetings. Now that the Computational Neuroscience meeting has taken place for 20 years, I can reflect on how far we have progressed since that time.

There were a number of changes in cultural styles from the 1980s to the 1990s. Neural modeling was dominated by connectionist models (Rumelhart et al. 1986; McClelland and Rumelhart 1988) and attractor dynamic models (Amit 1988; Amit and Treves 1989) in the 1980s. At the start of the 1990s, the excitement about connectionist models and attractor networks transitioned into a greater focus on biophysically detailed modeling of neural circuits. This type of work is essential to understanding the cellular and molecular mechanisms underlying behavior, which

M.E. Hasselmo (✉)
Department of Psychology and Graduate Program for Neuroscience, Center for Memory and Brain, Boston University, 2 Cummington Street, Boston, MA 02215, USA
e-mail: hasselmo@bu.edu

J.M. Bower (ed.), *20 Years of Computational Neuroscience*, Springer Series in Computational Neuroscience 9, DOI 10.1007/978-1-4614-1424-7_10,

will be essential to guiding the development of treatments for disorders such as schizophrenia and depression. Much of the influential work on biophysically detailed modeling was performed by founders and early participants of the Computational Neuroscience meeting, including John Rinzel, Bard Ermentrout, Jim Bower, Nancy Kopell, Matt Wilson, Erik DeSchutter, Charlie Wilson, David Golomb, Eve Marder, Todd Troyer, Francis Skinner, Alain Destexhe, Ron Calabrese, Orjan Ekeberg, and Christian Linster. There are too many names to provide a complete list here.

The growth of the field was facilitated tremendously by the dedicated work of Dennis Glanzman, as the program chief of the Theoretical and Computational Neuroscience program at NIMH. His program provided guidance toward funding for many of the influential modelers in those early years. The work was also facilitated by a later collaborative funding venture between Dennis Glanzman at NIMH, Yuan Liu at NINDS, and Ken Whang at NSF in the program for Collaborative Research in Computational Neuroscience (CRCNS).

In describing progress over the past 20 years, I will focus on the biological dynamics of memory function, with a particular emphasis on understanding how episodic memories are encoded. I will address the progress in three general areas: (1) associative memory function, (2) hippocampal function, and (3) theta rhythm and grid cells.

Associative Memory Function

The early days of the Computational Neuroscience meeting included presentations addressing biological mechanisms for associative memory function. The theory of associations has a long history in research on human cognition. A review can be found in Schacter (1982). These models received a more detailed mathematical treatment in early linear associative memory models (Anderson 1972; Kohonen 1972, 1984). In these models, vectors represented patterns of neural activity in the brain. An association was encoded by modification of synapses, represented mathematically by computing the outer product matrix between a presynaptic activity vector and the associated postsynaptic activity vector. Retrieval of the association was performed by allowing the presynaptic activity cue to spread across the modified synapses, represented mathematically by matrix multiplication of the presynaptic vector by the pattern of synaptic connections.

An important early paper by Marr proposed that the excitatory recurrent connections in hippocampal region CA3 could underlie autoassociative memory function (Marr 1971). This was expanded upon in subsequent papers by hippocampal researchers (McNaughton and Morris 1987) as described in more detail in the next section of the chapter. In addition, the primary olfactory cortex was also proposed by Haberly and Bower to function as an autoassociative memory (Haberly and Bower 1989). This proved an interesting model system. Early Computational Neuroscience meetings included presentations of detailed biophysical simulations of the olfactory cortex developed in the Bower laboratory (Bower 1990; Wilson and Bower 1992) and models of the olfactory cortex as an autoassociative memory (Hasselmo et al. 1990, 1994; Bergman et al. 1993). These biophysical simulations used the GENESIS simulation package initially written by Matt Wilson and developed further by many researchers within the Bower laboratory (Bower and Beeman 1995). The Bower laboratory provided an exciting environment where both biologically realistic modeling and intracellular recording experiments could be combined.

Excitatory recurrent connections will cause an explosion of activity unless the excitatory feedback is limited by the input–output function of individual neurons or by feedback inhibition. A dominant stream of research in the 1980s focused on fixed point attractor dynamics in associative memory function, in which activity converges to a stable fixed point. Mathematically, Lyapunov functions were used to show the stability of attractor states (Hopfield 1982, 1984; Cohen and Grossberg 1983). Many of these studies focused on relatively abstract representations of neurons and the computation of the storage capacity of attractor networks (Amit 1988). Initial models were highly unrealistic, for example, violating Dale's law by having both excitatory and inhibitory connections arise from the same neuron and driving neurons up to an asymptotic maximum activity. However, later studies addressed making these attractor networks more biologically realistic, for example, by modeling neurons with lower firing rates (Amit and Treves 1989; Amit et al. 1990).

Many of the early models used single neuron models that artificially limited the maximal output of neurons (i.e., using a step function or sigmoid function). This was justified as representing the maximal intrinsic firing rate of a neuron. However, recordings of cortical neurons in vivo almost never go above 100 Hz, whereas the maximal firing rate limited by intrinsic properties is usually higher. The intrinsic frequency–current (*f–I*) curve of a neuron is more accurately modeled with a threshold linear function. A more realistic way of limiting the maximal firing rate of modeled neurons is by use of feedback inhibition, for example as initially implemented by Wilson and Cowan (1972, 1973). In my own models, I used interactions of threshold linear excitatory and inhibitory neurons in attractor models of the hippocampus (Hasselmo et al. 1995; Kali and Dayan 2000). Carl van Vreeswijk wrote an unpublished paper with me on these types of models in my lab, and then went on to develop his model of balanced networks (van Vreeswijk and Sompolinsky 1996) in which chaotic activity involves a balance of excitatory and inhibitory activity.

Early associative memory models all used different dynamics during encoding and retrieval (Anderson 1972; Kohonen 1972, 1984; Hopfield 1982; Amit 1988). During encoding, activity in the network would be clamped to an external input pattern. The dynamics of retrieval were explicitly prevented during computation of an outer product for encoding of new input patterns. This was essential for the proper function of associative memory models, as retrieval during encoding would cause a build-up of interference between overlapping patterns (Hasselmo et al. 1992). However, there was no clear biological mechanism for this difference in dynamics during encoding and retrieval.

The effects of acetylcholine provide a potential biological mechanism for the difference in dynamics between encoding and retrieval in associative memory. Working on slices of the piriform cortex in the laboratory of Jim Bower, I studied differences between the properties of glutamatergic synaptic transmission at the afferent input

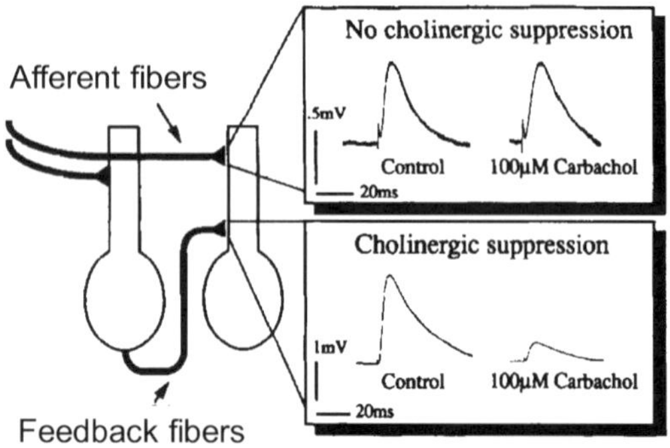

Fig. 10.1 Activation of acetylcholine receptors by the acetylcholine (ACh) agonist carbachol causes selective presynaptic inhibition of synaptic potentials evoked by stimulation of excitatory feedback synapses (*bottom*), with smaller change of synaptic potentials evoked by stimulation of excitatory afferent input (*top*) (Hasselmo and Bower 1992)

from the olfactory bulb in layer Ia and the glutamatergic excitatory recurrent connections in layer Ib arising from other piriform cortex pyramidal cells, extending previous work on the physiological properties of these synapses done by Jim Bower (Haberly and Bower 1984; Bower and Haberly 1986). I found a striking difference in the effects of acetylcholine on the glutamatergic transmission at these synapses (Fig. 10.1). Activation of muscarinic acetylcholine receptors caused much stronger presynaptic inhibition of glutamate release at excitatory recurrent synapses in layer Ib compared to afferent synapses in layer Ia (Hasselmo and Bower 1992, 1993).

The combined focus on modeling and physiology in the Bower lab gave me excellent tools for modeling the significance of this function. In the rooms on the top floor of the Beckman Behavioral Biology, I remember preparing piriform cortex slices, then starting simulations on a Sun workstation, then running a slice experiment, then checking on my simulation output and running a new batch file, in an interactive process throughout a 10 h experimental day. I found a clear effect of cholinergic modulation in abstract models of associative memory function in the piriform cortex. The selective suppression of excitatory recurrent connections clearly enhanced the encoding of new patterns by preventing interference from previously stored memories (Hasselmo et al. 1992; Hasselmo and Bower 1993). Later we simulated networks of piriform cortex neurons using the GENESIS simulation package for presentations at the Computational Neuroscience meeting (Bergman et al. 1993; Hasselmo et al. 1994), showing that encoding of new patterns was enhanced by these cholinergic effects. As shown in Fig. 10.2, interference from previously stored patterns was prevented by cholinergic suppression of synaptic transmission, and the rate of encoding was enhanced by cholinergic depolarization of pyramidal cells and the suppression of spike frequency accommodation (Barkai et al. 1994; Barkai and Hasselmo 1994).

These findings in the piriform cortex have been shown to generalize to other cortical structures in a wide range of subsequent studies. Research in my laboratory

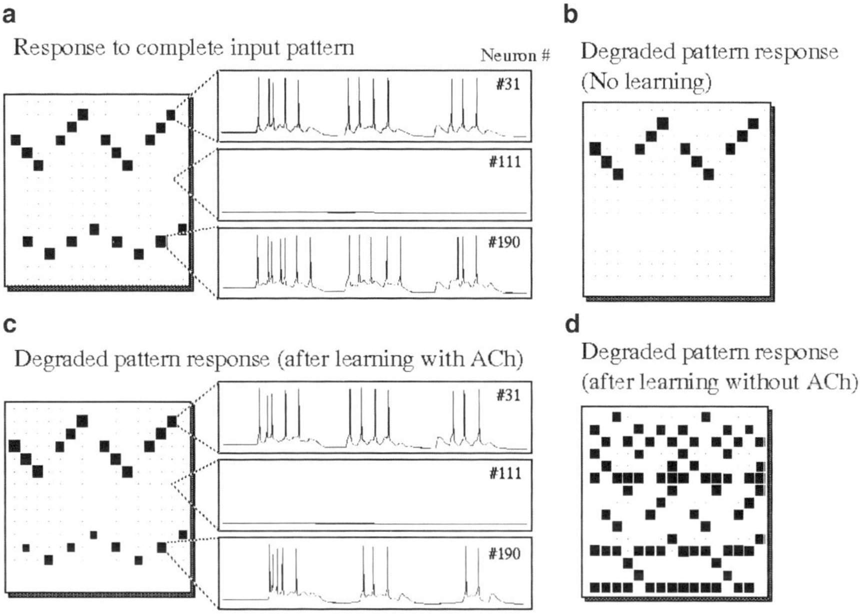

Fig. 10.2 (**a**) Biophysical simulation of spiking response to afferent input. Size of *black squares* indicates the amount of spiking activity (example membrane potential traces are shown). (**b**) With no synaptic modification (no learning), a degraded input pattern only activates a subset of neurons. (**c**) After learning with ACh, the network effectively completes missing components of the input pattern. (**d**) After learning without ACh, proactive interference results in retrieval of multiple different input patterns (Hasselmo et al. 1994; Barkai et al. 1994)

extended this work to subregions of the hippocampal formation. In region CA1 of the hippocampus, we showed that muscarinic presynaptic inhibition was stronger at excitatory connections arising from within the hippocampus (in region CA3) and terminating in stratum radiatum of region CA1 compared to afferent input from entorhinal cortex terminating in stratum lacunosum-moleculare (Hasselmo and Schnell 1994). Similarly, muscarinic presynaptic inhibition was stronger for synapses in stratum radiatum of region CA3 arising from CA3 pyramidal cells, compared to weaker presynaptic inhibition at afferent synapses in stratum lucidum, at synaptic input arising from the dentate gyrus (Hasselmo et al. 1995). This effect was later replicated in stratum lucidum (Vogt and Regehr 2001) and was extended to show less presynaptic inhibition in stratum lacunosum-moleculare of region CA3 (Kremin and Hasselmo 2007).

This principle of selective cholinergic suppression of excitatory feedback but not afferent input also proves to generalize to neocortical structures. In an early study, connections within somatosensory neocortex showed greater presynaptic inhibition than afferent input arising from the white matter (Hasselmo and Cekic 1996). This was subsequently shown in a study using thalamocortical slice preparations, showing muscarinic presynaptic inhibition of excitatory recurrent connections in neocortex and also showing nicotinic enhancement of afferent input (Gil et al. 1997).

In the visual cortex, optical imaging was used to show cholinergic suppression of the internal spread of activity along excitatory recurrent connections compared to afferent input (Kimura and Baughman 1997; Kimura 2000). This indicated that acetylcholine should reduce the functional spread of activity on excitatory recurrent connections in visual cortex. This was supported by in vivo experimental data showing that iontophoretic application of acetylcholine decreases the extent of spatial integration, assessed by measuring a neuron's tuning to length of visual stimuli (Roberts et al. 2005). These effects appear to contribute to the influence of top-down attention on the dynamics of visual cortex processing (Herrero et al. 2008). This work has been extended to human subjects in a study showing that the acetylcholinesterase blocker donepezil reduces the extent of the spread of activity in visual cortical areas associated with foveal stimulation (Silver et al. 2008). Thus, the physiological effects of muscarinic activation modeled in these early papers in the Computational Neuroscience meeting have proved to be a general principle of cortical function in subsequent studies.

The hippocampal data and modeling generated the prediction that blockade of muscarinic receptors by the muscarinic antagonist scopolamine should enhance proactive interference in a paired associate memory task (Hasselmo and Wyble 1997; Wyble and Hasselmo 1997). This was supported by experimental data on scopolamine effects in human subjects (Atri et al. 2004). Enhancement of proactive interference was also shown in studies on discrimination of pairs of odors in rats administered scopolamine (De Rosa and Hasselmo 2000) or after receiving selective lesions of the cholinergic innervation of the olfactory cortex (De Rosa et al. 2001). In computational models, the build-up of proactive interference causes runaway synaptic modification within cortical networks that can spread from one region to another. This mechanism was proposed to underlie the early appearance of Alzheimer's disease neuropathology in the form of neurofibrillary tangles in lateral entorhinal cortex and the progressive spread from lateral entorhinal cortex to other regions (Hasselmo 1994, 1997). This provides a computational framework that would predict reductions in Alzheimer's pathology with loss of fast hippocampal learning (e.g., in the most extreme case, patient HM would be expected to show absence of Alzheimer's pathology in his remaining temporal lobe structures). This framework could account for the beneficial effects of the NMDA blocker memantine on Alzheimer's disease (Reisberg et al. 2003) and supports the use of selective activation of presynaptic muscarinic receptors with M4 agonists to enhance presynaptic inhibition of glutamate release in treatment of Alzheimer's disease (Shirey et al. 2008).

The levels of acetylcholine change dramatically during different stages of waking and sleep. Acetylcholine levels are high during active waking, show decreases during quiet waking, and decrease to less than 1/3 of waking levels during slow wave sleep (Marrosu et al. 1995). The decrease in acetylcholine levels during slow wave sleep has been proposed to decrease the presynaptic inhibition of glutamatergic transmission

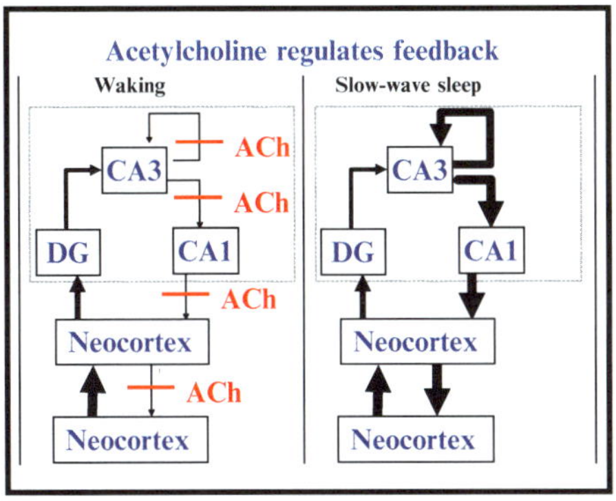

Fig. 10.3 (*Left*) During waking, high levels of ACh cause presynaptic inhibition of excitatory recurrent connections in CA3 as well as connections from region CA3 to region CA1 and feedback connections between neocortical structures. This allows a dominant influence of afferent input into the hippocampus during encoding. (*Right*) During slow wave sleep, lower levels of ACh allow stronger synaptic transmission at these connections. This results in a dominant influence of hippocampus on neocortex that could be appropriate for consolidation of previously encoded memories (Hasselmo 1999)

at connections from hippocampus back to neocortex, allowing activity based on recently formed associations in the hippocampus to spread back to the neocortex and drive consolidation of memories in the neocortex (for review see Hasselmo 1999). This proposal is consistent with the muscarinic cholinergic presynaptic inhibition shown at a number of stages of the feedback connections (Fig. 10.3), including the excitatory recurrent connections in region CA3 (Hasselmo et al. 1995; Vogt and Regehr 2001; Kremin and Hasselmo 2007), the connections from region CA3 to region CA1 (Hounsgaard 1978; Valentino and Dingledine 1981; Hasselmo and Schnell 1994; de Sevilla et al. 2002), and the feedback connections within neocortical structures (Hasselmo and Cekic 1996; Gil et al. 1997).

This model of the role of acetylcholine in consolidation led to some functional predictions that have been tested. If a reduction in cholinergic presynaptic inhibition enhances consolidation during slow wave sleep, then an increase in acetylcholine levels during slow wave sleep should impair consolidation. This was tested in a study in which subjects were administered physostigmine during slow wave sleep and showed reductions in subsequent tests of declarative memory consolidation performed after the subjects were awakened (Gais and Born 2004). On the other hand, the model predicts that reductions in acetylcholine modulation during waking should enhance consolidation. This was shown in a study in which scopolamine was administered to block muscarinic cholinergic receptors after encoding of information, and subjects showed an enhancement of consolidation on a later memory test (Rasch et al. 2006). Thus, computational modeling has provided an exciting link between cellular mechanisms of muscarinic presynaptic inhibition and behavioral studies in animals and humans.

This framework describes how the transitions between different levels of acetylcholine during waking and sleep can regulate the transition between encoding and consolidation. But this leaves the question of how more rapid transitions between encoding and retrieval could be regulated. Muscarinic presynaptic inhibition cannot change rapidly, as shown by studies in which 100 ms pressure pulse applications of acetylcholine cause changes in presynaptic inhibition that persist for 10–20 s (Hasselmo and Fehlau 2001). In contrast, rapid transitions between encoding and retrieval could be mediated by the change in dynamics during individual cycles of the theta rhythm oscillations in hippocampus (Hasselmo et al. 2002). These dynamical changes could be regulated by postsynaptic GABAA inhibition (Toth et al. 1997) and presynaptic GABAB inhibition (Molyneaux and Hasselmo 2002). Encoding could take place when entorhinal synaptic input is strongest at the trough of the EEG recorded at the hippocampal fissure (Hasselmo et al. 2002), and retrieval could be dominant when region CA3 input is strongest at the peak of fissure theta. The change in relative strength of synaptic input is supported by studies showing phasic changes in strength of evoked synaptic transmission on different pathways at different phases of the theta rhythm oscillation (Wyble et al. 2000; Villarreal et al. 2007). Consistent with the theorized role of these different phases in encoding and retrieval, the human EEG shows reset to different phases of theta rhythm during encoding versus during retrieval (Rizzuto et al. 2006), and spiking appears on different phases of hippocampal theta during match and nonmatch stimuli (Manns et al. 2007).

Hippocampus

In addition to these studies on associative memory in the piriform cortex, the early days of the Computational Neuroscience also included presentations of hippocampal models that have had a significant impact on subsequent research. These included papers on hippocampal models by Burgess, O'Keefe, and Recce; Idiart and Abbott; Redish and Touretzky; Holmes and Levy; Blum and Abbott; and Mehta and McNaughton. Modeling of the hippocampus has been very successful in guiding experimental work in this area. A number of experimental studies have tested specific predictions of computational models.

The phenomenon later described as spike-timing dependent plasticity was initially discovered by William B. "Chip" Levy (Levy and Steward 1983) and modeled extensively by Holmes and Levy (1990). The temporal asymmetry of synaptic modification modeled by Holmes and Levy was incorporated in a circuit model by Abbott and Blum (Abbott and Blum 1996; Blum and Abbott 1996). This model predicted that the potentiation of excitatory connections should cause a backward expansion of hippocampal place fields. An experimental test of the model was performed by Mayank Mehta in Bruce McNaughton's laboratory (Mehta and McNaughton 1997). They presented the experimental data from this test at the Computational Neuroscience meeting, showing the predicted backward expansion of the size of place fields of hippocampal place cells (Mehta et al. 1997; Mehta and McNaughton 1997). This phenomenon has been replicated extensively in subsequent studies (Mehta et al. 2000, 2002).

Some of the theories of hippocampal function had a slower time constant for their influence on experimental work in the field. For example, the early paper by Marr (1971) is extensively credited with proposing the principle of pattern completion on excitatory recurrent connections in region CA3 of hippocampus. Marr also proposed that interference between patterns stored in CA3 could be reduced by the process of pattern separation (orthogonalization) in the dentate gyrus (the codon hypothesis of Marr). Several papers in the late 1980s and early 1990s reviewed these basic ideas of pattern separation in the dentate gyrus (McNaughton and Morris 1987; McNaughton 1991; O'Reilly and McClelland 1994; Hasselmo and Wyble 1997) and pattern completion by autoassociative memory function in hippocampal region CA3 (McNaughton and Morris 1987; Treves and Rolls 1994; Hasselmo et al. 1995). These principles were also combined together in a simulation of the role of the hippocampus in human episodic memory function presented at the Computational Neuroscience meeting (Hasselmo and Wyble 1997; Wyble and Hasselmo 1997).

The basic principles proposed by Marr had an impact on experimental work over 20 years later. Selective genetic manipulations in mice allowed selective knockout of the NMDA receptor in hippocampal region CA3, and these mice showed an impairment of pattern completion based on learning a spatial response in an environment with multiple cues and being tested for their response in an environment with a single cue (Nakazawa et al. 2002). Similarly, selective expression of tetanus toxin in mouse region CA3 to block synaptic transmission from these neurons also impairs pattern completion in that task (Nakashiba et al. 2008). In contrast, selective knockout of NMDA receptors in the dentate gyrus caused impairment of responses that required distinguishing two separate but similar contextual environments (McHugh et al. 2007). In addition, selective lesions of the dentate gyrus impair the capacity of rats to encode and selectively respond to spatial locations that are close to each other (Gilbert et al. 2001).

Unit recording studies have also analyzed the response properties of the dentate gyrus versus other hippocampal subregions. Neurons in the dentate gyrus show sparser coding of the environment, with fewer responsive cells and smaller response fields for dentate place cells (Barnes et al. 1990). Minimal changes in the spatial environment can cause distinct responses of dentate gyrus granule cells (Leutgeb et al. 2007). Other unit recording studies have tested for the effect of partial shifts in the environment on neural responses in region CA3. In one study, the partial shift caused less change of neural response in CA3 compared to CA1 suggesting pattern completion (Lee et al. 2004), whereas in another study, region CA3 responded with distinct representations to partial changes in the environment (Leutgeb et al. 2004). These apparently conflicting results were unified by demonstration of a nonlinear transformation in region CA3 (Vazdarjanova and Guzowski 2004). Input patterns that are somewhat similar to each other induce very similar response patterns, whereas input patterns that are more different evoke more strongly differentiated patterns of neural activity (Guzowski et al. 2004; Vazdarjanova and Guzowski 2004).

Theta Rhythm, Theta Phase Precession, and Grid Cells

Another important body of modeling research has focused on the role of oscillations in cortical function. Here I will focus on models of the role of theta rhythm in hippocampal function. An early paper presented at the Computational Neuroscience meeting presented a model of goal-directed behavior in the hippocampus that used the phenomenon of theta phase precession to provide a more accurate spatial code (Burgess et al. 1994).

Theta phase precession was first discovered by O'Keefe and Recce (1993) and then replicated extensively in other studies (Skaggs et al. 1996; Huxter et al. 2003). In theta phase precession, the spiking response of hippocampal place cells changes relative to theta rhythm oscillations recorded simultaneously in the hippocampal EEG. When a rat first enters the place field of an individual place cell, the spikes occur predominantly at a relatively late phase of the theta rhythm. The spikes shift to progressively earlier phases as the rat traverses the field. In the original paper describing theta phase precession, the phenomenon was proposed to arise from a progressive phase shift between the network EEG oscillation frequency and the intrinsic spiking frequency of the neuron which was shown to have a higher frequency based on the autocorrelation of spiking activity (O'Keefe and Recce 1993). That paper presents a simple figure showing how the interaction of two oscillations of slightly different frequency will cause a precession of the summed oscillation relative to the lower frequency oscillation. This model makes an interesting additional prediction that there should be multiple firing fields, each showing the same precession. Since most place cells had a single firing field, this was perceived as a problem of the model, and later implementations kept the oscillations out of phase with each other until one was shifted to a higher frequency in the firing field (Lengyel et al. 2003). However, the later discovery of grid cells casts a different light on the original model, fulfilling the prediction of the model for multiple firing fields that was initially perceived as a problem of the model. Thus, the model by O'Keefe and Recce essentially predicted the existence of grid cells.

A number of other models have also simulated theta phase precession. For example, the oscillatory interference model has been presented in a variant involving inhibitory influences on pyramidal cells (Bose et al. 2000; Bose and Recce 2001). In another class of models, the replication of phase precession in the McNaughton laboratory was accompanied by a model of phase precession based on slow retrieval of a learned sequence of spatial locations during each theta cycle (Tsodyks et al. 1996). A similar sequence read-out model was presented that year by Jensen and Lisman (1996a). In the Jensen and Lisman model, the phase precession during encoding arose from a working memory buffer in which afterdepolarization allowed neurons to be played out in a sequence on each theta cycle (Jensen and Lisman 1996b). Both of these models required relatively slow read-out of the sequence across the full theta cycle, at a rate slower than the time constants of glutamatergic AMPA conductances. The following year a different model was presented (Wallenstein and Hasselmo 1997) in which read-out had the faster time course of

AMPA conductances, but the length of the read-out would shift across the theta cycle based on the level of presynaptic inhibition or the level of postsynaptic depolarization. This model was extended later to include the context-dependent retrieval of sequences, accounting for the reappearance of theta phase precession over initial trials on each new day (Hasselmo and Eichenbaum 2005).

Another class of models proposed that phase precession arose from progressive shifts in the postsynaptic depolarization of neurons, causing spikes to occur at different phases relative to network inhibitory oscillations (Kamondi et al. 1998; Magee 2001; Mehta et al. 2002). These different models have motivated a number of different experimental studies. The sequence retrieval models were supported by an initial study showing that reset of theta phase oscillations did not shift phase, spiking after reset would commence at the same phase as before the reset (Zugaro et al. 2005). However, a more recent study strongly supported the oscillatory interference model by showing that intracellularly recorded oscillations in membrane potential also show phase precession relative to network oscillations (Harvey et al. 2009), an effect not predicted by the sequence read-out model. The postsynaptic depolarization model did not predict this shift in phase of intracellular oscillations (Kamondi et al. 1998). In addition, the postsynaptic depolarization models predicted an asymmetrical sawtooth waveform for a depolarizing shift in the place field, whereas the data showed a symmetrical depolarization in the place field (Harvey et al. 2009).

As noted above, the original presentation of the oscillatory interference model of theta phase precession predicted the existence of neurons with multiple, regularly spaced firing fields (O'Keefe and Recce 1993). Though the authors initially saw this as a problem for the model, the generation of multiple firing fields by the model is explicitly shown in Fig. 10 of the O'Keefe and Recce (1993) paper. This initially undesired prediction of the model was validated by the later discovery of grid cells in the medial entorhinal cortex in the Moser laboratory. In the data from the Moser lab, the existence of repeating firing fields was first noted in the dorsal portion of medial entorhinal cortex (Fyhn et al. 2004), and subsequently the regular hexagonal arrangement of firing fields was noted and found to extend to more ventral regions of medial entorhinal cortex with larger spacing between the firing fields (Hafting et al. 2005). The systematic increase in spacing between firing fields for neurons in more ventral locations has been shown in great detail in subsequent papers (Sargolini et al. 2006), including very large and widely spaced firing fields in more ventral medial entorhinal cortex (Brun et al. 2008).

When the first paper on grid cells appeared, O'Keefe and Burgess immediately recognized the significance of the repeating nature of grid cell firing, as this had been a strong feature of the theta phase precession model. They rapidly pointed out how oscillatory interference could underlie the properties of grid cell firing (O'Keefe and Burgess 2005). In the Computational Cognitive Neuroscience meeting in Washington in 2005, Neil Burgess presented a poster with a detailed model using velocity modulation of firing frequency to generate realistic grid cell firing fields (Burgess et al. 2005). The oscillatory interference model of grid cells immediately generated a prediction about the mechanism for the difference in spacing of grid cells along the dorsal to ventral axis of medial entorhinal cortex (O'Keefe and Burgess 2005). To quote that paper directly: "The increasing spatial scale of the grid-like firing as you move from the postrhinal border of the medial entorhinal cortex would result from a gradually decreasing intrinsic frequency …" I saw Neil Burgess's poster in Washington and with graduate student Lisa Giocomo set out to test this explicit prediction of the model. Neil kindly sent us a copy of his poster with the model that he presented later in a full paper (Burgess et al. 2007).

To test the prediction, Lisa performed intracellular whole cell patch recording from stellate cells in slice preparations of medial entorhinal cortex (Giocomo et al. 2007). She used horizontal slices of entorhinal cortex and kept track of the dorsal to ventral position of the individual horizontal slices, so she could plot differences in intrinsic properties relative to anatomical position. We found a clear difference in the resonant frequency and the frequency of subthreshold membrane potential oscillations (Giocomo et al. 2007), with a gradual decrease in these intrinsic frequencies for slices more ventral relative to the postrhinal border. Thus, the prediction of the model was clearly supported by the data. The data on frequency membrane potential oscillations and resonance has been replicated by other groups (Boehlen et al. 2010) and by other researchers working in my laboratory (Heys et al. 2010).

In our initial presentation of the data on differences in intrinsic frequency (Giocomo et al. 2007), we illustrated the functional significance of the data by incorporating the difference in intrinsic frequency into the oscillatory interference model by Burgess (Fig. 10.4). Using a multiplicative version of the model, we showed that higher intrinsic frequency in dorsal cells could generate the narrower spacing between firing fields of grid cells recording in dorsal entorhinal cortex and the lower frequency in ventral cells could generate the wider spacing in more ventral cells. In a later paper, we showed that the data was more consistent with an additive model that could account for very wide spacings by having a shallower slope of change in frequency with velocity (Giocomo and Hasselmo 2008a).

The dorsal to ventral difference in intrinsic frequency was accompanied by a gradual slowing of the time constant of the depolarizing sag in stellate cells caused by hyperpolarizing current injections activating the H current and causing a depolarizing rebound (Giocomo et al. 2007). This suggested a role for H current in the dorsal to ventral difference in intrinsic frequency, which was supported by voltage clamp data suggesting a difference in the time constant of the H current as well as a trend toward differences in the magnitude of the H current (Giocomo and Hasselmo 2008b). Testing of intrinsic frequencies in mice with knockout of the H current showed a flattening of the gradients of intrinsic frequencies (Giocomo and Hasselmo 2009). These results were consistent with recordings in oocytes showing that homomeric H current channels using just HCN1 subunit had faster time constant than homomeric HCN2 channels, with an intermediate time constant for heteromeric channels combining HCN1 and HCN2 subunits (Chen et al. 2001). Thus, this model provided an exciting link between molecular and cellular properties of neurons in medial entorhinal cortex, and the functional coding of space by the grid cell firing properties of these neurons. This was beyond anything that I had dreamed of accomplishing when the Computational Neuroscience meeting started in the early 1990s.

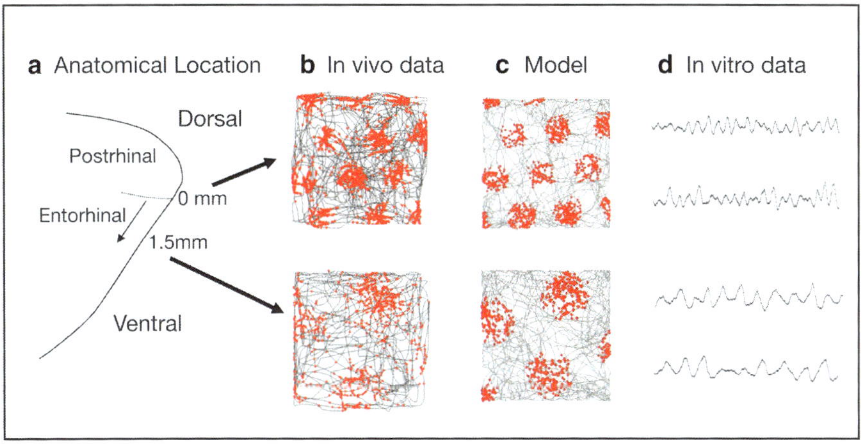

Fig. 10.4 (**a**) Anatomical location of grid cells with different spacing. (**b**) Dorsal cells near the postrhinal border have spacing between firing fields of about 40 cm (*top*). Cells recorded about 1.5 mm more ventral from the postrhinal border have spacing between firing fields of about 80 cm (*bottom*) (from Hafting et al. 2005). (**c**) The oscillatory interference model of grid cells can replicate these spacing properties based on a steeper slope of oscillation frequency to velocity in dorsal compared to ventral cells (Burgess et al. 2007; Hasselmo et al. 2007). (**d**) The prediction of the model for different intrinsic oscillation frequencies during depolarization is supported by whole cell patch recordings of stellate cells in slice preparations of medial entorhinal cortex from dorsal (*top*) versus ventral (*bottom*) anatomical locations (Giocomo et al. 2007)

After we published the Science paper, I felt that the next step would be simple. The model in the Science paper used interference of cosine functions. The next step would be to implement the model within a compartmental simulation of an entorhinal stellate cell as implemented in GENESIS by Fransén et al. (2004). I believed we could simulate subthreshold oscillations on different dendrites within a compartmental simulation (Hasselmo et al. 2007). However, in simulations run by Jim Heys in my laboratory, subthreshold oscillations on different dendrites tended to synchronize. The same result was obtained in work by Michiel Remme with Boris Gutkin and Mate Lengyel in extensive simulations and computational analysis (Remme et al. 2009, 2010). In addition, analysis of the variability of oscillation period showed that the membrane potential oscillations were too noisy to allow stable coding of location by phase (Giocomo and Hasselmo 2008a; Zilli et al. 2009). These points argued against a single cell implementation of the model and argued for a network implementation.

The effect of single cell resonance on spike timing is a topic of ongoing research. It is clear that resonance does not result in rhythmic spiking only at the resonant frequency, but allows a range of frequencies with only a small deflection at the resonant frequency (Giocomo and Hasselmo 2008a). In contrast, recordings of intrinsic persistent spiking mechanisms in medial entorhinal pyramidal cells show that cells tend to spike rhythmically at steady frequencies around theta rhythm (Egorov et al. 2002; Fransén et al. 2006; Tahvildari et al. 2007). Therefore, I developed a model of

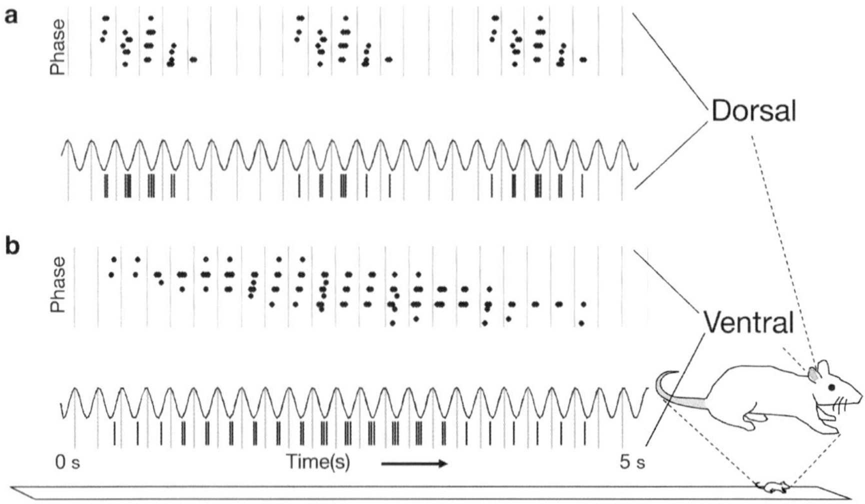

Fig. 10.5 Theta phase precession using the persistent spiking neuron model. (**a**) Simulation of neuron in dorsal entorhinal cortex with higher persistent firing frequency for a given velocity. *Black dots* show phase of spiking versus location during multiple passes through firing fields. *Blue trace* shows simulated EEG with *dashes* indicating spike times. (**b**) Ventral entorhinal cortex neuron with lower persistent spiking frequency for a given velocity, showing slower shift in phase with position in a larger grid cell firing field (Hasselmo 2008)

grid cells based on persistent spiking cells that could hold a steady baseline frequency. Cells with stable baseline frequencies have been shown in deep layers of medial entorhinal cortex (Egorov et al. 2002; Fransén et al. 2006; Tahvildari et al. 2007), in layer III of lateral entorhinal cortex (Tahvildari et al. 2007), and in the postsubiculum (Yoshida and Hasselmo 2009). These neurons tend to fire at the same stable baseline frequency regardless of the duration of the stimulation causing persistent spiking (Yoshida and Hasselmo 2009). A computational model of grid cells based on persistent spiking was developed using grid cells responding to the convergent input from different groups of persistent spiking cells that receive input from different sets of head direction cells (Hasselmo 2008). This effectively simulated grid cells based on shifts in the frequency of persistent spiking input (Hasselmo 2008), and as shown in Fig. 10.5, simulates theta phase precession in grid cells (Hasselmo 2008) consistent with experimental data showing theta phase precession in grid cells (Hafting et al. 2008).

Persistent spiking also shows variability in firing frequency that could interfere with the stability of phase coding. However, network level dynamics may overcome this variability, allowing cells that are intrinsically noisy and irregular in their firing to still participate in a network oscillation with frequency and phase sufficiently stable to generate grid cell firing (Zilli and Hasselmo 2010). This model can respond with different frequencies for different depolarizing inputs depending on the magnitude of the H current in individual neurons, though it is difficult to maintain a linear relationship between depolarizing input and magnitude of frequency change.

This model indicates the ongoing validity of the oscillatory interference model as a theory of the generation of grid cell firing responses and provides a framework for explaining the relationship between intrinsic resonance and the spacing of grid cell firing fields.

A number of alternate mechanisms have been proposed for the generation of grid cell firing properties, including attractor dynamics due to structured excitatory recurrent connectivity (Fuhs and Touretzky 2006; McNaughton et al. 2006; Burak and Fiete 2009) and self-organization of afferent input (Kropff and Treves 2008). The attractor dynamics models do not account for some data as well as oscillatory interference models, but they are better at accounting for the consistent orientation and spacing of grid cells within local regions of the medial entorhinal cortex (Hafting et al. 2005) and the apparent quantal transitions in the spacing between firing fields (Barry et al. 2007). However, most attractor dynamic models do not utilize theta frequency oscillations in spiking activity and do not account for theta phase precession. However, a recent model used attractor dynamics and simulated grid cell theta phase precession, while generating differences in spacing based on the time course of medium afterhyperpolarization (Navratilova et al. 2012). The importance of theta rhythm oscillations for grid cell generation has been demonstrated by local infusions into the medial septum that block network theta rhythm oscillations in the entorhinal cortex. Grid cell firing patterns do not appear during pharmacological blockade of theta rhythm oscillations (Brandon et al. 2011), whereas head direction responses are spared.

As described here, the discovery of grid cells and their relationship to the intrinsic resonance properties of entorhinal neurons provides fascinating clues to the function of the entorhinal cortex and hippocampus in human episodic memory. A theoretical framework based on the oscillatory interference model can perform the encoding and retrieval of complex trajectories as episodic memories. The data have not yet converged on a final model of the mechanism for generation of grid cells, but the ongoing interaction of computational modeling guiding experimental neurophysiology has provided insights beyond any that I imagined 20 years ago at the Computational Neuroscience meeting.

Acknowledgements Research supported by Silvio O. Conte Center grant NIMH MH71702, NIMH R01 MH61492; NSF Sciences of Learning Center CELEST SBE 0354378 and NIMH R01 60013 and ONR MURI N00014-10-1-0936.

References

Abbott LF, Blum KI (1996) Functional significance of long-term potentiation for sequence learning and prediction. Cereb Cortex 6(3):406–416

Amit DJ (1988) Modeling brain function: the world of attractor neural networks. Cambridge Univ Press, Cambridge, UK

Amit DJ, Treves A (1989) Associative memory neural networks with low temporal spiking rates. Proc Natl Acad Sci USA 86:7671–7673

Amit DJ, Evans MR, Abeles M (1990) Attractor neural networks with biological probe records. Network 1:381–405
Anderson JA (1972) A simple neural network generating an interactive memory. Math Biosci 14:197–220
Atri A, Sherman S, Norman KA, Kirchhoff BA, Nicolas MM, Greicius MD, Cramer SC, Breiter HC, Hasselmo ME, Stern CE (2004) Blockade of central cholinergic receptors impairs new learning and increases proactive interference in a word paired-associate memory task. Behav Neurosci 118(1):223–236
Barkai E, Hasselmo ME (1994) Modulation of the input/output function of rat piriform cortex pyramidal cells. J Neurophysiol 72(2):644–658
Barkai E, Bergman RE, Horwitz G, Hasselmo ME (1994) Modulation of associative memory function in a biophysical simulation of rat piriform cortex. J Neurophysiol 72(2):659–677
Barnes CA, McNaughton BL, Mizumori SJ, Leonard BW, Lin LH (1990) Comparison of spatial and temporal characteristics of neuronal activity in sequential stages of hippocampal processing. Prog Brain Res 83:287–300
Barry C, Hayman R, Burgess N, Jeffery KJ (2007) Experience-dependent rescaling of entorhinal grids. Nat Neurosci 10(6):682–684
Bergman RE, Vanier M, Horwitz G, Bower JM, Hasselmo ME (1993) Cholinergic modulation of associative memory function in a realistic computational model of piriform cortex. In: Eeckman F, Bower JM (eds) Computation and neural systems. Kluwer Academic, Norwell, MA, pp 273–280
Blum KI, Abbott LF (1996) A model of spatial map formation in the hippocampus of the rat. Neural Comput 8(1):85–93
Boehlen A, Heinemann U, Erchova I (2010) The range of intrinsic frequencies represented by medial entorhinal cortex stellate cells extends with age. J Neurosci 30(13):4585–4589
Bose A, Recce M (2001) Phase precession and phase-locking of hippocampal pyramidal cells. Hippocampus 11(3):204–215
Bose A, Booth V, Recce M (2000) A temporal mechanism for generating the phase precession of hippocampal place cells. J Comput Neurosci 9(1):5–30
Bower JM (1990) Reverse engineering the nervous system: an anatomical, physiological and computer based approach. In: Zornetzer S, Davis J, Lau C (eds) An introduction to neural and electronic networks. Academic, San Diego, pp 3–24
Bower JM, Beeman D (1995) The book of GENESIS: exploring realistic neural models with the GEneral NEural SImulation System. Springer, New York
Bower JM, Haberly LB (1986) Facilitating and nonfacilitating synapses on pyramidal cells: a correlation between physiology and morphology. Proc Natl Acad Sci USA 83:1115–1119
Brandon MP, Bogaard AR, Libby CP, Connerney MA, Gupta K, Hasselmo ME (2011) Reduction of theta rhythm dissociates grid cell spatial periodicity from directional tuning. Science 332:595–599
Brun VH, Solstad T, Kjelstrup KB, Fyhn M, Witter MP, Moser EI, Moser MB (2008) Progressive increase in grid scale from dorsal to ventral medial entorhinal cortex. Hippocampus 18(12):1200–1212
Burak Y, Fiete IR (2009) Accurate path integration in continuous attractor network models of grid cells. PLoS Comput Biol 5(2):e1000291
Burgess N, Barry C, Jeffery KJ, O'Keefe J (2005) A grid and place cell model of path integration utilizing phase precession versus theta. In: Computational cognitive neuroscience meeting, Washington, DC
Burgess N, Barry C, O'Keefe J (2007) An oscillatory interference model of grid cell firing. Hippocampus 17(9):801–812
Burgess N, O'Keefe J, Recce M (1994). Toward a mechanism for navigation by the rat hippocampus. In: Eeckman F, Bower JM (eds) Computation and neural systems. Kluwer, Norwell, MA, pp 257–262
Chen S, Wang J, Siegelbaum SA (2001) Properties of hyperpolarization-activated pacemaker current defined by coassembly of HCN1 and HCN2 subunits and basal modulation by cyclic nucleotide. J Gen Physiol 117(5):491–504

Cohen MA, Grossberg S (1983) Absolute stability of global pattern formation and parallel memory storage by competitive neural networks. IEEE Trans Syst Man Cybern 13:815–826
De Rosa E, Hasselmo ME (2000) Muscarinic cholinergic neuromodulation reduces proactive interference between stored odor memories during associative learning in rats. Behav Neurosci 114:32–41
De Rosa E, Hasselmo ME, Baxter MG (2001) Contribution of the cholinergic basal forebrain to proactive interference from stored odor memories during associative learning in rats. Behav Neurosci 115(2):314–327
de Sevilla DF, Cabezas C, de Prada AN, Sanchez-Jimenez A, Buno W (2002) Selective muscarinic regulation of functional glutamatergic Schaffer collateral synapses in rat CA1 pyramidal neurons. J Physiol 545(Pt 1):51–63
Egorov AV, Hamam BN, Fransen E, Hasselmo ME, Alonso AA (2002) Graded persistent activity in entorhinal cortex neurons. Nature 420(6912):173–178
Fransén E, Alonso AA, Dickson CT, Magistretti J, Hasselmo ME (2004) Ionic mechanisms in the generation of subthreshold oscillations and action potential clustering in entorhinal layer II stellate neurons. Hippocampus 14(3):368–384
Fransén E, Tahvildari B, Egorov AV, Hasselmo ME, Alonso AA (2006) Mechanism of graded persistent cellular activity of entorhinal cortex layer v neurons. Neuron 49(5):735–746
Fuhs MC, Touretzky DS (2006) A spin glass model of path integration in rat medial entorhinal cortex. J Neurosci 26(16):4266–4276
Fyhn M, Molden S, Witter MP, Moser EI, Moser MB (2004) Spatial representation in the entorhinal cortex. Science 305(5688):1258–1264
Gais S, Born J (2004) Low acetylcholine during slow-wave sleep is critical for declarative memory consolidation. Proc Natl Acad Sci USA 101(7):2140–2144
Gil Z, Conners BW, Amitai Y (1997) Differential regulation of neocortical synapses by neuromodulators and activity. Neuron 19:679–686
Gilbert PE, Kesner RP, Lee I (2001) Dissociating hippocampal subregions: double dissociation between dentate gyrus and CA1. Hippocampus 11(6):626–636
Giocomo LM, Hasselmo ME (2008a) Computation by oscillations: implications of experimental data for theoretical models of grid cells. Hippocampus 18(12):1186–1199
Giocomo LM, Hasselmo ME (2008b) Time constants of h current in layer II stellate cells differ along the dorsal to ventral axis of medial entorhinal cortex. J Neurosci 28(38):9414–9425
Giocomo LM, Hasselmo ME (2009) Knock-out of HCN1 subunit flattens dorsal-ventral frequency gradient of medial entorhinal neurons in adult mice. J Neurosci 29(23):7625–7630
Giocomo LM, Zilli EA, Fransen E, Hasselmo ME (2007) Temporal frequency of subthreshold oscillations scales with entorhinal grid cell field spacing. Science 315(5819):1719–1722
Guzowski JF, Knierim JJ, Moser EI (2004) Ensemble dynamics of hippocampal regions CA3 and CA1. Neuron 44(4):581–584
Haberly LB, Bower JM (1984) Analysis of association fiber system in piriform cortex with intracellular recording and staining techniques. J Neurophysiol 51:90–112
Haberly LB, Bower JM (1989) Olfactory cortex: model circuit for study of associative memory? Trends Neurosci 12:258–264
Hafting T, Fyhn M, Molden S, Moser MB, Moser EI (2005) Microstructure of a spatial map in the entorhinal cortex. Nature 436(7052):801–806
Hafting T, Fyhn M, Bonnevie T, Moser MB, Moser EI (2008) Hippocampus-independent phase precession in entorhinal grid cells. Nature 453(7199):1248–1252
Harvey CD, Collman F, Dombeck DA, Tank DW (2009) Intracellular dynamics of hippocampal place cells during virtual navigation. Nature 461(7266):941–946
Hasselmo ME (1994) Runaway synaptic modification in models of cortex: implications for Alzheimer's disease. Neural Netw 7:13–40
Hasselmo ME (1997) A computational model of the progression of Alzheimer's disease. MD Comput 14(3):181–191
Hasselmo ME (1999) Neuromodulation: acetylcholine and memory consolidation. Trends Cogn Sci 3(9):351–359

Hasselmo ME (2008) Grid cell mechanisms and function: contributions of entorhinal persistent spiking and phase resetting. Hippocampus 18(12):1213–1229
Hasselmo ME, Bower JM (1992) Cholinergic suppression specific to intrinsic not afferent fiber synapses in rat piriform (olfactory) cortex. J Neurophysiol 67(5):1222–1229
Hasselmo ME, Bower JM (1993) Acetylcholine and memory. Trends Neurosci 16(6):218–222
Hasselmo ME, Cekic M (1996) Suppression of synaptic transmission may allow combination of associative feedback and self-organizing feedforward connections in the neocortex. Behav Brain Res 79(1–2):153–161
Hasselmo ME, Eichenbaum H (2005) Hippocampal mechanisms for the context-dependent retrieval of episodes. Neural Netw 18(9):1172–1190
Hasselmo ME, Fehlau BP (2001) Differences in time course of ACh and GABA modulation of excitatory synaptic potentials in slices of rat hippocampus. J Neurophysiol 86(4):1792–1802
Hasselmo ME, Schnell E (1994) Laminar selectivity of the cholinergic suppression of synaptic transmission in rat hippocampal region CA1: computational modeling and brain slice physiology. J Neurosci 14(6):3898–3914
Hasselmo ME, Wyble BP (1997) Free recall and recognition in a network model of the hippocampus: simulating effects of scopolamine on human memory function. Behav Brain Res 89(1–2):1–34
Hasselmo ME, Wilson MA, Anderson BP, Bower JM (1990) Associative memory function in piriform (olfactory) cortex: computational modeling and neuropharmacology. Cold Spring Harb Symp Quant Biol 55:599–610
Hasselmo ME, Anderson BP, Bower JM (1992) Cholinergic modulation of cortical associative memory function. J Neurophysiol 67(5):1230–1246
Hasselmo ME, Barkai E, Horwitz G, Bergman RE (1994) Modulation of neuronal adaptation and cortical associative memory function. In: Eeckman F, Bower JM (eds) Computation and neural systems. Kluwer, Norwell, MA
Hasselmo ME, Schnell E, Barkai E (1995) Dynamics of learning and recall at excitatory recurrent synapses and cholinergic modulation in rat hippocampal region CA3. J Neurosci 15(7 Pt 2):5249–5262
Hasselmo ME, Bodelon C, Wyble BP (2002) A proposed function for hippocampal theta rhythm: separate phases of encoding and retrieval enhance reversal of prior learning. Neural Comput 14(4):793–817
Hasselmo ME, Giocomo LM, Zilli EA (2007) Grid cell firing may arise from interference of theta frequency membrane potential oscillations in single neurons. Hippocampus 17(12):1252–1271
Herrero JL, Roberts MJ, Delicato LS, Gieselmann MA, Dayan P, Thiele A (2008) Acetylcholine contributes through muscarinic receptors to attentional modulation in V1. Nature 454(7208):1110–1114
Heys JG, Giocomo LM, Hasselmo ME (2010) Cholinergic modulation of the resonance properties of stellate cells in layer II of medial entorhinal cortex. J Neurophysiol 104(1):258–270
Holmes WR, Levy WB (1990) Insights into associative long-term potentiation from computational models of NMDA receptor-mediated calcium influx and intracellular calcium concentration changes. J Neurophysiol 63(5):1148–1168
Hopfield JJ (1982) Neural networks and physical systems with emergent selective computational abilities. Proc Natl Acad Sci USA 79:2554–2559
Hopfield JJ (1984) Neurons with graded response have collective computational properties like those of two-state neurons. Proc Natl Acad Sci USA 81(10):3088–3092
Hounsgaard J (1978) Presynaptic inhibitory action of acetylcholine in area CA1 of the hippocampus. Exp Neurol 62(3):787–797
Huxter J, Burgess N, O'Keefe J (2003) Independent rate and temporal coding in hippocampal pyramidal cells. Nature 425(6960):828–832
Jensen O, Lisman JE (1996a) Hippocampal CA3 region predicts memory sequences: accounting for the phase precession of place cells. Learn Mem 3:279–287

Jensen O, Lisman JE (1996b) Novel lists of 7+/−2 known items can be reliably stored in an oscillatory short-term memory network: interaction with long-term memory. Learn Mem 3:257–263
Kali S, Dayan P (2000) The involvement of recurrent connections in area CA3 in establishing the properties of place fields: a model. J Neurosci 20(19):7463–7477
Kamondi A, Acsady L, Wang XJ, Buzsaki G (1998) Theta oscillations in somata and dendrites of hippocampal pyramidal cells in vivo: activity-dependent phase-precession of action potentials. Hippocampus 8(3):244–261
Kimura F (2000) Cholinergic modulation of cortical function: a hypothetical role in shifting the dynamics in cortical network. Neurosci Res 38(1):19–26
Kimura F, Baughman RW (1997) Distinct muscarinic receptor subtypes suppress excitatory and inhibitory synaptic responses in cortical neurons. J Neurophysiol 77(2):709–716
Kohonen T (1972) Correlation matrix memories. IEEE Trans Comput C-21:353–359
Kohonen T (1984) Self-organization and associative memory. Springer, Berlin
Kremin T, Hasselmo ME (2007) Cholinergic suppression of glutamatergic synaptic transmission in hippocampal region CA3 exhibits laminar selectivity: implication for hippocampal network dynamics. Neuroscience 149(4):760–767
Kropff E, Treves A (2008) The emergence of grid cells: intelligent design or just adaptation? Hippocampus 18(12):1256–1269
Lee I, Yoganarasimha D, Rao G, Knierim JJ (2004) Comparison of population coherence of place cells in hippocampal subfields CA1 and CA3. Nature 430(6998):456–459
Lengyel M, Szatmary Z, Erdi P (2003) Dynamically detuned oscillations account for the coupled rate and temporal code of place cell firing. Hippocampus 13(6):700–714
Leutgeb S, Leutgeb JK, Treves A, Moser MB, Moser EI (2004) Distinct ensemble codes in hippocampal areas CA3 and CA1. Science 305(5688):1295–1298
Leutgeb JK, Leutgeb S, Moser MB, Moser EI (2007) Pattern separation in the dentate gyrus and CA3 of the hippocampus. Science 315(5814):961–966
Levy WB, Steward O (1983) Temporal contiguity requirements for long-term associative potentiation/depression in the hippocampus. Neuroscience 8(4):791–797
Magee JC (2001) Dendritic mechanisms of phase precession in hippocampal CA1 pyramidal neurons. J Neurophysiol 86(1):528–532
Manns JR, Zilli EA, Ong KC, Hasselmo ME, Eichenbaum H (2007) Hippocampal CA1 spiking during encoding and retrieval: relation to theta phase. Neurobiol Learn Mem 87(1):9–20
Marr D (1971) Simple memory: a theory for archicortex. Philos Trans R Soc B 262:23–81
Marrosu F, Portas C, Mascia MS, Casu MA, Fa M, Giagheddu M, Imperato A, Gessa GL (1995) Microdialysis measurement of cortical and hippocampal acetylcholine release during sleep-wake cycle in freely moving cats. Brain Res 671(2):329–332
McClelland JL, Rumelhart DE (1988) Explorations in parallel distributed processing. MIT Press, Cambridge, MA
McHugh TJ, Jones MW, Quinn JJ, Balthasar N, Coppari R, Elmquist JK, Lowell BB, Fanselow MS, Wilson MA, Tonegawa S (2007) Dentate gyrus NMDA receptors mediate rapid pattern separation in the hippocampal network. Science 317(5834):94–99
McNaughton BL (1991) Associative pattern completion in hippocampal circuits: new evidence and new questions. Brain Res Rev 16:193–220
McNaughton BL, Morris RGM (1987) Hippocampal synaptic enhancement and information storage within a distributed memory system. Trends Neurosci 10:408–415
McNaughton BL, Battaglia FP, Jensen O, Moser EI, Moser MB (2006) Path integration and the neural basis of the 'cognitive map'. Nat Rev Neurosci 7(8):663–678
Mehta MR, McNaughton BL (1997) Expansion and shift of hippocampal place fields: evidence for synaptic potentiation during behavior. In: Bower JM (ed) Computational neuroscience. Plenum Press, New York, pp 741–744
Mehta MR, Barnes CA, McNaughton BL (1997) Experience-dependent, asymmetric expansion of hippocampal place fields. Proc Natl Acad Sci USA 94(16):8918–8921

Mehta MR, Quirk MC, Wilson MA (2000) Experience-dependent asymmetric shape of hippocampal receptive fields. Neuron 25(3):707–715
Mehta MR, Lee AK, Wilson MA (2002) Role of experience and oscillations in transforming a rate code into a temporal code. Nature 417(6890):741–746
Molyneaux BJ, Hasselmo ME (2002) GABA(B) presynaptic inhibition has an in vivo time constant sufficiently rapid to allow modulation at theta frequency. J Neurophysiol 87(3):1196–1205
Nakashiba T, Young JZ, McHugh TJ, Buhl DL, Tonegawa S (2008) Transgenic inhibition of synaptic transmission reveals role of CA3 output in hippocampal learning. Science 319(5867):1260–1264
Nakazawa K, Quirk MC, Chitwood RA, Watanabe M, Yeckel MF, Sun LD, Kato A, Carr CA, Johnston D, Wilson MA et al (2002) Requirement for hippocampal CA3 NMDA receptors in associative memory recall. Science 297(5579):211–218
Navratilova Z, Giocomo LM, Fellous JM, Hasselmo ME, McNaughton BL (2012) Phase precession and variable spatial scaling in a periodic attractor map model of medial entorhinal grid cells with realistic after-spike dynamics. Hippocampus 22:772–789
O'Keefe J, Burgess N (2005) Dual phase and rate coding in hippocampal place cells: theoretical significance and relationship to entorhinal grid cells. Hippocampus 15(7):853–866
O'Keefe J, Recce ML (1993) Phase relationship between hippocampal place units and the EEG theta rhythm. Hippocampus 3:317–330
O'Reilly RC, McClelland JL (1994) Hippocampal conjunctive encoding, storage, and recall: avoiding a trade-off. Hippocampus 4(6):661–682
Rasch BH, Born J, Gais S (2006) Combined blockade of cholinergic receptors shifts the brain from stimulus encoding to memory consolidation. J Cogn Neurosci 18(5):793–802
Reisberg B, Doody R, Stoffler A, Schmitt F, Ferris S, Mobius HJ (2003) Memantine in moderate-to-severe Alzheimer's disease. N Engl J Med 348(14):1333–1341
Remme MW, Lengyel M, Gutkin BS (2009) The role of ongoing dendritic oscillations in single-neuron dynamics. PLoS Comput Biol 5(9):e1000493
Remme MW, Lengyel M, Gutkin BS (2010) Democracy-independence trade-off in oscillating dendrites and its implications for grid cells. Neuron 66(3):429–437
Rizzuto DS, Madsen JR, Bromfield EB, Schulze-Bonhage A, Kahana MJ (2006) Human neocortical oscillations exhibit theta phase differences between encoding and retrieval. Neuroimage 31(3):1352–1358
Roberts MJ, Zinke W, Guo K, Robertson R, McDonald JS, Thiele A (2005) Acetylcholine dynamically controls spatial integration in marmoset primary visual cortex. J Neurophysiol 93(4):2062–2072
Rumelhart RE, Hinton GE, Williams RJ (1986) Learning representations by back-propagating errors. Nature 323:533–536
Sargolini F, Fyhn M, Hafting T, McNaughton BL, Witter MP, Moser MB, Moser EI (2006) Conjunctive representation of position, direction, and velocity in entorhinal cortex. Science 312(5774):758–762
Schacter D (1982) Stranger behind the engram: theories of memory and the psychology of science. Lawrence, Hillsdale, NJ
Shirey JK, Xiang Z, Orton D, Brady AE, Johnson KA, Williams R, Ayala JE, Rodriguez AL, Wess J, Weaver D et al (2008) An allosteric potentiator of M4 mAChR modulates hippocampal synaptic transmission. Nat Chem Biol 4(1):42–50
Silver MA, Shenhav A, D'Esposito M (2008) Cholinergic enhancement reduces spatial spread of visual responses in human early visual cortex. Neuron 60(5):904–914
Skaggs WE, McNaughton BL, Wilson MA, Barnes CA (1996) Theta phase precession in hippocampal neuronal populations and the compression of temporal sequences. Hippocampus 6:149–172

Tahvildari B, Fransen E, Alonso AA, Hasselmo ME (2007) Switching between "On" and "Off" states of persistent activity in lateral entorhinal layer III neurons. Hippocampus 17(4):257–263

Toth K, Freund TF, Miles R (1997) Disinhibition of rat hippocampal pyramidal cells by GABAergic afferent from the septum. J Physiol 500:463–474

Treves A, Rolls ET (1994) Computational analysis of the role of the hippocampus in memory. Hippocampus 4(3):374–391

Tsodyks MV, Skaggs WE, Sejnowski TJ, McNaughton BL (1996) Population dynamics and theta rhythm phase precession of hippocampal place cell firing: a spiking neuron model. Hippocampus 6(3):271–280

Valentino RJ, Dingledine R (1981) Presynaptic inhibitory effect of acetylcholine in the hippocampus. J Neurosci 1:784–792

van Vreeswijk C, Sompolinsky H (1996) Chaos in neuronal networks with balanced excitatory and inhibitory activity. Science 274(5293):1724–1726

Vazdarjanova A, Guzowski JF (2004) Differences in hippocampal neuronal population responses to modifications of an environmental context: evidence for distinct, yet complementary, functions of CA3 and CA1 ensembles. J Neurosci 24(29):6489–6496

Villarreal DM, Gross AL, Derrick BE (2007) Modulation of CA3 afferent inputs by novelty and theta rhythm. J Neurosci 27(49):13457–13467

Vogt KE, Regehr WG (2001) Cholinergic modulation of excitatory synaptic transmission in the CA3 area of the hippocampus. J Neurosci 21(1):75–83

Wallenstein GV, Hasselmo ME (1997) GABAergic modulation of hippocampal population activity: sequence learning, place field development, and the phase precession effect. J Neurophysiol 78(1):393–408

Wilson MA, Bower JM (1992) Cortical oscillations and temporal interactions in a computer simulation of piriform cortex. J Neurophysiol 67:981–995

Wilson HR, Cowan JD (1972) Excitatory and inhibitory interactions in localized populations of model neurons. Biophys J 12(1):1–24

Wilson HR, Cowan JD (1973) A mathematical theory of the functional dynamics of cortical and thalamic nervous tissue. Kybernetik 1973(13):2

Wyble BP, Hasselmo ME (1997) A model of the effects of scopolamine on human memory performance. In: Bower JM (ed) Computational neuroscience: trends in research 1997. Plenum Press, New York, pp 891–896

Wyble BP, Linster C, Hasselmo ME (2000) Size of CA1-evoked synaptic potentials is related to theta rhythm phase in rat hippocampus. J Neurophysiol 83(4):2138–2144

Yoshida M, Hasselmo ME (2009) Persistent firing supported by an intrinsic cellular mechanism in a component of the head direction system. J Neurosci 29(15):4945–4952

Zilli EA, Hasselmo ME (2010) Coupled noisy spiking neurons as velocity-controlled oscillators in a model of grid cell spatial firing. J Neurosci 30(41):13850–13860

Zilli EA, Yoshida M, Tahvildari B, Giocomo LM, Hasselmo ME (2009) Evaluation of the oscillatory interference model of grid cell firing through analysis and measured period variance of some biological oscillators. PLoS Comput Biol 5(11):e1000573

Zugaro MB, Monconduit L, Buzsaki G (2005) Spike phase precession persists after transient intrahippocampal perturbation. Nat Neurosci 8(1):67–71

Chapter 11
Spatiotemporal Coding in the Olfactory System

Christiane Linster and Thomas A. Cleland

Abstract Beginning in 1994, Gilles Laurent and colleagues published a series of studies describing odor-induced field potential oscillations in the locust olfactory system. While field oscillations had been described in the olfactory system previously—beginning with the work of Lord Adrian in the 1940s and including the extensive studies performed by Walter Freeman and colleagues and the later work of Gelperin and colleagues—the Laurent laboratory's work emerged at a time in which oscillations and spike synchronization in the visual system were attracting substantial attention, such that the emergence of this work triggered a renewed interest in the temporal properties of olfactory system activation and what it implied for the representation of odor stimuli.

Introduction

Beginning in 1994, Gilles Laurent and colleagues published a series of studies describing odor-induced field potential oscillations in the locust olfactory system (Laurent 1996a, b; Laurent and Davidowitz 1994; Laurent and Naraghi 1994; Laurent et al. 1996). While field oscillations had been described in the olfactory

C. Linster (✉)
Department of Neurobiology and Behavior, Computational Physiology Laboratory, Cornell University, 1 Mudd Hall, Ithaca, NY 14853, USA
e-mail: CL243@cornell.edu

T.A. Cleland
Department of Psychology, Computational Physiology Laboratory, Cornell University, Ithaca, NY, USA

J.M. Bower (ed.), *20 Years of Computational Neuroscience*, Springer Series in Computational Neuroscience 9, DOI 10.1007/978-1-4614-1424-7_11,

system previously—beginning with the work of Lord Adrian in the 1940s (Adrian 1942, 1950, 1957) and including the extensive studies performed by Walter Freeman and colleagues (Di Prisco and Freeman 1985; Freeman 1978, 1979a, b; Freeman and Schneider 1982; Freeman and Skarda 1985), and the later work of Gelperin and colleagues (Delaney et al. 1994; Gelperin et al. 1993; Gelperin and Tank 1990; Kleinfeld et al. 1994)—the Laurent laboratory's work emerged at a time in which oscillations and spike synchronization in the visual system were attracting substantial attention (Engel et al. 1990; Gray et al. 1990, 1992), such that the emergence of this work triggered a renewed interest in the temporal properties of olfactory system activation and what it implied for the representation of odor stimuli.

The work of Freeman and colleagues showed that odor stimulation triggers odor-specific patterns of oscillatory activity in the olfactory bulb and piriform cortex of rabbits. These evoked patterns reflected the identity (or quality) of the odorant and also were modulated by (1) the behavioral relevance of the odor to the animal, (2) the activity of neuromodulatory and feedback inputs arising from other brain areas, and (3) olfactory learning (Di Prisco and Freeman 1985; Freeman and Grajski 1987; Grajski and Freeman 1989; Gray et al. 1986, 1987). Interestingly, these results directly coincided with similar conclusions reached by other laboratories based on the odor-specific activation of characteristic populations of neurons; these odor-specific spatial patterns comprised *identity codes* in that odor quality was represented by the identities of activated neurons (Kauer 1988; Kauer et al. 1987; Lancet et al. 1982; Stewart et al. 1979). These identity codes also were found to be modulated by (1) the behavioral relevance of the odor to the animal, (2) the activity of neuromodulatory inputs to olfactory regions, and (3) olfactory learning (Coopersmith et al. 1986; Coopersmith and Leon 1986; Salcedo et al. 2005; Sullivan and Leon 1986; Sullivan et al. 1988). While continued exploration of both the identity-code ("spatial") and temporal approaches revealed substantial complexities and mechanisms of regulation, no clear division of labor between the two became apparent.

Given that both spatial and temporal activity patterns in the olfactory system exhibit specificity for odors as well as dependence on learning, experience, and behavioral state, researchers in the field have sought to determine the relationship between the two as well as the relative importance of each. Recent years have seen a substantial increase in research focusing on the relationship between dynamics, spatial activity patterns, and odor perception. A parallel line of research, mostly theoretical, has emphasized study of the cellular and network mechanisms underlying field oscillations in the olfactory system. We here review the function and mechanisms of the olfactory bulb as it is understood today, emphasizing both spatial and dynamical odor representations and the behavioral evidence pertaining to each; for reasons of space, we omit discussion of the equally important work on piriform cortex conducted by Haberly, Bower, Hasselmo, D. Wilson, and others. We conclude by reviewing recent research illustrating how dynamical and spatial activity patterns build upon one another to establish an informative and flexible code.

Olfactory Bulb Circuitry

The main olfactory bulb in rodents has been extensively described in a number of review articles (Cleland 2010; Linster and Cleland 2009) which we reiterate here briefly. Distributed patterns of activity evoked in primary olfactory sensory neurons (OSNs) by volatile chemical stimuli (odorants) are transmitted to the olfactory bulb via OSN axons. The axons of OSNs that express the same receptors, and hence exhibit the same chemoreceptive fields, converge together to form the glomeruli of the olfactory bulb input layer (Fig. 11.1a); hence, each glomerulus effectively inherits the chemoreceptive field of the OSN population that converges upon it. The olfactory bulb is believed to both filter and actively transform these incoming sensory data, performing operations such as normalization, contrast enhancement, signal-to-noise regulation, and other state-dependent operations before conveying the processed olfactory information to multiple secondary olfactory structures via the axons of mitral cells. These transformations are performed by the interaction of OSN arbors and mitral cells with multiple classes of local interneurons, notably including the periglomerular cells, external tufted cells, and superficial short-axon cells of the glomerular layer as well as the more deeply positioned granule cells, which reciprocally interact with the lateral dendrites of mitral cells (Fig. 11.1a). The olfactory bulb also receives extensive ascending inputs from other brain areas, including piriform cortex, and noradrenergic, serotonergic, and cholinergic nuclei.

Bulbar Processing of Spatial Activation Patterns

Spatially distributed neuronal activity patterns specific to individual odorants were described as early as the 1970s and are present in all species that have been investigated. Each olfactory stimulus activates a specific subset of olfactory receptor types, and hence glomeruli, that is uniquely defined by stimulus quality and concentration and can be presented as an activity map of the olfactory bulb surface (Fig. 11.1b). These bulbar activity maps have been thoroughly analyzed by Michael Leon and colleagues, who have measured the glomerular activation responses to hundreds of different odor stimuli in rats and mice and shown not only that each evokes a characteristic pattern of activity, but also that these patterns, under certain circumstances, are predictive of perceptual qualities (Cleland et al. 2002, 2007; Johnson and Leon 2007; Leon and Johnson 2003, 2006; Linster et al. 2001; Youngentob et al. 2006) (Fig. 11.1c). Similar results have been obtained by other groups using different methods or species (Carlsson et al. 2002; Galizia and Menzel 2000; Guerrieri et al. 2005; Laska and Galizia 2001; Rubin and Katz 2001).

The spatial activation patterns measured in the glomerular layer are thought to represent the average afferent activity conveyed to the glomeruli by OSNs and hence to heavily influence the activation of the postsynaptic mitral cells and glomerular interneurons that innervate each glomerulus. Any computations performed

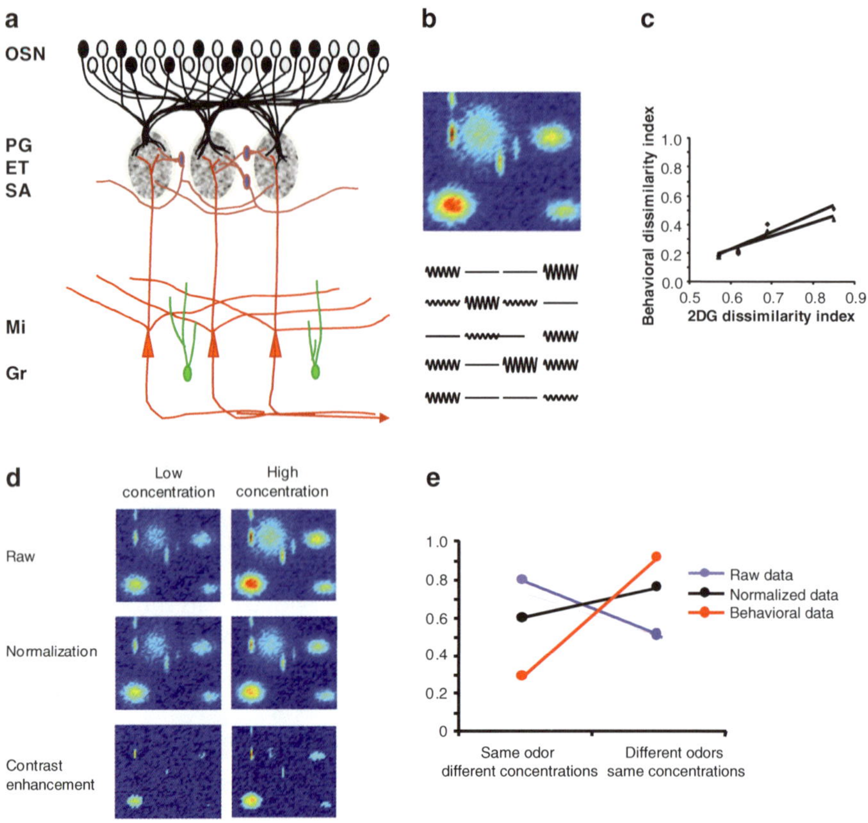

Fig. 11.1 Olfactory bulb processing. (**a**) Schematic diagram of olfactory bulb circuitry. Olfactory sensory neurons (OSNs), each exhibiting a specific receptive field for odorant stimuli, project to the olfactory bulb glomerular layer where they form excitatory synapses with mitral (Mi), external tufted (ET), and a subclass of periglomerular (PG) cells. Within the glomerular layer of the olfactory bulb, local interneurons (PG, ET, and superficial short-axon cells (SA)) interact with one another and with the principal output cells (Mi) to construct odor representations that are conveyed to the deeper layers of the olfactory bulb by Mi activity patterns. In the deeper bulb, Mi cells interact via their widespread lateral dendrites with another major class of local interneurons known as granule cells (Gr). The olfactory bulb also receives extensive inputs from other brain areas such as piriform cortex and noradrenergic, serotonergic, and cholinergic nuclei. (**b**) *Top panel.* Schematic depiction of an odor-evoked spatial activation pattern on the surface of olfactory bulb. Various methods of neuronal activity mapping, both histological (e.g., 2-deoxyglucose, c-*fos*, *Zif268*) and physiological (e.g., calcium imaging), enable visualization of the odor-specific spatial activity patterns conveyed to olfactory bulb by incoming OSNs. *Bottom panel.* Schematic illustration of odor-evoked field oscillations measured in different physical locations across olfactory bulb. The distribution of oscillatory amplitudes reflects odor quality and concentration. (**c**) Spatial activation patterns measured in the olfactory bulb glomerular layer are predictive of the perceptual similarity of odorants. Adapted from Cleland et al. (2002). (**d**) The relatively concentration-invariant representations of mitral cells are believed to be generated by computations in the OB glomerular layer that normalize incoming activation patterns (Cleland et al. 2007); other glomerular circuits perform contrast enhancement functions to decorrelate similar odor representations (Cleland and Sethupathy 2006). The figures represent odor-evoked spatial activation patterns at two different

on these spatial patterns would alter the relative activation levels of mitral cells and consequently alter the pattern of activity that is conveyed to the piriform cortex and other postbulbar structures. Indeed, the spatial activation pattern across OB mitral cells changes in response to different types of learning and has been shown to depend on the behavioral relevance of the odor stimulation (Coopersmith et al. 1986; Coopersmith and Leon 1986; Faber et al. 1999; Fernandez et al. 2009; Salcedo et al. 2005; Sullivan and Leon 1986; Sullivan et al. 1988). Mechanistically, a number of transformations have been proposed to be performed on these spatial activation patterns in the glomerular layer, including contrast enhancement and concentration invariance functions. These functions rely on interactions between mitral cells and local interneurons and, despite differences in detail, are thought to be substantially similar across species.

Concentration invariance, or normalization, of odor representations is clearly observable in the concentration profiles of mitral cells. That is, while activity in mitral cells is altered by changes in odor concentration, it does not monotonically increase with concentration as does activity in OSNs (Wellis et al. 1989) (Fig. 11.1d); indeed, on average, higher odor concentrations probably evoke slightly lower total activity levels across the mitral cell population. The network mechanism underlying this partial implementation of concentration invariance is not confirmed, though it has been proposed (Cleland et al. 2007) to rely on the deep glomerular networks of external tufted cells, superficial short-axon cells, and periglomerular cells first described by Shipley and colleagues (Aungst et al. 2003), and also likely involves feedback inhibition of OSN presynaptic arbors (McGann et al. 2005; Wachowiak et al. 2002). Notably, behavioral data collected in rats demonstrate that glomerular activation patterns normalized with respect to mean bulbar activity levels are better predictors of odor perception than raw patterns (Fig. 11.1e), as would be predicted if mitral cell activation levels were comparably normalized across the bulbar population (Cleland et al. 2007). A recent experimental study proposed a similar computational scheme in the *Drosophila* antennal lobe, concluding that relative concentration invariance is implemented in this structure as well (Olsen et al. 2010).

Contrast enhancement is a common property of sensory systems that narrows (sharpens) sensory representations by specifically inhibiting neurons on the periphery of the representation, thus enhancing the contrast between signal and

Fig. 11.1 (continued) concentrations; hot colors correspond to higher activation levels. Raw, normalized, and contrast-enhanced patterns are represented. The details of these functions and their underlying neural mechanisms have been previously reviewed (Cleland 2010; Cleland and Linster 2005; Linster and Cleland 2009). (**e**) Normalized activity patterns across OB are better predictors of odorant perceptual similarity than are raw activity patterns. The graph illustrates the pairwise perceptual dissimilarity between two different concentrations of the same odor compared to the dissimilarity between that odor and a second odor presented at the same concentration (*Behavior*), compared to the dissimilarities predicted from calculations of the overlap between their raw (*Raw*) and normalized (*Normalized*) glomerular activation patterns. The important feature is that the slopes of the *Behavior* and *Normalized* plots are both positive, in contrast to the *Raw* plot. Adapted from Cleland et al. (2007)

background (Fig. 11.1d). Contrast enhancement of spatial odor representations in the olfactory bulb is thought to be mediated by inhibitory interneurons both in the glomerular layer (*periglomerular cells*, Cleland and Sethupathy 2006; Linster and Cleland 2004, 2009; Linster and Gervais 1996; Linster and Hasselmo 1997; Linster et al. 2005)) and in the granule cell layer/external plexiform layer (*granule cells*, Arevian et al. 2008; Urban 2002; Urban and Arevian 2009)). A number of computational models have proposed solutions for this important function in the mammalian OB and insect antennal lobe, including lateral interactions between glomeruli (Linster and Gervais 1996; Linster and Hasselmo 1997), computations local to each glomerulus (Cleland and Sethupathy 2006; Cleland 2010 #50), and local and lateral interactions between mitral and granule cells (Urban and Arevian 2009).

Field Oscillations and Temporal Activity Patterns in Olfactory Bulb

In the deeper layers of the olfactory bulb, or among global interneurons in the antennal lobe, the modulation of mitral cell *spike timing and synchronization* rather than the modulation of absolute response magnitude (numbers of action potentials) is thought to be the dominant means by which interneuronal interactions affect the content of odor representations (Fig. 11.1b, *lower panel*). Degrees of synchronization among OB or antennal lobe outputs were proposed to contribute to odor processing and learning by Laurent and colleagues in a long series of studies in locust and honeybee (MacLeod et al. 1998; Stopfer et al. 2003; Stopfer and Laurent 1999; Wehr and Laurent 1999). These studies showed that the patterns of synchronization among principal neuron activation patterns, rather than the gross patterns of all odor-responsive cells, best identified specific odorants (Fig. 11.2a) and that these patterns of synchronization changed as a function of experience (Stopfer et al. 2003). Earlier studies in rabbits also had shown the odor-specificity and sensitivity to learning of dynamical activity patterns in olfactory bulb (Freeman and Schneider 1982; Gray et al. 1986), first demonstrating a potential functional role for bulbar dynamics. More recent studies have shown that olfactory bulb dynamics are modulated by behavioral demands and that behavioral performance in olfactory perceptual tasks is correlated with these dynamics (Beshel et al. 2007; Kay 2003; Kay et al. 2009; Nusser et al. 2001; Rojas-Libano and Kay 2008). For example, Nusser and colleagues showed, using genetically modified mice in which fast OB field oscillations in the gamma range were stronger than in their wild-type littermates, a robust relationship between oscillatory power and odor discrimination performance (Nusser et al. 2001). Data from Ravel and colleagues have shown that oscillations in the beta band are modulated during a behavioral experiment, strongly correlating with the animal's task performance (Martin et al. 2004a, b, 2006; Ravel et al. 2003). In these experiments, strong oscillations in the beta band (15–30 Hz) appeared in the OB field potential while the animal was first learning to discriminate between a rewarded and a non-rewarded odorant; during this same epoch,

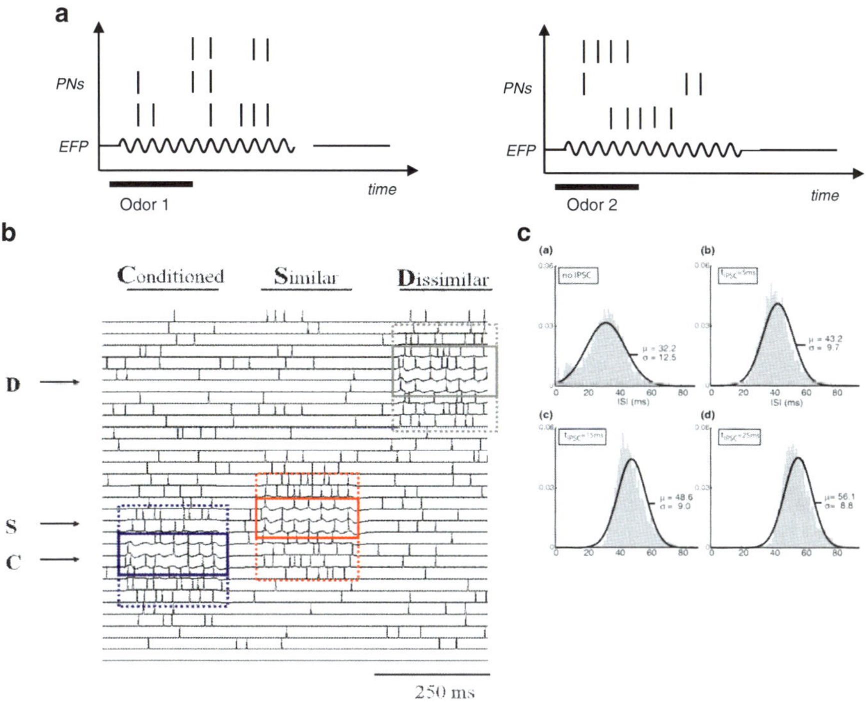

Fig. 11.2 Temporal processing of OB activation patterns. (**A**) In locusts and honeybees, patterns of synchronized spikes, rather than coarse firing rates, have been proposed to represent odor quality (Cleland 2010; Laurent and Davidowitz 1994; Wehr and Laurent 1999). In this schematic depiction, a stimulus-evoked oscillation is present in the field potential recording (*EFP*) while several projection neurons (*PNs*) fire action potentials in response to odorant presentation. While the overall PN firing rate does not enable discrimination of Odor 1 (*left panel*) and Odor 2 (*right panel*), the temporal organization of the action potentials and their synchronization patterns do enable discrimination of the two odors. (**B**) Contrast enhancement using synchronization properties. The neural responses to three odor stimuli C, S, and D are schematically depicted. Stimuli C and S evoke highly overlapping responses when coarse firing rates are used to determine their similarity (enclosed in *dotted boxes*), whereas stimuli C and D evoke very different response patterns under the same conditions. In contrast, if only synchronized action potentials are considered relevant (enclosed in *solid boxes*), the patterns evoked by stimuli C and S become nearly nonoverlapping and hence easily differentiated. While odorant D is affected in the same way, nothing is gained by the consideration of temporal information because the spatial patterns alone were already entirely nonoverlapping. (**C**) Regulation of the temporal precision of action potentials by inhibitory dendritic inputs. (**a**) Distribution of mitral cell interstimulus intervals (ISIs) under baseline conditions in the absence of incoming inhibitory postsynaptic currents (IPSCs) on the lateral dendrites. (**b–d**) Distribution of ISIs when shunting inhibitory currents were opened 5, 15, or 25 ms after a mitral cell spike. Inhibitory inputs on the lateral dendrites increased the temporal precision of mitral spiking. Adapted from David et al. (2009)

oscillations in the faster gamma band (40–80 Hz) were reduced in power. The occurrence of this phenomenon depended strongly on intact ascending projections to the olfactory bulb from other brain areas (Martin et al. 2006).

Evidence that field oscillations and spike synchronization patterns play a role in odor perception had previously been gathered in honeybees in a study showing that bees in which oscillatory dynamics and synchronization patterns were pharmacologically impaired were more prone to confuse chemically similar odorants (Stopfer et al. 1997). This study drew a lot of attention to the importance of synchronization for odor representations but said little regarding the role of spatial activation patterns. Nevertheless, by the late 1990s, the role of dynamical patterning in the olfactory bulb and antennal lobe had been widely accepted. Many laboratories began working on related questions, notably on the underlying mechanisms by which these field oscillations were generated. Presently, OB field oscillations are usually ascribed to the feedback loop between principal neurons and inhibitory interneurons (Bathellier et al. 2008; David et al. 2009; Davison et al. 2003; Lagier et al. 2004; Li and Hopfield 1989) or to intrinsic oscillatory properties of principal neurons synchronized by common inhibitory inputs (Brea et al. 2009; Ermentrout et al. 2007; Galan et al. 2006). Interestingly, to date little more has been learned about the function of these dynamical processes beyond their suggested role in further sharpening odor representations so as to improve olfactory discrimination.

Contrast enhancement by synchronization. Given that the regulation of field oscillations and mitral cell spike synchronization by dynamical interactions in the deep bulb affects perception, how might these subtle modifications of neuronal activity be interpreted by downstream structures? Theoretical models have established some mechanisms by which patterns of neuronal synchronization can be regulated by bulbar circuitry to effect arbitrary patterns of contrast enhancement and subsequently interpreted by postbulbar neurons (Cleland and Linster 2002; Linster and Cleland 2001, 2010). Interestingly, such models suggest that only spikes that are relatively synchronized with others are read out by downstream neurons, with asynchronously firing neurons effectively becoming excluded from the odor representation. This could be an important coding principle for systems in which principal neurons exhibit substantial input-independent baseline activity, as mitral cells do. Among synchronized neurons, the degree of contrast can be manipulated by changing synchronization properties (Fig. 11.2b), e.g., by changes in neuromodulatory input activity mediating attentive processes or changes in stimulus salience. Computational models using this approach have been able to explain behavioral results demonstrating that changes in synchronization properties correspond to changes in the perceptual discrimination of odors (Cleland and Linster 2002).

Signal-to-noise ratio. Muscarinic cholinergic neuromodulation, the receptors for which are expressed in the deeper layers of the OB, enhances response precision in granule and mitral cells in OB slices (Pressler et al. 2007). Simulations of mitral–granule cell interactions, in conjunction with experimental data, show that inhibitory inputs along the secondary dendrites affect spike timing in mitral cells and enhance the temporal precision of spikes occurring in response to odor stimuli (David et al. 2009) (Fig. 11.2c). While more thorough study is necessary to explore the implications, these data in conjunction suggest a role in signal-to-noise modulation for the deeper layers of bulbar processing.

Spatiotemporal Activity Patterns and Odor Perception

As reviewed above, the evidence to date clearly demonstrates that both spatial and temporal activation patterns reflect odor identity, predict perceptual qualities to a certain degree, and are modified by learning. Combined experiments in honeyebees (Stopfer et al. 1997) and rats (Beshel et al. 2007) have begun to explain their relationship to one another. In the honeybee antennal lobe, odor stimulation evokes stimulus-locked oscillations in the 15–30 Hz frequency range that are accompanied by synchronization of action potentials among output neurons (Stopfer et al. 1997) (Fig. 11.2a). Blocking fast GABAergic transmission in the antennal lobe abolished the stimulus-evoked field oscillations without evoking clearly observable changes in the odor response properties of output neurons (MacLeod and Laurent 1996). According to the Laurent group, only the synchronization properties of these neurons changed, and not their individual responses to odors. In a parallel honeybee behavioral experiment, blockade of GABAergic transmission was shown (1) to have no effect on the acquisition of an odor-reward association, (2) to have no effect on the discrimination of a chemically dissimilar odorant from the conditioned odorant, but (3) to impair the discrimination of chemically and perceptually *similar* odorants from the conditioned odorant (Stopfer et al. 1997). Subsequent calcium imaging experiments established that these chemically similar odorants evoked highly overlapping spatial patterns in the antennal lobe (Sachse and Galizia 2002). It is clear from these data that spike synchronization in olfaction becomes functionally important specifically when structurally similar odorants must be discriminated, since the perceptual discrimination of dissimilar odorants was not affected by the impairment of synchronization (Fig. 11.3).

In related work in rats, Kay and colleagues (Beshel et al. 2007) have shown that oscillatory synchrony in the olfactory bulb is systematically affected by the difficulty of an odor discrimination task. Specifically, when discriminating between highly similar odorants in a forced-choice task, the power of OB gamma oscillations was significantly increased in comparison to the oscillatory power recorded when the same rats were discriminating dissimilar odorants. These results strongly suggest that oscillatory dynamics are functionally utilized during odor discrimination in proportion to task difficulty and that behavioral demands can regulate oscillatory dynamics (Fig. 11.3). As in the honeybee experiments described above, prior knowledge and understanding of the role of odor-specific spatial activity patterns was crucial to the success of these experiments.

Together, these two data sets demonstrate that temporal dynamics and spatial activation patterns both play important roles in odor perception. Specifically, temporal properties appear to serve a secondary role, modulating and fine-tuning the basic spatial activation patterns evoked by odor stimuli in response to behavioral demands and neuromodulatory state. While much remains to be studied, the integration to date of these sophisticated mechanisms of perception has helped support

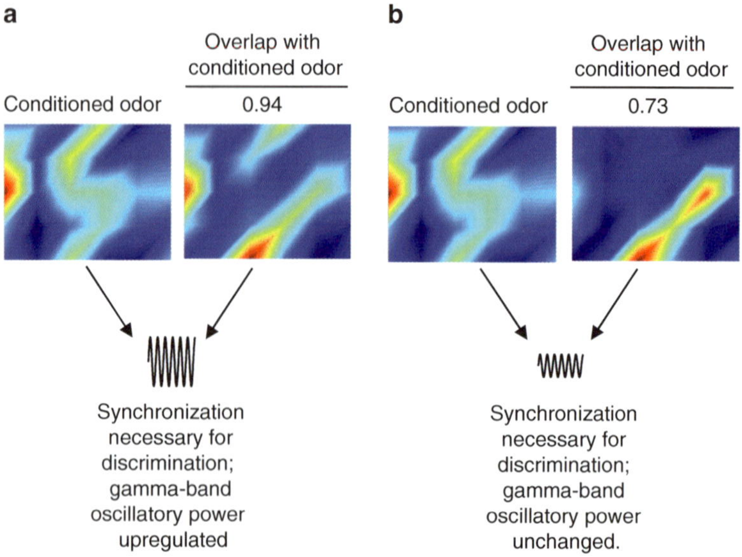

Fig. 11.3 Relationship between temporal and spatial representations of odorants. Experiments in honeybees showed that pharmacological manipulations which led to a reduction in synchrony among AL neurons also led to a reduction in perceptual odor discrimination (Stopfer et al. 1997). Specifically, this reduction in odor discrimination could only be observed in odorants that activated highly overlapping spatial representations, as schematically depicted in these color-coded images (courtesy of G. Galizia). During olfactory discrimination tasks in rats, oscillatory power in the gamma range increased when more highly similar, overlapping odorants were presented (i.e., when task difficulty was increased) (Beshel et al. 2007). Numbers above the *right panels* indicate the proportion of spatial pattern overlap with the conditioned odor (*left panels*)

a substantial revival in computational olfaction over the last 2 decades, facilitating increasingly comprehensive analyses of both spatial and temporal processing capabilities.

References

Adrian ED (1942) Olfactory reactions in the brain of the hedgehog. J Physiol 100:459–473

Adrian ED (1950) The electrical activity of the mammalian olfactory bulb. Electroencephalogr Clin Neurophysiol 2:377–388

Adrian ED (1957) Electrical oscillations recorded from the olfactory organ. J Physiol 136:29P–30P

Arevian AC, Kapoor V, Urban NN (2008) Activity-dependent gating of lateral inhibition in the mouse olfactory bulb. Nat Neurosci 11:80–87

Aungst JL, Heyward PM, Puche AC, Karnup SV, Hayar A, Szabo G, Shipley MT (2003) Centre-surround inhibition among olfactory bulb glomeruli. Nature 426:623–629

Bathellier B, Carleton A, Gerstner W (2008) Gamma oscillations in a nonlinear regime: a minimal model approach using heterogeneous integrate-and-fire networks. Neural Comput 20:2973–3002

Beshel J, Kopell N, Kay LM (2007) Olfactory bulb gamma oscillations are enhanced with task demands. J Neurosci 27:8358–8365
Brea JN, Kay LM, Kopell NJ (2009) Biophysical model for gamma rhythms in the olfactory bulb via subthreshold oscillations. Proc Natl Acad Sci USA 106:21954–21959
Carlsson MA, Galizia CG, Hansson BS (2002) Spatial representation of odours in the antennal lobe of the moth Spodoptera littoralis (Lepidoptera: Noctuidae). Chem Senses 27:231–244
Cleland TA (2010) Early transformations in odor representation. Trends Neurosci 33:130–139
Cleland TA, Linster C (2002) How synchronization properties among second-order sensory neurons can mediate stimulus salience. Behav Neurosci 116:212–221
Cleland TA, Linster C (2005) Computation in the olfactory system. Chem Senses 30:801–813
Cleland TA, Sethupathy P (2006) Non-topographical contrast enhancement in the olfactory bulb. BMC Neurosci 7:7
Cleland TA, Morse A, Yue EL, Linster C (2002) Behavioral models of odor similarity. Behav Neurosci 116:222–231
Cleland TA, Johnson BA, Leon M, Linster C (2007) Relational representation in the olfactory system. Proc Natl Acad Sci USA 104:1953–1958
Coopersmith R, Leon M (1986) Enhanced neural response by adult rats to odors experienced early in life. Brain Res 371:400–403
Coopersmith R, Lee S, Leon M (1986) Olfactory bulb responses after odor aversion learning by young rats. Brain Res 389:271–277
David FO, Hugues E, Cenier T, Fourcaud-Trocme N, Buonviso N (2009) Specific entrainment of mitral cells during gamma oscillation in the rat olfactory bulb. PLoS Comput Biol 5:e1000551
Davison AP, Feng J, Brown D (2003) Dendrodendritic inhibition and simulated odor responses in a detailed olfactory bulb network model. J Neurophysiol 90:1921–1935
Delaney KR, Gelperin A, Fee MS, Flores JA, Gervais R, Tank DW, Kleinfeld D (1994) Waves and stimulus-modulated dynamics in an oscillating olfactory network. Proc Natl Acad Sci USA 91:669–673
Di Prisco GV, Freeman WJ (1985) Odor-related bulbar EEG spatial pattern analysis during appetitive conditioning in rabbits. Behav Neurosci 99:964–978
Engel AK, Konig P, Gray CM, Singer W (1990) Stimulus-dependent neuronal oscillations in cat visual cortex: inter-columnar interaction as determined by cross-correlation analysis. Eur J Neurosci 2:588–606
Ermentrout GB, Galan RF, Urban NN (2007) Relating neural dynamics to neural coding. Phys Rev Lett 99:248103
Faber T, Joerges J, Menzel R (1999) Associative learning modifies neural representations of odors in the insect brain. Nat Neurosci 2:74–78
Fernandez PC, Locatelli FF, Person-Rennell N, Deleo G, Smith BH (2009) Associative conditioning tunes transient dynamics of early olfactory processing. J Neurosci 29:10191–10202
Freeman WJ (1978) Spatial properties of an EEG event in the olfactory bulb and cortex. Electroencephalogr Clin Neurophysiol 44:586–605
Freeman WJ (1979a) EEG analysis gives model of neuronal template-matching mechanism for sensory search with olfactory bulb. Biol Cybern 35:221–234
Freeman WJ (1979b) Nonlinear dynamics of paleocortex manifested in the olfactory EEG. Biol Cybern 35:21–37
Freeman WJ, Grajski KA (1987) Relation of olfactory EEG to behavior: factor analysis. Behav Neurosci 101:766–777
Freeman WJ, Schneider W (1982) Changes in spatial patterns of rabbit olfactory EEG with conditioning to odors. Psychophysiology 19:44–56
Freeman WJ, Skarda CA (1985) Spatial EEG patterns, non-linear dynamics and perception: the neo-Sherringtonian view. Brain Res 357:147–175
Galan RF, Fourcaud-Trocme N, Ermentrout GB, Urban NN (2006) Correlation-induced synchronization of oscillations in olfactory bulb neurons. J Neurosci 26:3646–3655
Galizia CG, Menzel R (2000) Odour perception in honeybees: coding information in glomerular patterns. Curr Opin Neurobiol 10:504–510

Gelperin A, Tank DW (1990) Odour-modulated collective network oscillations of olfactory interneurons in a terrestrial mollusc. Nature 345:437–440
Gelperin A, Rhines LD, Flores J, Tank DW (1993) Coherent network oscillations by olfactory interneurons: modulation by endogenous amines. J Neurophysiol 69:1930–1939
Grajski KA, Freeman WJ (1989) Spatial EEG correlates of nonassociative and associative olfactory learning in rabbits. Behav Neurosci 103:790–804
Gray CM, Freeman WJ, Skinner JE (1986) Chemical dependencies of learning in the rabbit olfactory bulb: acquisition of the transient spatial pattern change depends on norepinephrine. Behav Neurosci 100:585–596
Gray CM, Freeman WJ, Skinner JE (1987) Induction and maintenance of epileptiform activity in the rabbit olfactory bulb depends on centrifugal input. Exp Brain Res 68:210–212
Gray CM, Engel AK, Konig P, Singer W (1990) Stimulus-dependent neuronal oscillations in cat visual cortex: receptive field properties and feature dependence. Eur J Neurosci 2:607–619
Gray CM, Engel AK, Konig P, Singer W (1992) Synchronization of oscillatory neuronal responses in cat striate cortex: temporal properties. Vis Neurosci 8:337–347
Guerrieri F, Schubert M, Sandoz JC, Giurfa M (2005) Perceptual and neural olfactory similarity in honeybees. PLoS Biol 3:e60
Johnson BA, Leon M (2007) Chemotopic odorant coding in a mammalian olfactory system. J Comp Neurol 503:1–34
Kauer JS (1988) Real-time imaging of evoked activity in local circuits of the salamander olfactory bulb. Nature 331:166–168
Kauer JS, Senseman DM, Cohen LB (1987) Odor-elicited activity monitored simultaneously from 124 regions of the salamander olfactory bulb using a voltage-sensitive dye. Brain Res 418:255–261
Kay LM (2003) Two species of gamma oscillations in the olfactory bulb: dependence on behavioral state and synaptic interactions. J Integr Neurosci 2:31–44
Kay LM, Beshel J, Brea J, Martin C, Rojas-Libano D, Kopell N (2009) Olfactory oscillations: the what, how and what for. Trends Neurosci 32:207–214
Kleinfeld D, Delaney KR, Fee MS, Flores JA, Tank DW, Gelperin A (1994) Dynamics of propagating waves in the olfactory network of a terrestrial mollusk: an electrical and optical study. J Neurophysiol 72:1402–1419
Lagier S, Carleton A, Lledo PM (2004) Interplay between local GABAergic interneurons and relay neurons generates gamma oscillations in the rat olfactory bulb. J Neurosci 24:4382–4392
Lancet D, Greer CA, Kauer JS, Shepherd GM (1982) Mapping of odor-related neuronal activity in the olfactory bulb by high-resolution 2-deoxyglucose autoradiography. Proc Natl Acad Sci USA 79:670–674
Laska M, Galizia CG (2001) Enantioselectivity of odor perception in honeybees (Apis mellifera carnica). Behav Neurosci 115:632–639
Laurent G (1996a) Dynamical representation of odors by oscillating and evolving neural assemblies. Trends Neurosci 19:489–496
Laurent G (1996b) Odor images and tunes. Neuron 16:473–476
Laurent G, Davidowitz H (1994) Encoding of olfactory information with oscillating neural assemblies. Science 265:1872–1875
Laurent G, Naraghi M (1994) Odorant-induced oscillations in the mushroom bodies of the locust. J Neurosci 14:2993–3004
Laurent G, Wehr M, Davidowitz H (1996) Temporal representations of odors in an olfactory network. J Neurosci 16:3837–3847
Leon M, Johnson BA (2003) Olfactory coding in the mammalian olfactory bulb. Brain Res Brain Res Rev 42:23–32
Leon M, Johnson B (2006) Functional units in the olfactory system. Proc Natl Acad Sci USA 103:14985–14986
Li Z, Hopfield JJ (1989) Modeling the olfactory bulb and its neural oscillatory processings. Biol Cybern 61:379–392

Linster C, Cleland TA (2001) How spike synchronization among olfactory neurons can contribute to sensory discrimination. J Comput Neurosci 10:187–193
Linster C, Cleland TA (2004) Configurational and elemental odor mixture perception can arise from local inhibition. J Comput Neurosci 16:39–47
Linster C, Cleland TA (2009) Glomerular microcircuits in the olfactory bulb. Neural Netw 22:1169–1173
Linster C, Cleland TA (2010) Decorrelation of odor representations via spike timing-dependent plasticity. Front Comput Neurosci 4:157
Linster C, Gervais R (1996) Investigation of the role of interneurons and their modulation by centrifugal fibers in a neural model of the olfactory bulb. J Comput Neurosci 3:225–246
Linster C, Hasselmo M (1997) Modulation of inhibition in a model of olfactory bulb reduces overlap in the neural representation of olfactory stimuli. Behav Brain Res 84:117–127
Linster C, Johnson BA, Yue E, Morse A, Xu Z, Hingco EE, Choi Y, Choi M, Messiha A, Leon M (2001) Perceptual correlates of neural representations evoked by odorant enantiomers. J Neurosci 21:9837–9843
Linster C, Sachse S, Galizia CG (2005) Computational modeling suggests that response properties rather than spatial position determine connectivity between olfactory glomeruli. J Neurophysiol 93:3410–3417
MacLeod K, Laurent G (1996) Distinct mechanisms for synchronization and temporal patterning of odor-encoding neural assemblies. Science 274:976–979
MacLeod K, Backer A, Laurent G (1998) Who reads temporal information contained across synchronized and oscillatory spike trains? Nature 395:693–698
Martin C, Gervais R, Chabaud P, Messaoudi B, Ravel N (2004a) Learning-induced modulation of oscillatory activities in the mammalian olfactory system: the role of the centrifugal fibres. J Physiol Paris 98:467–478
Martin C, Gervais R, Hugues E, Messaoudi B, Ravel N (2004b) Learning modulation of odor-induced oscillatory responses in the rat olfactory bulb: a correlate of odor recognition? J Neurosci 24:389–397
Martin C, Gervais R, Messaoudi B, Ravel N (2006) Learning-induced oscillatory activities correlated to odour recognition: a network activity. Eur J Neurosci 23:1801–1810
McGann JP, Pirez N, Gainey MA, Muratore C, Elias AS, Wachowiak M (2005) Odorant representations are modulated by intra- but not interglomerular presynaptic inhibition of olfactory sensory neurons. Neuron 48:1039–1053
Nusser Z, Kay LM, Laurent G, Homanics GE, Mody I (2001) Disruption of GABA(A) receptors on GABAergic interneurons leads to increased oscillatory power in the olfactory bulb network. J Neurophysiol 86:2823–2833
Olsen SR, Bhandawat V, Wilson RI (2010) Divisive normalization in olfactory population codes. Neuron 66:287–299
Pressler RT, Inoue T, Strowbridge BW (2007) Muscarinic receptor activation modulates granule cell excitability and potentiates inhibition onto mitral cells in the rat olfactory bulb. J Neurosci 27:10969–10981
Ravel N, Chabaud P, Martin C, Gaveau V, Hugues E, Tallon-Baudry C, Bertrand O, Gervais R (2003) Olfactory learning modifies the expression of odour-induced oscillatory responses in the gamma (60–90 Hz) and beta (15–40 Hz) bands in the rat olfactory bulb. Eur J Neurosci 17:350–358
Rojas-Libano D, Kay LM (2008) Olfactory system gamma oscillations: the physiological dissection of a cognitive neural system. Cogn Neurodyn 2:179–194
Rubin BD, Katz LC (2001) Spatial coding of enantiomers in the rat olfactory bulb. Nat Neurosci 4:355–356
Sachse S, Galizia CG (2002) Role of inhibition for temporal and spatial odor representation in olfactory output neurons: a calcium imaging study. J Neurophysiol 87:1106–1117
Salcedo E, Zhang C, Kronberg E, Restrepo D (2005) Analysis of training-induced changes in ethyl acetate odor maps using a new computational tool to map the glomerular layer of the olfactory bulb. Chem Senses 30:615–626

Stewart WB, Kauer JS, Shepherd GM (1979) Functional organization of rat olfactory bulb analysed by the 2-deoxyglucose method. J Comp Neurol 185:715–734
Stopfer M, Laurent G (1999) Short-term memory in olfactory network dynamics. Nature 402:664–668
Stopfer M, Bhagavan S, Smith BH, Laurent G (1997) Impaired odour discrimination on desynchronization of odour-encoding neural assemblies. Nature 390:70–74
Stopfer M, Jayaraman V, Laurent G (2003) Intensity versus identity coding in an olfactory system. Neuron 39:991–1004
Sullivan RM, Leon M (1986) Early olfactory learning induces an enhanced olfactory bulb response in young rats. Brain Res 392:278–282
Sullivan RM, Wilson DA, Kim MH, Leon M (1988) Behavioral and neural correlates of postnatal olfactory conditioning: I. Effect of respiration on conditioned neural responses. Physiol Behav 44:85–90
Urban NN (2002) Lateral inhibition in the olfactory bulb and in olfaction. Physiol Behav 77:607–612
Urban NN, Arevian AC (2009) Computing with dendrodendritic synapses in the olfactory bulb. Ann N Y Acad Sci 1170:264–269
Wachowiak M, Cohen LB, Ache BW (2002) Presynaptic inhibition of olfactory receptor neurons in crustaceans. Microsc Res Tech 58:365–375
Wehr M, Laurent G (1999) Relationship between afferent and central temporal patterns in the locust olfactory system. J Neurosci 19:381–390
Wellis DP, Scott JW, Harrison TA (1989) Discrimination among odorants by single neurons of the rat olfactory bulb. J Neurophysiol 61:1161–1177
Youngentob SL, Johnson BA, Leon M, Sheehe PR, Kent PF (2006) Predicting odorant quality perceptions from multidimensional scaling of olfactory bulb glomerular activity patterns. Behav Neurosci 120:1337–1345

Chapter 12
20 Years of Learning About Vision: Questions Answered, Questions Unanswered, and Questions Not Yet Asked

Bruno A. Olshausen

Abstract I have been asked to review the progress that computational neuroscience has made over the past 20 years in understanding how vision works. In reflecting on this question, I come to the conclusion that perhaps the most important advance we have made is in gaining a deeper appreciation of the magnitude of the problem before us. While there has been steady progress in our understanding—and I will review some highlights here—we are still confronted with profound mysteries about how visual systems work. These are not just mysteries about biology, but also about the *general principles* that enable vision in any system whether it be biological or machine. I devote much of this chapter to examining these open questions, as they are crucial in guiding and motivating current efforts. Finally, I shall argue that the biggest mysteries are likely to be ones we are not currently aware of, and that bearing this in mind is important as it encourages a more exploratory, as opposed to strictly hypothesis-driven, approach.

Introduction

I am both honored and delighted to speak at this symposium. The CNS meetings were pivotal to my own coming of age as a scientist in the early 1990s, and today they continue to constitute an important part of my scientific community. Now that 20 years have passed since the first meeting, we are here today to ask, what have we

---

B.A. Olshausen (✉)
Helen Wills Neuroscience Institute and School of Optometry, University of California—Berkeley, Berkeley, CA, USA

Redwood Center for Theoretical Neuroscience, UC Berkeley, 575A Evans Hall, MC 3198, Berkeley, CA 94720-3198, USA
e-mail: baolshausen@berkeley.edu

J.M. Bower (ed.), *20 Years of Computational Neuroscience*, Springer Series in Computational Neuroscience 9, DOI 10.1007/978-1-4614-1424-7_12,

learned? I have been tasked with addressing the topic of vision, which is of course a huge field, and so before answering I should disclose my own biases and the particular lens through which I view things: I began as an engineer wanting to build robotic vision systems inspired by biology, and I evolved into a neuroscientist trying to understand how brains work inspired by principles from mathematics and engineering. Along the way, I was fortunate to have worked and trained with some of the most creative and pioneering scientists of our field: Pentti Kanerva, David Van Essen, Charlie Anderson, Mike Lewicki, David Field, and Charlie Gray. Their own way of thinking about computation and the brain has shaped much of my own outlook, and the opinions expressed below stem in large part from their influence. I also benefited enormously from my fellow students in the Computation and Neural Systems program at Caltech in the early 1990s and the interdisciplinary culture that flourished there. They impressed upon me that the principles of vision are not owned by biology, nor by engineering—they are universals that transcend discipline, and they will be discovered by thinking outside the box.

Now to begin our journey into the past 20 years, let us first gain some perspective by looking back nearly half a century, to a time when it was thought that vision would be a fairly straightforward problem. In 1966, the MIT AI Lab assigned their summer students the task of building an artificial vision system (Papert 1966). This effort came on the heels of some early successes in artificial intelligence in which it was shown that computers could solve simple puzzles and prove elementary theorems. There was a sense of optimism among AI researchers at the time that they were conquering the foundations of intelligence (Dreyfus and Dreyfus 1988). Vision it seemed would be a matter of feeding the output of a camera to the computer, extracting edges, and performing a series of logical operations. They were soon to realize however that the problem is orders of magnitude more difficult. David Marr summarized the situation as follows:

> …in the 1960s almost no one realized that machine vision was difficult. The field had to go through the same experience as the machine translation field did in its fiascoes of the 1950s before it was at last realized that here were some problems that had to be taken seriously. … the idea that extracting edges and lines from images might be at all difficult simply did not occur to those who had not tried to do it. It turned out to be an elusive problem. Edges that are of critical importance from a three-dimensional point of view often cannot be found at all by looking at the intensity changes in an image. Any kind of textured image gives a multitude of noisy edge segments; variations in reflectance and illumination cause no end of trouble; and even if an edge has a clear existence at one point, it is as likely as not to fade out quite soon, appearing only in patches along its length in the image. The common and almost despairing feeling of the early investigators like B.K.P. Horn and T.O. Binford was that practically anything could happen in an image and furthermore that practically everything did. (Marr 1982)

The important lesson from these early efforts is that it was from *trying to solve the problem* that these early researchers learned what were the difficult computational problems of vision, and thus what were the important questions to ask. This is still true today: Reasoning from first principles and introspection, while immensely valuable, can only go so far in forming hypotheses that guide our study of the visual system. *We will learn what questions to ask by trying to solve the problems of vision.* Indeed,

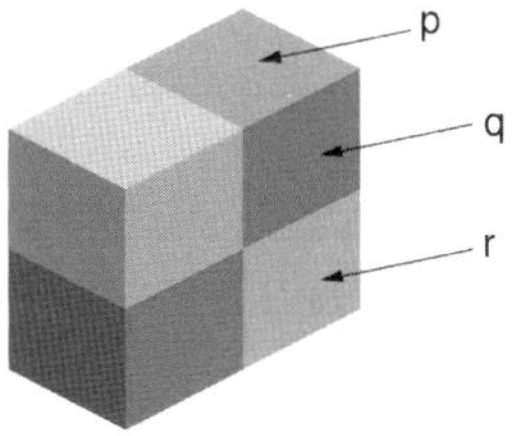

Fig. 12.1 Image of a block painted in two shades of *gray* (from Adelson 2000). The edges in this image are easy to extract, but understanding what they mean is far more difficult

this is one of the most important contributions that computational neuroscience can make to the study of vision.

A decade after the AI Lab effort, David Marr began asking very basic questions about information processing in the visual system that had not yet been asked. He sought to develop a computational theory of biological vision, and he stressed the importance of *representation* and the different types of information that need to be extracted from images. Marr envisioned the problem being broken up into a series of processing stages: a primal sketch in which features and tokens are extracted from the image, a 2.5D sketch that begins to make explicit aspects of depth and surface structure, and finally an object-centered, 3D model representation of objects (Marr 1982). He attempted to specify the types of computations involved in each of these steps as well as their neural implementations.

One issue that appears to have escaped Marr at the time is the importance of *inferential computations* in perception. Marr's framework centered around a mostly feedforward chain of processing in which features are extracted from the image and progressively built up into representations of objects through a logical chain of computations in which information flows from one stage to the next. After decades of research following Marr's early proposals, it is now widely recognized (though still not universally agreed upon) by those in the computational vision community that the features of the *world* (not images) that we care about can almost never be computed in a purely bottom-up manner. Rather, they require inferential computation in which data is combined with prior knowledge in order to estimate the underlying causes of a scene (Mumford 1994; Knill and Richards 1996; Rao et al. 2002; Kersten et al. 2004). This is due to the fact that natural images are full of ambiguity. The causal properties of images—illumination, surface geometry, reflectance (material properties), and so forth—are entangled in complex relationships among pixel values. In order to tease these apart, aspects of scene structure must be estimated simultaneously, and the inference of one variable affects the other. This area of research is still in its infancy and models for solving these types of problems are just beginning to emerge (Tappen et al. 2005; Barron and Malik 2012; Cadieu and Olshausen 2012). As they do, they prompt us to ask new questions about how visual systems work.

To give a concrete example, consider the simple image of a block painted in two shades of gray, as shown in Fig. 12.1 (Adelson 2000). The edges in this image are easy to extract, but understanding what they mean is far more difficult. Note that there are three different types of edges: (1) those due to a change in reflectance (the boundary between q and r), (2) those due to a change in object shape (the boundary between p and q), and (3) those due to the boundary between the object and background. Obviously it is impossible for any computation based on purely local image analysis to tell these edges apart. It is the context that informs us what these different edges mean, but how exactly? More importantly, *how are these different edges represented in the visual system and at what stage of processing do they become distinct?*

As one begins asking these questions, an even more troubling question arises: How can we not have the answers after a half century of intensive investigation of the visual system? By now there are literally mounds of papers examining how neurons in the retina, LGN, and V1 respond to test stimuli such as isolated spots, white noise patterns, gratings, and gratings surrounded by other gratings. We know much—perhaps too much—about the orientation tuning of V1 neurons. Yet we remain ignorant of how this very basic and fundamental aspect of scene structure is represented in the system. The reason for our ignorance is not that many have looked and the answer proved to be too elusive. Surprisingly, upon examining the literature one finds that, other than a handful of studies (Rossi et al. 1996; Lee et al. 2002; Boyaci et al. 2007), no one has bothered to ask the question.

Vision, though a seemingly simple act, presents us with profound computational problems. Even stating what these problems are has proven to be a challenge. One might hope that we could gain insight from studying biological vision systems, but this approach is plagued with its own problems: Nervous systems are composed of many tiny, interacting devices that are difficult to penetrate. The closer one looks, the more complexity one is confronted with. The solutions nature has devised will not reveal themselves easily, but as we shall see the situation is not hopeless.

Here I begin by reviewing some of the areas where our field has made remarkable progress over the past 20 years. I then turn to the open problems that lie ahead, where I believe we have the most to learn over the next several decades. Undoubtedly though there are other problems lurking that we are not even aware of, questions that have not yet been asked. I conclude by asking how we can best increase our awareness of these questions, as these will drive the future paths of investigation.

Questions Answered

Since few questions in biology can be answered with certainty, I cannot truly claim that we have fully answered any of the questions below. Nevertheless these are areas where our field has made concrete progress over the past 20 years, both in terms of theory and in terms of empirical findings that have changed the theoretical landscape.

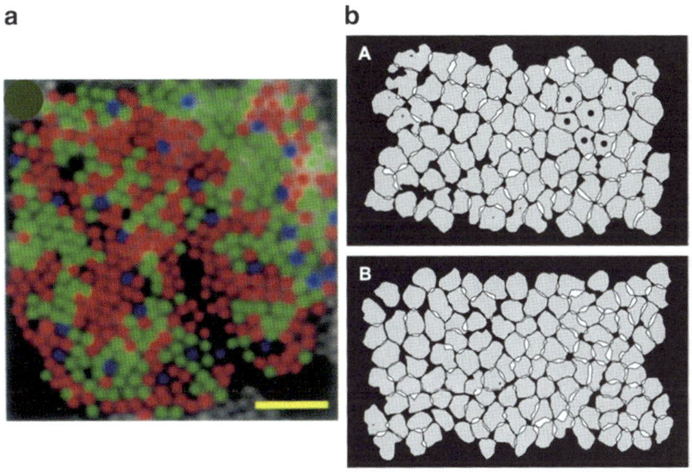

Fig. 12.2 Tiling in the retina. (**a**) Tiling of L, M, S cones; scale bar = 5 arcmin (from Roorda and Williams 1999). (**b**) Tiling of parasol retinal ganglion cell receptive fields; *A*, on cells; *B*, off cells (from Gauthier et al. 2009a, b)

Tiling in the Retina

A long-standing challenge facing computational neuroscience, especially at the systems level, is that the data one is constrained to work with are often sparse or incomplete. Recordings from one or a few units out of a population of thousands of interconnected neurons, while suggestive, cannot help but leave one unsatisfied when attempting to test or form hypotheses about what the system is doing as a whole. In recent years, however, a number of advances have made it possible to break through this barrier in the retina.

The retina contains an array of photoreceptors of different types, and the output of the retina is conveyed by an array of ganglion cells which come in even more varieties. How these different cell types tile the retina—that is, how a complete population of cells of each type cover the two-dimensional image through the spatial arrangement of their receptive fields—has until recently evaded direct observation. As the result of advances in adaptive optics and multielectrode recording arrays, we now have a more complete and detailed picture of tiling in the retina which illuminates our understanding of the first steps in visual processing.

Adaptive optics corrects for optical aberrations of the eye by measuring and compensating for wavefront distortions (Roorda 2011). With this technology, it is now possible to resolve individual cones within the living human eye, producing breathtakingly detailed pictures of how L, M, and S cones tile the retina (Fig. 12.2a) (Roorda and Williams 1999). Surprisingly, L and M cones appear to be spatially clustered beyond what one would expect from a strictly stochastic positioning according to density (Hofer et al. 2005). New insights into the mechanism of color perception have been obtained by stimulating individual cones and looking at how subjects report the corresponding color (Hofer and Williams 2005). Through computational modeling studies, one can show that an individual cone's response is interpreted according to a Bayesian estimator that is attempting to infer the actual color present in the scene in the face of subsampling by the cone mosaic, not simply the cone's "best color" (Brainard et al. 2008). It is also possible to map out receptive fields of LGN neurons cone by cone, providing a more direct picture of how these neurons integrate across space and wavelength (Sincich et al. 2009).

Another important question that can be addressed with adaptive optics is the effect of fixational drifts and microsaccades on perception. It is now possible to track movements of the retina in real-time with single-cone precision, allowing one to completely stabilize retinal images or even introduce artificially generated drifts (Vogel et al. 2006; Arathorn et al. 2007). These studies strongly suggest the presence of internal mechanisms that compensate for drifts during fixation to produce stable percepts (Austin Roorda, personal communication).

At the level of retinal ganglion cells, large-scale neural recording arrays have enabled the simultaneous mapping of receptive fields over an entire local population (Litke et al. 2004). These studies reveal a beautifully ordered arrangement not only in how receptive fields are positioned but also in how they are shaped so as to obtain optimal coverage of the image for each of the four major cell types (i.e., each of the different combinations of on/off and midget/parasol) (Gauthier et al. 2009a, b). Although the position of receptive fields can be somewhat irregular, the shape of each receptive field is morphed so as to fill any gaps in coverage, as shown in Fig. 12.2b. Remarkably, despite the irregular spacing, the receptive field overlap with nearest neighbors is fairly constant, which is a further testament to the degree of precision that is present in retinal image encoding.

Together, these developments provide a solid picture of retinal organization and resolve questions regarding the completeness of coverage that were unresolved just a decade ago. Importantly, these developments also open a new door in allowing us to ask more detailed questions about the link between neural mechanisms and perception.

The Relation Between Natural Image Statistics and Neural Coding

Twenty years ago, most people (myself included) thought of neurons at early stages of the visual system in terms of feature detection. For example, Marr had proposed that retinal ganglion cells function as edge detectors by computing zero crossings of the Laplacian operator (which indicates extrema in the first derivative) and this became a fairly popular idea. Similarly, the oriented receptive fields of V1 neurons were thought to operate as oriented edge detectors that encode the boundaries or geometric shape of objects. However, in the early 1990s it became clear there is another way to think about what these neurons are doing in terms of *efficient coding principles*. Here the goal is to consider how information about the image can be encoded and represented in a complete manner that is adapted to the input statistics. In contrast to detection, which is typically a lossy process designed for a specific purpose, the goal of efficient coding is to form a generic representation that could be used for myriad tasks, but which nevertheless exploits and makes explicit the structure contained in natural images.

Although the efficient coding hypothesis was first proposed by Barlow more than 50 years ago (Barlow 1961), it was not until decades later that investigators such as Laughlin and Srinivasan began making serious quantitative connections between the statistics of natural scenes and neural coding (Srinivasan et al. 1982). David Field subsequently showed that the power spectrum of natural images follows a characteristic $1/f^2$ power law, and he pointed out how the scale-invariant structure of cortical receptive fields is well matched to encode this structure (Field 1987). Atick and Redlich formulated the whitening theory of retinal coding, which proposed that the purpose of the circularly symmetric, center-surround receptive fields of retinal ganglion cells is not to detect edges as Marr claimed, but rather to remove redundancies in natural images so as to make maximal use of channel capacity in the optic nerve (Atick and Redlich 1992). Subsequent neurophysiological experiments in the LGN seemed to support this assertion (Dan et al. 1996). Around the same time, David Field and I showed through computer simulation that the localized, oriented, and multiscale receptive fields of V1 neurons could be accounted for in terms of a sparse coding strategy adapted to natural images (Olshausen and Field 1996). These theories and findings have drawn considerable interest because they offer an intimate, quantitative link between theories of neural coding and experimental data. Moreover it is not just a theory of vision, but a general theory of sensory coding that could be applied to other modalities or subsequent levels of representation, and indeed there has been much work investigating these directions (Geisler et al. 2001; Hyvarinen and Hoyer 2001; Schwartz and Simoncelli 2001; Karklin and Lewicki 2003, 2005, 2009; Hyvarinen et al. 2005; Smith and Lewicki 2006).

A related theoretical framework that has been used to make connections between natural scene statistics and neural representation is that of *Bayesian inference*. Here the goal is to go beyond coding to consider how the properties of scenes are inferred from image data. As mentioned above, making inferences about the world depends upon strong prior knowledge. Often this knowledge is probabilistic in nature. For example, in the simple scene of Fig. 12.1, we could choose to interpret it either as a flat scene created entirely by paint (which it is), as a scene created entirely by structured light, or as a three-dimensional object in two shades of paint (Adelson 2000). All three are valid interpretations when judged purely in terms of the image data. Our visual system chooses the latter interpretation because it is the most parsimonious or *probable* interpretation that is consistent not only with the data but also with our experience in interacting with the world. A goal of many modeling efforts over the past 20 years has been to show how probabilistic information about the world can be learned from visual experience and how inferential computations can be performed in neural systems (Dayan et al. 1995; Rao et al. 2002; Ma et al. 2006). Some of these models make predictions about higher level visual representations beyond V1, in addition to providing a possible account for the role of feedback connections from higher areas to lower areas (Lee and Mumford 2003; Karklin and Lewicki 2005; Cadieu and Olshausen 2012). An important property of these models is the manner in which different hypotheses compete to explain the data—termed "explaining away" (Pearl 1988)—which provides an account for the nonlinear, suppressive effects of context upon the responses of visual neurons (Vinje and Gallant 2000; Murray et al. 2002; Zhu and Rozell 2011).

The Nature of Intermediate-Level Vision

For many years intermediate-level vision was the *terra incognita* of our field. It is the murkiest territory because unlike low-level vision its neural substrates cannot be directly identified or characterized, and unlike high-level phenomena such as object recognition and attention we have no well-established terms or conceptual frameworks for what goes on at this stage. In fact, it is difficult even to define what "intermediate-level vision" means. Processes such as grouping or segmentation are often ascribed to this stage, but the range of other things that could be going on is so broad and ill-defined that it is semi-seriously referred to as "everything between low-level and high-level vision." Over the past 20 years however this area has become progressively less murky through insightful and penetrating psychophysical experiments.

In particular, Nakayama and colleagues have provided compelling evidence that intermediate-level representations are organized around *surfaces* in the 3D environment, and that these representations serve as a basis for high-level processes such as visual search and attention (Nakayama et al. 1995). This view stands in contrast to previous theories postulating 2D features such as orientation and motion energy as the basis of perceptual grouping that underlies texture segmentation, search, and attention (Treisman and Gelade 1980; Julesz 1981). Nakayama's experiments suggest that representations of 3D surface structure are formed prior to this stage, and that perceptual grouping operates primarily on surface representations rather than 2D features. For example, when colored items are arranged on surfaces in different depth planes, detection of an odd-colored target is facilitated when pre-cued to the depth plane containing the target; but if the items are arranged so as to appear attached to a common surface receding in depth, then pre-cueing to a specific depth has little effect. Thus, it would appear that attention spreads within surfaces in 3D coordinates in the environment, not within 2D proximity or a simple disparity measure.

Another contribution of Nakayama's work is in pointing out the importance of *occlusion* in determining how features group within a scene. Once again, they show that simple grouping rules based on 2D proximity or similarity do not suffice. This should not be surprising, because under natural viewing conditions the 2D image

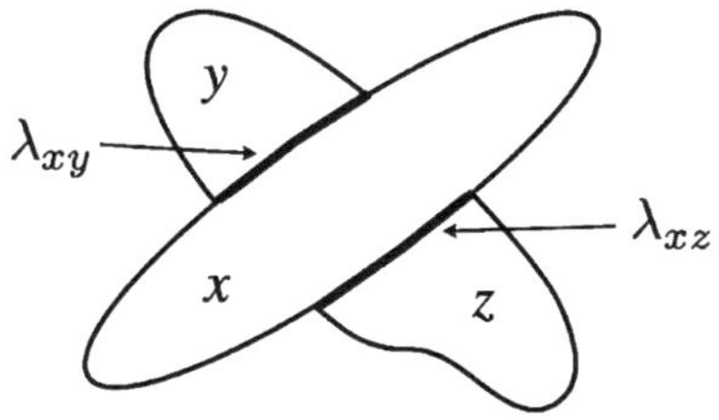

Fig. 12.3 Occlusion and border ownership. When image regions corresponding to different surfaces meet in the projection of a scene, the region corresponding to the surface in front "owns" the border between them. A region that does not own a border is essentially unbounded and can group together with other unbounded regions. Here, surface x owns the borders λ_{xy} and λ_{xz}. Thus, regions y and z are unbounded at these borders and they are free to group with each other, but not with region x because it owns these borders and is therefore bounded by them (adapted from Nakayama et al. 1995)

arises from the projection of 3D surfaces in the environment. When these surfaces overlap in the projection, the one nearest the observer "overwrites" or occludes the other. Thus, a proper grouping of features would need to take this aspect of scene composition into account in determining what goes together with what, as shown in Fig. 12.3. By manipulating disparity cues so as to reverse figure ground relationships in a scene, they show that the visual system groups features in a way that obeys the rules of 3D scene composition. Features are grouped within surfaces, even when parts of the surface are not visible, but not beyond the boundary of a surface. Thus, the neural machinery mediating this grouping would seem to require an explicit representation of border ownership, such as described by von der Heydt (Zhou et al. 2000; Qiu and von der Heydt 2005), or some other variable that expresses the boundaries and ordinal relationship of surfaces.

Nakayama's work is not the only in this realm, there are many others (Adelson 1993; Mamassian et al. 1998; Knill and Saunders 2003). It is a body of work that suggests what to look for at the neural level. Much as color psychophysics preceded the discovery of its neural mechanisms, these psychophysical experiments suggest the existence of certain neural representations at the intermediate level of vision.

Functional Organization of Human Visual Cortex

In 1991, Felleman and Van Essen published their now famous diagram of connections between visual cortical areas in the macaque monkey (Felleman and Van Essen 1991). This diagram and the detailed information about laminar patterns of connections that went alongside it shed new light on the hierarchical organization and division of labor in visual cortex. In the years since, we have seen an almost equally detailed picture of the functional organization of human visual cortex emerge from fMRI studies (Wandell et al. 2007). The significance of having these areas mapped out in humans is that it enables a more direct connection to perception, since one can tie the amount of activity in a given brain area to variations in both stimulus space and psychophysical performance (Heeger 1999; Grill-Spector et al. 2000; Ress and Heeger 2003). This has made it possible to identify areas involved in the representation of three-dimensional form, such as the lateral occipital complex (Kourtzi and Kanwisher 2001). It has also enabled us for the first time to see evidence of "explaining away," in which top-down signals originating from high-level areas appear to decrease the activity in lower level areas when subjects perceive an entire 3D object or scene layout as opposed to its individual parts (Murray et al. 2002).

Some visual areas and neurons exhibit a striking degree of specificity, such as those responsive to faces. Tsao and Livingston used fMRI to localize areas in macaque cortex that are selectively activated by faces and then subsequently recorded in those areas with microelectrodes to characterize responses of individual neurons (Tsao et al. 2006). These studies have revealed a complex of areas that appear to specialize for different aspects of faces such as identity vs. pose (Freiwald et al. 2009). There is now evidence for corresponding areal specializations in humans (Tsao et al. 2008). In addition, Izhak Fried's recordings from the medial temporal lobes in humans have revealed neurons that appear every bit as selective as "grandmother cells," an idea which for years was the subject of theoretical speculation but usually regarded with great skepticism (Quiroga et al. 2005).

Another method that is providing new insights about cortical organization in humans is *neural decoding*. In contrast to traditional approaches that attempt to characterize which class of stimuli a neuron or cortical region responds to, here the goal is to find out what those neurons tell you about the stimulus. When applied to BOLD signals measured over a wide swath of human visual cortex in response to natural images, one finds that lower level areas do a reasonable job at reconstructing image properties such as color and texture, whereas higher level areas reconstruct information about the semantic content of the scene (Naselaris et al. 2009, 2011; Nishimoto et al. 2011). While these particular findings are not surprising given our current understanding of visual cortex, they are nevertheless a testament to the rich, multidimensional information provided by fMRI. Rather than testing specific hypotheses about selected regions of interest, this approach treats the entire 3D volume of BOLD signals as a multielectrode recording array and lets the data speak for itself. Importantly, these studies are most informative when the visual system is presented with complex natural scenes or movies, since these stimuli contain the rich, multidimensional forms of information that are most likely to evoke patterns of activity revealing the functional significance of different brain regions.

How to Infer Scene Geometry from Multiple Views

In parallel with these achievements in neuroscience and psychophysics, the field of computer vision has undergone a number of dramatic advances. Chief among these is the ability to infer three-dimensional scene structure from multiple views, termed *multiple-view geometry* (Hartley and Zisserman 2003). This has been enabled in part by the discovery of stable and unique keypoint detectors and invariant feature descriptors which allow for solving the correspondence problem efficiently (Lowe 2004). It is now possible, given an unordered set of images of the same three-dimensional scene taken from different viewpoints, to simultaneously recover a representation of the 3D scene structure as well as the positions in the scene from which the images were taken (Brown and Lowe 2005). This technology has enabled commercial products such as *Photosynth* which assimilate information from the many thousands of photographs stored on repositories such as Flickr into a unified scene model (Snavely et al. 2006).

While many computer vision algorithms are divorced from biology, there has long been a productive interchange of ideas between the fields of computer vision and biological vision. I believe the advances in multiple-view geometry tell us something important about vision, and that they open the door to a new area of investigation in visual neuroscience—namely, how do animals assimilate the many views they obtain of their environment into a unified representation of the 3D scene? The ability to navigate one's surroundings, to remember where food is, and how to get home is fundamental to the survival of nearly all animals. It would seem to demand an allocentric representation of the 3D environment. However, there has been considerable debate among cognitive psychologists as to whether humans or other animals actually build 3D models as opposed to simply storing 2D views. It is often tacitly assumed that storing 2D views is the simpler, cheaper strategy. But from the standpoint of efficient coding it actually makes the most sense to combine the images acquired while moving through the environment into a single 3D representation, since that is the lowest entropy explanation of the incoming data stream. Now the mathematics and algorithms of multiple-view geometry show us that the computations needed to do this are really quite feasible. In fact these algorithms can run in real-time from video camera input (Newcombe and Davison 2010). The challenge for theorists and modelers now is to figure out how these computations can be performed in a more holistic manner (drawing upon all the data rather than just keypoints), how to exploit the continuity in images over time, and in what format 3D scene information should be represented.

Questions Unanswered

There is little doubt that we are closer to understanding how visual systems work than we were 20 years ago. But how much remains to be understood? Here I shall review areas in which there are still gaping holes in our knowledge. As we shall see, the scope of our ignorance is vast. It is not simply a matter of filling in holes here and there; rather we are missing something fundamental.

How Is Sophisticated Vision Possible in Tiny Nervous Systems?

Much effort in neuroscience is expended to understand how neural circuits in the visual cortex of cats and monkeys enable their perceptual abilities. An often unstated assumption behind these studies is that mammalian cortex is uniquely suited for gaining insight into the neural mechanisms of perception. But one must begin questioning this assumption when confronted with the highly sophisticated visual capabilities found in nervous systems that are smaller by several orders of magnitude.

Consider for example the jumping spider (Fig. 12.4). Unlike other spiders that use a web to extend their sensory space, this animal relies entirely upon vision to localize prey, identify potential mates, and navigate complex terrain. It does so

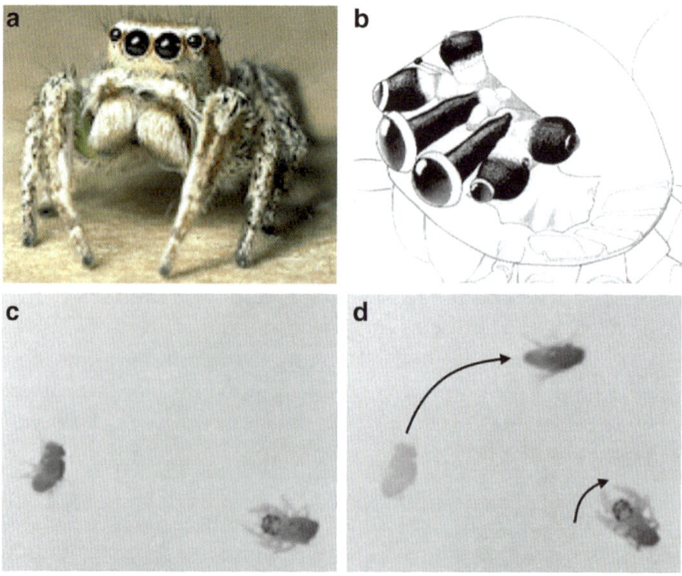

Fig. 12.4 (**a**) Jumping spider (*Habronattus*). (**b**) Jumping spider visual system showing antero-median, antero-lateral, and posterior-lateral eyes. (**c**, **d**) Orienting behavior of a 1-day-old jumping spider (*lower right*) during prey capture. (**a**, **b**) From Wayne Maddison's Tree of Life; (**c**, **d**) video frames filmed by Bruno Olshausen and Wyeth Bair in the Bower lab (Caltech 1991)

using a highly elaborate visual system comprising four pairs of eyes: one pair of frontal facing principal eyes (antero-median eyes) provide a high-resolution image over a narrow field of view, while the other three pairs provide lower resolution images over a wide field of view and are mounted on different parts of the head so as to provide 360° coverage of the entire visual field (Land 1985). Interestingly, the retinae of the antero-median eyes are highly elongated in the vertical direction so as to essentially form a one-dimensional array of photoreceptors. These retinae move from side to side within the head in a smooth (approximately 1 Hz) scanning motion to perform pattern analysis (Land 1969). The jumping spider uses its low resolution system to detect targets or objects of interest, and then orients its body to position the target within the field of view of the high-resolution antero-median eyes for more detailed spatial analysis via scanning (Land 1971).

The jumping spider exhibits a number of striking visual behaviors. Figure 12.4c, d illustrates the tracking and pursuit behavior involved in hunting. The spider initially follows the target (in this case, a fruit fly) with its eye and head movements. It then stalks the fly in a crouching motion before pouncing on it. Mediating this behavior demands the ability to maintain attention on a target, to track the target via appropriate motor commands, and to perform distance estimation. In this case the spider happens to be only 1 day old, so these abilities are largely innate. Another striking visual behavior of the jumping spider is exhibited during courtship, in which the male performs an elaborate dance for the female. During these dances the female visually inspects and attends to the male. Complex pattern recognition via scanning is utilized by both parties during this interaction. Courtship dances may be elicited by presenting a video image of a female (Clark and Uetz 1990), or even a line drawing depicting a jumping spider, to the male (Drees 1952), which further testifies to the role of vision in mediating this behavior. Vision also plays an important role in 3D path planning and navigation. One particular species, *Portia fimbriata*, appears to use its visual system to survey the 3D visual environment before embarking on a path that requires a complex detour to obtain a prey item beyond jumping range (Tarsitano and Jackson 1997; Tarsitano and Andrew 1999).

Thus it would seem that the jumping spider performs complex pattern recognition, visual attention, motion analysis and tracking, distance estimation via stereopsis, and 3D path planning. These are all abilities that most would consider the hallmark of visual cortical function, yet in the jumping spider they are being carried out by a visual system that is no larger than a single hypercolumn of V1, and requiring little or no visual experience during development. There seems to be a huge explanatory gap here between our conventional wisdom and reality.

Another small animal that challenges our conventional wisdom is the sand wasp, *Philanthus triangulum*. The navigational abilities of this animal were intensely studied and described by Tinbergen (1974). He demonstrated that the wasp finds its nest, consisting of a small burrow in the sand, by memorizing the spatial arrangement of debris that happen to immediately surround the nest such as twigs, rocks, or other items. If these items are displaced by a meter or so while the wasp is away hunting, keeping the relative spatial positions of the items intact, it returns to a point in the center of this new arrangement rather than the actual location of its nest. Initially stunned, the animal eventually finds its nest. However, when it next emerges to go out hunting it makes an extra set of circular flights over its nest, as though recommitting to memory the pattern of landmarks surrounding the nest. What is perhaps most astonishing here is that the sand wasp does all of this utilizing only a compound eye, which has very low spatial-resolution. Thus, the complex spatial layout of the environment must somehow be accumulated over time from the dynamic pattern of activity coming from the ommatidia during flight.

It is often tempting to explain away these abilities as the result of simple but clever tricks. To those who try I challenge them to prove such strategies are actually viable by building an autonomous system by these rules that exhibits the same degree of robust, visually guided behavior. Such systems do not exist, and I contend they are still far away from being realized because *we do not understand the fundamental principles governing robust, autonomous behavior in complex environments*. Evolution has discovered these principles and they are embodied in the nervous systems of insects and spiders. There are valuable lessons to be learned from studying them.

The fact that sophisticated visual abilities are present in simpler animals also raises a disturbing question: *If so much can be done with a tiny brain, what more can be done with a large brain?* Perhaps the vast cortical circuits of mammals are carrying out a more complex set of functions than we are currently considering. Perhaps we lack the intellectual maturity needed to ask the right questions about what cortex is doing.

I do not suggest that we must fully understand invertebrate vision as a prerequisite to studying vision in mammals. But I do think that our field is guilty of taking a cortico-centric approach, and that simpler animals have been prematurely dismissed and unjustly neglected in the quest to understand intelligent behavior. One often hears the argument that invertebrates are likely to utilize highly specialized or idiosyncratic neural processing strategies that will not generalize to mammals. But biology is teeming with examples of molecular and cellular mechanisms that are recapitulated across the animal kingdom. Those who study fly genetics are not just interested in flies, they want to know how genes work. At this point there are astonishingly few examples of computations in the nervous system that anyone truly understands. Thus, gaining a solid understanding of neural computation as it occurs in *any* animal would give us much needed insight into the space of possible solutions.

How Do Cortical Microcircuits Contribute to Vision?

Not long after the discovery of orientation selectivity and columnar structure in visual cortex, the view began to emerge that V1 operates as a filter bank in which the image is analyzed in terms of oriented features at different spatial scales (Blakemore and Campbell 1969; De Valois et al. 1982), now often modeled with Gabor functions (Marcelja 1980; Daugman 1985). Others further elaborated on this idea by building hierarchical models composed of successive stages of feature detection and spatial pooling (Fukushima 1980), inspired by Hubel and Wiesel's early proposals (Hubel and Wiesel 1962, 1965). In the ensuing decades, this conceptual framework has come to dominate the theoretical landscape. It has had a profound impact in shaping how neuroscientists form and test hypotheses regarding visual cortical function, and it has influenced the development of computer vision algorithms. It is even referred to as the "standard model" (Riesenhuber and Poggio 2004), and theories that strongly deviate from this framework are often dismissed as biologically implausible. However, this view begins to clash with reality as one takes a closer look at the detailed structure of cortical circuits.

As all students of neuroanatomy know, mammalian neocortex is a layered structure. By convention it has been subdivided into six laminar zones according to various histological criteria such as cell density and morphology. Underlying this overt structure is a detailed microcircuit that connects neurons in a specific way according to the layer they reside in (Douglas et al. 1989; Thomson and Bannister 2003; Douglas and Martin 2004). Inputs from thalamus terminate principally on neurons in layer 4. These neurons in turn project to neurons in layers 2 and 3, which then project back down to layers 5 and 6. Neurons within each layer are recurrently connected by horizontal fibers, with the most extensive of these networks found in layers 2 and 3. Inhibitory interneurons have their own specialized cell types and circuits, and some are interconnected by gap junctions and exhibit synchronous, high gamma oscillations (Mancilla et al. 2007). Layer 1 is mostly composed of the distal tufts of pyramidal cell apical dendrites and the axonal fibers of neurons in other layers. On top of all this, we are beginning to appreciate the "deep molecular diversity" of cortical synapses, which increases the potential complexity of synaptic transmission and plasticity (O'Rourke et al. 2012).

To those who subscribe to the Gabor filter model of V1 I ask, where are these filters? In which layers do they reside, and why do you need such a complex circuit to assemble them? In 1 mm^2 of macaque V1 there are 100,000 neurons, yet the number of LGN afferents innervating this same amount of cortex amounts to the equivalent of only a 14×14 sample node array within the retinal image (Van Essen and Anderson 1995). Why so many neurons for such a small patch of image? To complicate matters further, each neuron is a highly nonlinear device with inputs combining in a multiplicative or "and-like" manner within local compartments of the dendritic tree (Poirazi et al. 2003; Polsky et al. 2004). Such nonlinearities are notably absent from the L-N cascade models commonly utilized within the neural coding community. What are the consequences of these nonlinearities when large numbers of such devices are densely interconnected with one another in a recurrent circuit? It is well known that recurrent networks composed of perceptron-type neurons (linear sum followed by point-wise nonlinearity) can have attractor dynamics, but what are the consequences of dendritic nonlinearities? Is such complexity compatible with the simple notion of a filter or a receptive field? Moreover, why have different layers of processing, and how do the computations and formatting of visual information differ between these layers?

There are numerous hand-wavy explanations and ad hoc models that can be (and have been) constructed to account for all of these things. At the end of the day we are faced with this simple truth: *No one has yet spelled out a detailed model of V1 that incorporates its true biophysical complexity and exploits this complexity to process visual information in a meaningful or useful way*. The problem is not just that we lack the proper data, but that we don't even have the right conceptual framework for thinking about what is happening.

In light of the strong nonlinearities and other complexities of neocortical circuits, one should view the existing evidence for filters or other simple forms of feature extraction in V1 with great skepticism. The vast majority of experiments that claim to measure and characterize "receptive fields" were conducted assuming a linear systems identification framework. We are now discovering that for many V1 neurons these receptive field models perform poorly in predicting responses to complex, time-varying natural images (David et al. 2004; Frégnac et al. 2005; Khosrowshahi et al. 2007). Some argue that with the right amount of tweaking and by including proper gain control mechanisms and other forms of contextual modulation that you can get these models to work (Carandini et al. 2005; Rust and Movshon 2005). My own view is that the standard model is not just in need of revision, *it is the wrong starting point and needs to be discarded altogether.* What is needed in its place is a model that embraces the true biophysical complexity and structure of cortical microcircuits, especially dendritic nonlinearities. The ultimate test of such a model will be in how well it accounts for neural population activity in response to dynamic natural scenes (as opposed to simple test stimuli), and the extent to which it can begin to account for our robust perceptual abilities.

How Does Feedback Contribute to Vision?

At nearly every stage of processing in the visual system, one finds feedback loops in which information flows from one set of neurons to another and then back again. At the very first stage, photoreceptors provide input to a network of horizontal cells which in turn provide negative feedback onto photoreceptors. Hence a photoreceptor does not report a veridical measurement of the amount of light falling upon it, but rather a signal that is modified by context. At later stages, LGN relay neurons provide input to the reticular nucleus which in turn provides negative feedback to LGN relay neurons; LGN projects to V1 and V1 projects back to LGN; V1 projects to V2 which projects back to V1, and so on. What are these feedback loops doing and how do they help us see?

In some cases, such as horizontal cells in the retina, we have fairly good models to suggest what feedback is doing and what it might be good for (i.e., mediating lateral inhibition among photoreceptors to reduce redundancy and increase dynamic range). But in other cases, such as in the thalamo-cortical loop or cortico-cortical loops, there has yet to emerge a clear conceptual model, supported by the data, that tells us what function is being served. There have been numerous experimental attempts to uncover what feedback is doing, for example, by cooling or disabling the neurons in a higher area that feedback onto a lower area and characterizing how response properties in the lower area change (Hupé et al. 2001; Angelucci and Bullier 2003; Andolina et al. 2007). One sees a variety of modulatory effects, but so far there has not emerged a clear consensus or framework for how to incorporate these findings into a larger theory. Indeed there is considerable doubt among neuroscientists as to whether feedback plays any role in dynamically shaping information processing (Lennie 1998).

Perhaps the most striking sign of our conceptual ignorance here is the fact that modern computer vision systems are still largely based on feedforward processing pipelines: image data is preprocessed, features are extracted and then pooled and fed to another layer of processing, or histogrammed and fed to a classifier. One does not typically see algorithms that use the outputs of a higher stage of processing to modify the input coming from a lower stage (though see Arathorn 2005 for a notable exception). In other areas of engineering, such as in the design of control systems or electronic amplifiers, the advantages of feedback are well understood and it is exploited to build robust, stable systems that work in practice. But currently, other than automatic gain control or other early forms of preprocessing, researchers have not discovered how to exploit feedback for more advanced forms of processing that support recognition or other perceptual tasks.

One rationale that is offered in support of feedforward models is that visual recognition occurs so exceedingly fast that there is little time for the iterative type of processing that feedback loops would entail (Thorpe and Imbert 1989). EEG signals correlated with visual recognition in humans arise 150 ms after stimulus onset (Thorpe et al. 1996). In macaque monkey cortex, the earliest neural signals in inferotemporal cortical areas that are discriminative for objects occur ca. 125 ms after stimulus onset (Oram and Perrett 1992; Hung et al. 2005). Given the number of stages of processing and axonal and synaptic delays, it is argued, there is precious little time for any feedback loops to play a significant role in supporting these signals. But this reasoning is based upon overly simplistic and dour assumptions about how feedback works. The conduction velocities of feedforward and feedback axons between V1 and V2 are on the order of 2–4 ms (Angelucci and Bullier 2003). Even between thalamus and V1 the round trip travel time can be as short as 9 ms (Briggs and Usrey 2007). Most importantly though, vision does not work in terms of static snapshots but rather as a dynamical system operating on a continuous, time-varying input stream. Axonal and synaptic delays simply mean that sensory information arriving at the present moment is processed in the context of past information that has gone through a higher level of processing.

Given the space and resource constraints faced by the brain, it seems unlikely that such vast amounts of white matter would be devoted to feedback pathways unless they were serving a useful purpose in shaping information processing. Over the past decade two promising theoretical ideas have been advanced. One is based on the idea of *predictive coding*, in which higher levels send their predictions to lower levels where they are compared, and the residual or degree of mismatch is sent forward (Rao and Ballard 1999). Such a coding scheme would be useful to reduce redundancy and detect novelty. The other is based on *perceptual inference* (or Bayesian inference, as described above) (Lee and Mumford 2003). Here, higher levels also send their predictions to lower levels, but rather than computing differences, the parts where the predictions agree are amplified and the parts where they disagree are suppressed. This type of processing is most useful when lower levels of representation are ambiguous (such as the aperture problem in the computation of motion). Higher level knowledge and context are used to adjudicate between different interpretations and resolve ambiguity. Formally this may be cast in terms of probabilistic inference in graphical models or "belief propagation." To validate either of these hypotheses one would need to investigate the effects of feedback during the viewing of natural images or other complex, structured images where prediction can play a role, or the need for disambiguation arises. Indeed this may explain why the findings of previous experiments using simplified test stimuli have been rather inconclusive.

What Is the Role of Neuronal Oscillations in Visual Processing?

Since Hans Berger's first EEG measurements in the 1920s it has been known that the brain oscillates. Early investigators ascribed the terms *alpha*, *beta*, and *gamma* to oscillations occurring in different frequency bands, and they attempted to relate these oscillations to various states of arousal, perception, cognition, or clinical pathologies. Later, when neurophysiologists such as Barlow, Kuffler, Hubel, and Wiesel began achieving success with single-unit recordings, attention turned to the activity of individual neurons. Interest in oscillations dissipated, and the focus instead shifted to studying how the *stimulus-driven* firing rate of neurons encodes features of the visual world. Against this backdrop in 1989, Gray and Singer showed that the activity of single neurons in V1 is phase-locked to gamma oscillations in the local field potential (LFP), and furthermore that the degree of synchrony between neurons depends on whether the features they encode belong to a common object (Gray and Singer 1989). This finding reignited interest in oscillations, especially among theorists who speculated that they may serve as a mechanism for feature binding and attention, or even consciousness. Experimentalists argued among themselves as to whether oscillations or synchrony were actually present. Sides were taken and debates were staged (e.g., at the Society for Neuroscience 1993 annual meeting), and each side argued passionately for their point of view.

Now almost 20 years later the debate has mostly subsided. Few doubt the existence of oscillations—they have withstood the test of time and have been shown to be a ubiquitous property of sensory systems, from the locust olfactory system to the mammalian retina and visual cortex. One senses that the field has settled into taking a more dispassionate approach to investigate what causes these oscillations, under

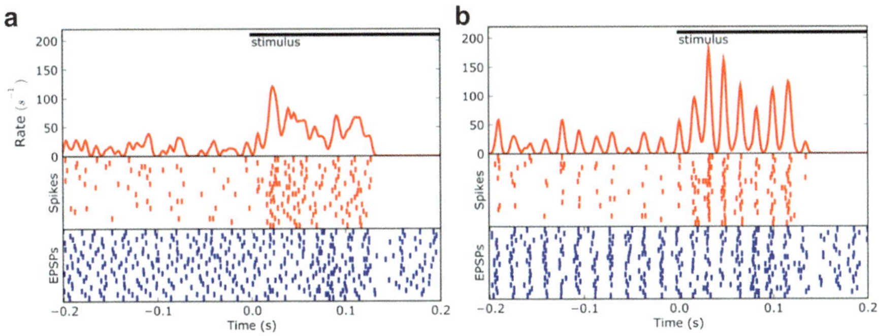

Fig. 12.5 LGN neurons synchronize to 50 Hz retinal oscillations. (**a**) PSTH and spike rasters in response to repeated presentations of a stimulus. Note the apparent variability in the latency of the LGN neuron's response. (**b**) When the LGN spikes are realigned to the instantaneous phase of retinal oscillations extracted from the EPSPs for each trial, the variability in response latency is vastly reduced (from Koepsell et al. 2009)

what conditions they arise, and how they relate to perception. However, there is still little concrete evidence that suggests what they are doing and how they help us see.

One recent finding that I believe points to an important role for oscillations in vision comes from recordings from cat LGN neurons in Judith Hirsch's laboratory (Koepsell et al. 2009). These data reveal that the spiking activity of some neurons in the LGN is phase-locked to the 50 Hz oscillations arising from the retina. These oscillations are readily apparent in the electro-retinogram and have been observed in recordings from retinal neurons, but their effect on downstream processing was previously unknown. Koepsell et al. showed that when the phase of these ongoing oscillations is taken into account, the apparent variability in the response latency of LGN neurons—commonly attributed to "noise"—is vastly reduced (Fig. 12.5). In other words, LGN neurons exhibit a much higher degree of temporal precision—and hence information carrying capacity—when the phase of ongoing oscillations is included in reading out their activity (as opposed to considering the stimulus-driven component only). What could this extra information be used for? Koepsell and Sommer propose that oscillations propagating through distributed networks in the retina could be used to compute "graph cuts," an effective method of image segmentation that is widely used in computer vision (Koepsell et al. 2010). In their model, the firing rate of a neuron encodes contrast and the phase of oscillation encodes region membership. While highly speculative, the theory nevertheless demonstrates how oscillations could be leveraged in a profound and elegant way to carry out computations requiring the rapid and global spread of information across an image to solve a difficult problem in vision.

When considering oscillation-based theories it is important to bear in mind that the prevailing rate-based, stimulus-driven view of neural function, while often portrayed as fact, is itself a theory. Though there are countless examples where firing rate correlates with perceptual variables, this in itself does not demonstrate that information is actually encoded and read out this way. So little is known at this point that there is much room for alternative theories. But if one accepts that neural activity is an information-bearing signal in the brain, then oscillations and other forms of ongoing activity must be included in a full account of neural function.

How to Build Robust, Autonomous Vision Systems?

In 1973 Sir James Lighthill issued a report to the British Parliament that condemned AI for failing to achieve its grandiose objectives and recommended that its funding be cut off (the recommendation was subsequently adopted, killing AI research in the UK for nearly a decade). At the center of his argument was that robotic systems were only capable of operating in restricted domains, and that scaling up to general purpose intelligence that could deal with real world conditions would require a combinatorial explosion in computational resources. The idea that we might someday build general purpose robots, he claimed, was a "mirage." A debate was held between Lighthill and three leading AI researchers, Donald Michie, John McCarthy, and Richard Gregory, who defended their aims and work as realistic and worthwhile (BBC 1973). State-of-the-art robots of the day such as SRI's *Shakey* and Edinburgh's *Freddy* took center stage to illustrate the promising achievements of AI. These robots could perceive the world through cameras that extracted the outlines of objects and could guide an actuator to grasp or manipulate the objects. They could execute complex tasks, such as assembling a toy car from parts randomly arranged on a table, in a completely autonomous manner without human intervention.

Now almost 40 years later, with all of the participants of that debate gone, it is almost too painful to ask this, but … was Lighthill right? Consider that over this span of time Moore's law has brought us an increase of *six orders of magnitude* in available computational resources. Can we claim that robots have similarly advanced compared to their predecessors in the early 1970s? *Stanley*, the robot that won DARPA's Grand Challenge desert road race in 2005, is heralded as a triumph of AI. But upon closer examination it would seem to exemplify exactly the sort of domain-specific limitations that Lighthill railed against—it was preprogrammed with a map of the entire route and 3000 GPS waypoints, and it followed a road with few major obstacles on a bright sunny day. As such, it was primarily a test of high-speed road finding, obstacle detection, and avoidance in desert terrain (Thrun et al. 2006). Its success in navigating the course was mainly the result of clever engineering—Kalman filters to compute robust, optimal estimates of position, and combining LIDAR and image data to find drivable terrain and stay in the center of the road. These are notable achievements, but it is difficult to imagine that this is the level of visual intelligence that Michie, McCarthy, and Gregory would have hoped to see emerge by the early twenty-first century.

Now consider these robots in comparison to the jumping spider or sand wasp. To survive they must navigate unfamiliar, complex terrain that is filled with obstacles, variable illumination from shadows, and potentially unstable surfaces. They have no GPS way points or roads to provide guidance. Rather, they must acquire and store

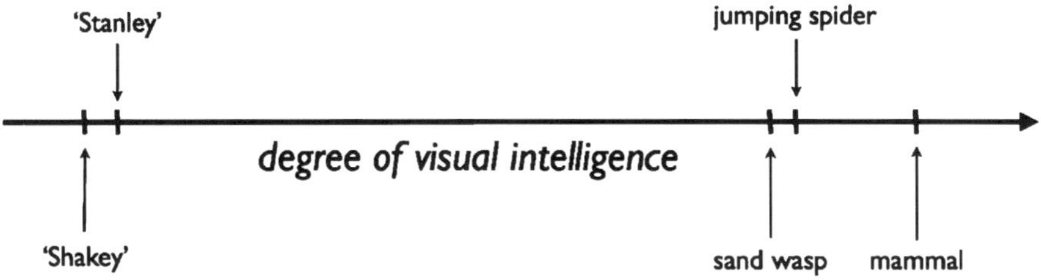

Fig. 12.6 When measured in terms of visual intelligence, there is still a wide gulf separating robots such as Shakey and Stanley from biological visual systems

information about the environment as they go so as to remember where they have been, where the food is, and how to get home. They must detect, localize, track, and successfully capture prey, even when seen against complex backgrounds. They must deal with unforeseen events such as getting knocked off course by wind or debris. They must continue to function 24/7 in the face of the elements such as rain or dust or changes in lighting conditions. And they do all of this while consuming only minuscule amounts of power in comparison to their robotic counterparts.

While *Stanley* unquestionably represents an advance over *Shakey*, both of these systems would seem equally far removed from the jumping spider or sand wasp, let alone humans, when measured in terms of the level of robust, autonomous behavior they exhibit (Fig. 12.6). That we stand at this impasse after 40 years I believe tells us something important. It suggests that the problem we face is not just technological but rather due to a scientific gap in our knowledge. *We are missing something fundamental about the principles of vision and how it enables autonomous behavior.* Computing optic flow or building a depth map of a scene, while useful, is not sufficient to robustly navigate, interact with, and survive in the natural three-dimensional environment. What exactly *is* needed is of course difficult to say—that is the problem we are up against. But I would point to two things. One is a richer representation of surface layout in the surrounding environment that expresses not only its 3D geometry but also its *affordances*—that is, the actions that are possible (Gibson 1986). The other is to move beyond the Turing machine, procedural framework that today's robots are trapped in—that is, an infinite loop of "acquire data," "make decisions," and "execute actions." What is needed is a more fluid, dynamic interaction between perception and action. Theories for how to do this are now beginning to emerge but it is a field still in its infancy (Gordon et al. 2011).

Questions Not Yet Asked

The answers we get from experiments are only as useful as the questions we ask. The key is to ask the right questions to begin with. But how do we know what these are? Most of the questions described in the preceding section are ones that scientists are already keenly aware of and which drive current research efforts. Undoubtedly though there are other important questions that no one working in the field today has even thought to ask yet, just as computer vision researchers in the 1960s never thought to ask how you find the edges of an object in a scene. This points to the importance of another process of discovery beyond answering questions—that is, discovering the questions that need to be asked.

Here I will suggest two ways that we can accelerate the process of discovering what these questions are. One is to take an *exploratory approach* that casts a wide net and seeks to reveal interesting phenomena. The other is to educate ourselves about the problems of vision by attempting to *build* neuromorphic visual systems that enable autonomous behavior.

The Need for Exploratory Approaches

Scientists by their nature are eager to test hypotheses or to tell a story about how a given set of facts or findings fit together and explain perceptual phenomena. But as we have seen, vision presents us with deep computational problems, and nervous systems confront us with stunning complexity. Most of the hypotheses we test and the stories we tell are far too simple minded by comparison, and ultimately they turn out to be wrong. Worse yet, they can be misleading and stifling because they encourage one to look at the data through a narrow lens. When one carefully designs a set of experiments to test a specific set of hypotheses, the data obtained are often of little value for looking at other issues. In some cases this may be warranted, but when the hypothesis landscape is not well formed to begin with it may be more worthwhile to take an exploratory approach.

The exploratory approach is more observational in nature. The goal is to document how the system works in its natural state—for example, what are the distributions of firing rates among neurons in different layers, and in different cortical areas, during natural vision? Such experiments do not test any particular hypothesis, and the outcome may simply be a large table of numbers. But such data would be of immense value in helping us to understand what kind of a system we are dealing with, and they are of pivotal importance in shaping theories.

Another goal of the exploratory approach is discover new phenomena that surprise us and defy conventional wisdom. These can then provide clues about what we *should* be looking for. A notable example is the discovery of orientation selectivity. The idea that visual neurons might be selective to lines or edges at different orientations did not occur to Hubel and Wiesel a priori. Rather, they were probing the visual cortex with spots of light using a slide projector, and in the process of moving slides in and out of the projector they noticed that the edge of the slide moving over the receptive field happened to elicit a robust neural response (Hubel 1982). This observation in turn led to a revolution in visual neuroscience. Tinkering is often frowned upon in scientific circles, especially by study sections and review panels of the major scientific funding bodies. But when one is mostly in the dark to begin with—as I would argue we are in our understanding of the visual cortex—a certain amount of tinkering seems warranted.

I do not advocate that we abandon the hypothesis-based approach—it has formed the bedrock of modern science because in many cases it has been a fruitful and productive path to knowledge. But we should recognize when this approach is appropriate and when it is not. Storytelling makes science interesting, and it often makes a finding seem more compelling, but it can also lead to a false sense of complacency, a feeling that we have understood something when in fact the real story is orders of magnitude more complicated. We should be more inclined to take these stories with a grain of salt and instead be on the lookout for something deeper lurking beneath the surface. And no one should feel ashamed to report a complete, unfiltered set of findings without a story to envelop them. After all, one person's untidy finding may provide the missing piece in another person's theory.

Learning About Vision by Building Autonomous Systems

There is very little that neuroscience per se has taught us about the principles of vision. That we know there is a ventral and dorsal stream, a hierarchy of visual areas, and neurons that selectively respond to certain visual features in these areas does not tell us *what* problems are being solved and *how*. They provide strong hints and tantalizing clues to be sure, but trying to build a functional vision system by directly mimicking these attributes in a computer chip is like trying to build a flying machine out of flapping wings and feathers.

By contrast, the failures of robot vision in the 1960s were a transformative learning experience in the study of vision. They set the stage for people like David Marr to intensely study the computational problems of vision and to theorize how biological vision systems work. The field thus made an advance by trying to solve an important and unsolved problem, the depth of which was previously unappreciated. I believe this will continue to be the case in the future—we will learn the most about the principles of vision by attempting to build autonomous vision systems, learning what works and what does not, and then drawing upon these insights in studying the visual systems of humans and other animals.

To some extent this is a role that computer vision already plays. However, mainstream computer vision is focused on solving a prescribed set of problems that have been defined by computer scientists and engineers. Algorithms for shape from shading, optic flow, and stereo are judged by how well they perform on standard benchmarks, where the correct representation is assumed to be known. Object recognition is distilled down to a problem of classification, one of converting pixels to labels, again with benchmark datasets for judging performance. If we wish to gain insight into the principles of biological vision, or autonomous visual behavior in general, it will require a different approach.

What is needed is an approach that, like computer vision, attempts to solve problems, but where more attention is paid to how we define those problems, and the computational architectures we draw upon to solve them. The choice of problems should be guided by animal behavior and psychophysics: What are the tasks that animals need to solve in order to survive in the natural environment? What are the performance characteristics of human or other animal observers in these tasks? In addition, it is important to take into account and exploit the unique computational properties of neural systems, what Carver Mead called "neuromorphic engineering." The only functional vision systems we know of today are built out of nonlinear recurrent networks, they compute with analog values, and they run in continuous time. They are not Turing machines. Thus, in considering the space of solutions to visual problems this needs to be taken into account.

Finally, it is important to bear in mind that vision did not evolve as a stand-alone function, but rather as part of the perception–action cycle. As philosopher Robert Cummins put it, "Why don't plants have eyes?" We have much to gain by building vision systems with tight sensorimotor loops and learning what problems need to be overcome in doing so. This area remains vastly under investigated, and is likely to uncover to many questions that have yet to be asked.

Acknowledgments I thank Jim Bower for encouraging me to write this article and for his patience in giving me the time to complete it, and Jim DiCarlo for providing the MIT AI Memo. Supported by funding from NSF (IIS-1111654), NIH (EY019965), NGA (HM1582-08-1-0007), and the Canadian Institute for Advanced Research.

References

Adelson EH (1993) Perceptual organization and the judgment of brightness. Science 262(5142):2042–2044

Adelson EH (2000) Lightness perception and lightness illusions. In: Gazzaniga M (ed) The new cognitive neurosciences, 2nd edn. MIT, Cambridge, MA, pp 339–351

Andolina IM, Jones HE et al (2007) Corticothalamic feedback enhances stimulus response precision in the visual system. Proc Natl Acad Sci USA 104(5):1685–1690

Angelucci A, Bullier J (2003) Reaching beyond the classical receptive field of V1 neurons: horizontal or feedback axons? J Physiol Paris 97(2–3):141–154

Arathorn DW (2005) Computation in the higher visual cortices: map-seeking circuit theory and application to machine vision. In: Proceedings of the 33rd applied imagery pattern recognition workshop (AIPR 2004), Washington, DC, 1–6

Arathorn DW, Yang Q et al (2007) Retinally stabilized cone-targeted stimulus delivery. Opt Express 15(21):13731–13744

Atick J, Redlich A (1992) What does the retina know about natural scenes? Neural Comput 4:196–210

Barlow H (1961) Possible principles underlying the transformation of sensory messages. In: Rosenblith WA (ed) Sensory communications. MIT, Cambridge, MA, pp 217–234

Barron JT, Malik J (2012) Shape, albedo, and illumination from a single image of an unknown object. In: Conference on computer vision and pattern recognition, Washington, DC, 1–8

BBC (1973) Controversy. http://www.aiai.ed.ac.uk/events/lighthill1973/1973-BBC-Lighthill-Controversy.mov

Blakemore C, Campbell FW (1969) On the existence of neurones in the human visual system selectively sensitive to the orientation and size of retinal images. J Physiol 203(1):237–260

Boyaci H, Fang F, Murray SO, Kersten D (2007) Responses to lightness variations in early human visual cortex. Curr Biol 17:989–993

Brainard DH, Williams DR et al (2008) Trichromatic reconstruction from the interleaved cone mosaic: Bayesian model and the color appearance of small spots. J Vis 8(5):15
Briggs F, Usrey WM (2007) A fast, reciprocal pathway between the lateral geniculate nucleus and visual cortex in the macaque monkey. J Neurosci 27(20):5431–5436
Brown M, Lowe DG (2005) Unsupervised 3D object recognition and reconstruction in unordered datasets. In: Fifth international conference on 3-D digital imaging and modeling, 2005, 3DIM 2005, Ottawa, 56–63
Cadieu CF, Olshausen BA (2012) Learning intermediate-level representations of form and motion from natural movies. Neural Comput 24(4):827–866
Carandini M, Demb JB et al (2005) Do we know what the early visual system does? J Neurosci 25(46):10577–10597
Clark D, Uetz G (1990) Video image recognition by the jumping spider, Maevia inclemens(Araneae: Salticidae). Anim Behav 40:884–890
Dan Y, Atick JJ et al (1996) Efficient coding of natural scenes in the lateral geniculate nucleus: experimental test of a computational theory. J Neurosci 16(10):3351–3362
Daugman J (1985) Uncertainty relation for resolution in space, spatial frequency, and orientation optimized by two-dimensional visual cortical filters. J Opt Soc Am 2:1160–1169
David SV, Vinje WE et al (2004) Natural stimulus statistics alter the receptive field structure of v1 neurons. J Neurosci 24(31):6991–7006
Dayan P, Hinton GE et al (1995) The Helmholtz machine. Neural Comput 7(5):889–904
De Valois RL, Albrecht DG et al (1982) Spatial frequency selectivity of cells in macaque visual cortex. Vision Res 22(5):545–559
Douglas RJ, Martin KAC (2004) Neuronal circuits of the neocortex. Annu Rev Neurosci 27(1):419–451
Douglas RJ, Martin KAC et al (1989) A canonical microcircuit for neocortex. Neural Comput 1(4):480–488
Drees O (1952) Untersuchungen über die angeborenen verhaltensweisen bei springspinnen (salticidae). Z Tierpsychol 9(2):169–207
Dreyfus HL, Dreyfus SE (1988) Making a mind versus modeling the brain: artificial intelligence back at a branchpoint. Daedalus 117(1):15–43
Felleman DJ, Van Essen DC (1991) Distributed hierarchical processing in the primate cerebral cortex. Cereb Cortex 1(1):1–47
Field D (1987) Relations between the statistics of natural images and the response properties of cortical-cells. J Opt Soc Am A 4:2379–2394
Frégnac Y, Baudot P, Levy M, Marre O (2005) An intracellular view of time coding and sparseness in V1 during virtual oculomotor exploration of natural scenes. In: 2nd International Cosyne conference in computational and systems neuroscience, Salt Lake City, UT, 17
Freiwald WA, Tsao DY et al (2009) A face feature space in the macaque temporal lobe. Nat Neurosci 12(9):1187–1196
Fukushima K (1980) Neocognitron: a self-organizing neural network model for a mechanism of pattern recognition unaffected by shift in position. Biol Cybern 36(4):193–202
Gauthier JL, Field GD et al (2009a) Receptive fields in primate retina are coordinated to sample visual space more uniformly. PLoS Biol 7(4):e1000063
Gauthier JL, Field GD et al (2009b) Uniform signal redundancy of parasol and midget ganglion cells in primate retina. J Neurosci 29(14):4675–4680
Geisler WS, Perry JS et al (2001) Edge co-occurrence in natural images predicts contour grouping performance. Vision Res 41(6):711–724
Gibson J (1986) The ecological approach to visual perception—James Jerome Gibson—Google books. Erlbaum, Hillsdale, NJ
Gordon G, Kaplan DM, Lankow B, Little DY, Sherwin J, Suter BA, Thaler L (2011) Toward an integrated approach to perception and action: conference report and future directions. Front Syst Neurosci 5:20
Gray CM, Singer W (1989) Stimulus-specific neuronal oscillations in orientation columns of cat visual cortex. Proc Natl Acad Sci USA 86:1698–1702

Grill-Spector K, Kushnir T et al (2000) The dynamics of object-selective activation correlate with recognition performance in humans. Nat Neurosci 3(8):837–843
Hartley R, Zisserman A (2003) Multiple view geometry in computer vision. Cambridge University Press, Cambridge, MA
Heeger DJ (1999) Linking visual perception with human brain activity. Curr Opin Neurobiol 9(4):474–479
Hofer H, Williams D (2005) Different sensations from cones with the same photopigment. J Vis 5(5):444–454
Hofer H, Carroll J et al (2005) Organization of the human trichromatic cone mosaic. J Neurosci 25(42):9669–9679
Hubel DH (1982) Exploration of the primary visual cortex, 1955–78. Nature 299(5883):515–524
Hubel DH, Wiesel TN (1962) Receptive fields, binocular interaction and functional architecture in the cat's visual cortex. J Physiol 160:106–154
Hubel DH, Wiesel TN (1965) Receptive fields and functional architecture in two nonstriate visual areas (18 and 19) of the cat. J Neurophysiol 28:229–289
Hung CP, Kreiman G et al (2005) Fast readout of object identity from macaque inferior temporal cortex. Science 310(5749):863–866
Hupé JM, James AC et al (2001) Feedback connections act on the early part of the responses in monkey visual cortex. J Neurophysiol 85(1):134–145
Hyvarinen A, Hoyer PO (2001) A two-layer sparse coding model learns simple and complex cell receptive fields and topography from natural images. Vision Res 41(18):2413–2423
Hyvarinen A, Gutmann M et al (2005) Statistical model of natural stimuli predicts edge-like pooling of spatial frequency channels in V2. BMC Neurosci 6(1):12
Julesz B (1981) Textons, the elements of texture perception, and their interactions. Nature 290(5802):91–97
Karklin Y, Lewicki M (2003) Learning higher-order structures in natural images. Network 14(3):483–499
Karklin Y, Lewicki M (2005) A hierarchical Bayesian model for learning nonlinear statistical regularities in nonstationary natural signals. Neural Comput 17(2):397–423
Karklin Y, Lewicki MS (2009) Emergence of complex cell properties by learning to generalize in natural scenes. Nature 457(7225):83–85
Kersten D, Mamassian P et al (2004) Object perception as Bayesian inference. Annu Rev Psychol 55:271–304
Khosrowshahi A, Baker J et al (2007) Predicting responses of V1 neurons to natural movies. Society for Neuroscience, San Diego, CA, p 33
Knill D, Richards W (1996) Perception as Bayesian inference. Cambridge University Press, Cambridge, MA
Knill DC, Saunders JA (2003) Do humans optimally integrate stereo and texture information for judgments of surface slant? Vision Res 43(24):2539–2558
Koepsell K, Wei Y, Wang Q, Rathbun DL, Usrey WM, Hirsch JA, Sommer FT (2009) Retinal oscillations carry visual information to cortex. Front Syst Neurosci 3
Koepsell K, Wang X et al (2010) Exploring the function of neural oscillations in early sensory systems. Front Neurosci 4:53
Kourtzi Z, Kanwisher N (2001) Representation of perceived object shape by the human lateral occipital complex. Science 293(5534):1506–1509
Land MF (1969) Movements of the retinae of jumping spiders (Salticidae: dendryphantinae) in response to visual stimuli. J Exp Biol 51(2):471–493
Land MF (1971) Orientation by jumping spiders in the absence of visual feedback. J Exp Biol 54(1):119–139
Land MF (1985) Fields of view of the eyes of primitive jumping spiders. J Exp Biol 119:381–384
Lee TS, Mumford D (2003) Hierarchical Bayesian inference in the visual cortex. J Opt Soc Am A Opt Image Sci Vis 20(7):1434–1448
Lee TS, Yang CF et al (2002) Neural activity in early visual cortex reflects behavioral experience and higher-order perceptual saliency. Nat Neurosci 5(6):589–597

Lennie P (1998) Single units and visual cortical organization. Perception 27(8):889–935
Litke AM, Bezayiff N et al (2004) What does the eye tell the brain? Development of a system for the large-scale recording of retinal output activity. IEEE Trans Nucl Sci 51(4):1434–1440
Lowe DG (2004) Distinctive image features from scale-invariant keypoints. Int J Comput Vis 60(2):91–110
Ma WJ, Beck JM et al (2006) Bayesian inference with probabilistic population codes. Nat Neurosci 9(11):1432–1438
Mamassian P, Knill DC et al (1998) The perception of cast shadows. Trends Cogn Sci 2(8):288–295
Mancilla JG, Lewis TJ et al (2007) Synchronization of electrically coupled pairs of inhibitory interneurons in neocortex. J Neurosci 27(8):2058–2073
Marcelja S (1980) Mathematical description of the responses of simple cortical cells. J Opt Soc Am 70(11):1297–1300
Marr D (1982) Vision: a computational investigation into the human representation and processing of visual information. WH Freeman, San Francisco, CA
Mumford D (1994) Neuronal architectures for pattern-theoretic problems. Large-scale neuronal theories of the brain. MIT, Cambridge, MA
Murray SO, Kersten D et al (2002) Shape perception reduces activity in human primary visual cortex. Proc Natl Acad Sci USA 99(23):15164–15169
Nakayama K, He Z et al (1995) Visual surface representation: a critical link between lower-level and higher-level vision. In: Kosslyn SM, Osherson DN (eds) An invitation to cognitive science: visual cognition, vol 2. MIT, Cambridge, MA, pp 1–70
Naselaris T, Prenger RJ et al (2009) Bayesian reconstruction of natural images from human brain activity. Neuron 63(6):902–915
Naselaris T, Kay KN et al (2011) Encoding and decoding in fMRI. Neuroimage 56(2):400–410
Newcombe RA, Davison AJ (2010) Live dense reconstruction with a single moving camera. In: 2010 IEEE conference on computer vision and pattern recognition (CVPR), San Francisco, CA, 1498–1505
Nishimoto S, Vu AT et al (2011) Reconstructing visual experiences from brain activity evoked by natural movies. Curr Biol 21(19):1641–1646
O'Rourke NA, Weiler NC, Micheva KD, Smith SJ (2012) Deep molecular diversity of mammalian synapses: why it matters and how to measure it. Nat Rev Neurosci 13:365–379
Olshausen BA, Field DJ (1996) Emergence of simple-cell receptive field properties by learning a sparse code for natural images. Nature 381(6583):607–609
Oram MW, Perrett DI (1992) Time course of neural responses discriminating different views of the face and head. J Neurophysiol 68(1):70–84
Papert S (1966) The summer vision project. MIT Artificial Intelligence Group, Vision Memo No. 100, 1–6
Pearl J (1988) Probabilistic reasoning in intelligent systems: networks of plausible inference—Judea Pearl—Google books. Morgan Kaufmann Publishers, San Francisco, CA
Poirazi P, Brannon T et al (2003) Pyramidal neuron as two-layer neural network. Neuron 37(6):989–999
Polsky A, Mel BW et al (2004) Computational subunits in thin dendrites of pyramidal cells. Nat Neurosci 7(6):621–627
Qiu FT, von der Heydt R (2005) Figure and ground in the visual cortex: v2 combines stereoscopic cues with gestalt rules. Neuron 47(1):155–166
Quiroga RQ, Reddy L et al (2005) Invariant visual representation by single neurons in the human brain. Nature 435(7045):1102–1107
Rao RP, Ballard DH (1999) Predictive coding in the visual cortex: a functional interpretation of some extra-classical receptive-field effects. Nat Neurosci 2(1):79–87
Rao RPN, Olshausen BA et al (2002) Probabilistic models of the brain: perception and neural function. MIT, Cambridge, MA
Ress D, Heeger DJ (2003) Neuronal correlates of perception in early visual cortex. Nat Neurosci 6(4):414–420

Riesenhuber M, Poggio T (2004) How the visual cortex recognizes objects: the tale of the standard model. In: Chalupa L, Werner J (eds) The visual neurosciences. MIT, Cambridge, MA, pp 1–14
Roorda A (2011) Adaptive optics for studying visual function: a comprehensive review. J Vis 11(7)
Roorda A, Williams DR (1999) The arrangement of the three cone classes in the living human eye. Nature 397(6719):520–522
Rossi AF, Rittenhouse CD et al (1996) The representation of brightness in primary visual cortex. Science 273(5278):1104–1107
Rust N, Movshon J (2005) In praise of artifice. Nat Neurosci 8(12):1647–1650
Schwartz O, Simoncelli EP (2001) Natural signal statistics and sensory gain control. Nat Neurosci 4:819–825
Sincich LC, Zhang Y et al (2009) Resolving single cone inputs to visual receptive fields. Nat Neurosci 12(8):967–969
Smith E, Lewicki M (2006) Efficient auditory coding. Nature 439(7079):978–982
Snavely N, Seitz SM et al (2006) Photo tourism: exploring photo collections in 3D. ACM, New York, NY
Srinivasan MV, Laughlin SB et al (1982) Predictive coding: a fresh view of inhibition in the retina. Proc R Soc Lond B Biol Sci 216(1205):427–459
Tappen MF, Freeman WT et al (2005) Recovering intrinsic images from a single image. IEEE Trans Pattern Anal Mach Intell 27(9):1459–1472
Tarsitano M, Andrew R (1999) Scanning and route selection in the jumping spider Portia labiata. Anim Behav 58(2):255–265
Tarsitano M, Jackson RR (1997) Araneophagic jumping spiders discriminate between detour routes that do and do not lead to prey. Anim Behav 53:257–266
Thomson AM, Bannister AP (2003) Interlaminar connections in the neocortex. Cereb Cortex 13(1):5–14
Thorpe SJ, Imbert M (1989) Biological constraints on connectionist models. In: Pfeifer R, Schreter Z, Fogelman-Soulie F, Steels L (eds) Connectionism in perspective. Elsevier Inc., Amsterdam, pp 63–92
Thorpe S, Fize D et al (1996) Speed of processing in the human visual system. Nature 381:520–522
Thrun S, Montemerlo M et al (2006) Stanley: the robot that won the DARPA grand challenge. J Field Robot 23(9):661–692
Tinbergen N (1974) Curious naturalists (revised edition). University of Massachusetts Press, Amherst, MA
Treisman AM, Gelade G (1980) A feature-integration theory of attention. Cogn Psychol 12(1):97–136
Tsao DY, Freiwald WA et al (2006) A cortical region consisting entirely of face-selective cells. Science 311(5761):670–674
Tsao DY, Moeller S et al (2008) Comparing face patch systems in macaques and humans. Proc Natl Acad Sci 105(49):19514–19519
Van Essen DC, Anderson CH (1995) Information processing strategies and pathways in the primate visual system. In: Zornetzer SF, Davis JL, Lau C, McKenna T (eds) An introduction to neural and electronic networks. Academic, San Diego, CA, pp 45–76
Vinje WE, Gallant JL (2000) Sparse coding and decorrelation in primary visual cortex during natural vision. Science 287(5456):1273–1276
Vogel CR, Arathorn DW et al (2006) Retinal motion estimation in adaptive optics scanning laser ophthalmoscopy. Opt Express 14(2):487–497
Wandell BA, Dumoulin SO et al (2007) Visual field maps in human cortex. Neuron 56(2):366–383
Zhou H, Friedman HS et al (2000) Coding of border ownership in monkey visual cortex. J Neurosci 20(17):6594–6611
Zhu M, Rozell C (2011) Population characteristics and interpretations of nCRF effects emerging from sparse coding. In: Computational and Systems Neuroscience (COSYNE), Salt Lake City, UT

Chapter 13
Reinforcement Learning Models Then-and-Now: From Single Cells to Modern Neuroimaging

P. Read Montague

Abstract Although much ignored in some intellectual circles today, behaviorism and its models from the early to mid parts of the twentieth century provided the basis for some of the first computational accounts of reward learning. The best expression of this work emerged in the early 1970s with the Rescorla–Wagner model of Pavlovian conditioning. This model accounted for a range of behavioral data about how animals learn about cues that predict rewarding outcomes. The step forward in this account was that learning was depicted as being driven by failed predictions—that is, some system collected information, formed expectations about how much reward to expect (associated with "conditioned stimuli" or cs), and generated learning updates that were proportional to the size and sign of the error. While successful in describing a large body of data, the Rescorla–Wagner model failed at one critical aspect of simple learning—the capacity to "chain" important cues together into a trajectory of learned associations—a feature called secondary conditioning: "A predicts B predicts food," for example.

Although much ignored in some intellectual circles today, behaviorism and its models from the early to mid parts of the twentieth century provided the basis for some of the first computational accounts of reward learning. The best expression of this work emerged in the early 1970s with the Rescorla–Wagner model of Pavlovian conditioning (Rescorla and Wagner 1972). This model accounted for a range of behavioral data about how animals learn about cues that predict rewarding outcomes. The step forward in this account was that learning was depicted as being

P.R. Montague (✉)
Human Neuroimaging Lab, The Virginia Tech Carilion Research Institute,
2 Riverside Circle, Roanoke, VA 24016, USA

Department of Physics, Virginia Tech, Roanoke, VA, USA

Wellcome Trust Centre for Neuroimaging, University College London, London, UK
e-mail: read@vtc.vt.edu

J.M. Bower (ed.), *20 Years of Computational Neuroscience*, Springer Series in Computational Neuroscience 9, DOI 10.1007/978-1-4614-1424-7_13,

driven by failed predictions—that is, some system collected information, formed expectations about how much reward to expect (associated with "conditioned stimuli" or cs), and generated learning updates that were proportional to the size and sign of the error. While successful in describing a large body of data, the Rescorla–Wagner model failed at one critical aspect of simple learning—the capacity to "chain" important cues together into a trajectory of learned associations—a feature called secondary conditioning: "A predicts B predicts food," for example.

$$V^i_{\text{new}} = V^i_{\text{old}} + \lambda\left(r^{\text{US}} - \sum_i V^i_{\text{old}}\right) \tag{13.1}$$

Here, we have written the Rescorla–Wagner rule in a form that relates to more modern "rewardese." The V's represent the predicted value of each conditioned stimulus and the designation "old" and "new" refer, respectively, to previous values and updated values for the V's. This expression remains agnostic to the timescale over which such changes take place. λ is a learning rate that scales the difference between the reward r and the summed predictions of this reward contributed by all the conditioned stimuli in a learning trial. The main point is that learning was framed as a problem of prediction and the signals for learning to occur were the errors in those predictions.

Over roughly the same time period, there was developing an independent line of research in the area of *optimal control* where the central problems shared many features of learning studied in animal experiments by a more psychologically minded community. This literature was large then and is even larger now and so no attempt will be made to summarize it here. We focus here on a family of computational methods that solve sequential decision problems (Bellman 1957). In these problems, there is a learning agent situated in some state space and this agent has available to it (a) the ability to emit actions and thus transition from one state to another and (b) a signal to criticize the value to the agent of the transition, but with the proviso that this criticism might be delayed. This latter feature distinguishes reinforcement learning (RL) approaches from other forms of machine learning.

One of the first fruitful fusions of this optimal control arena and psychological learning theory was proposed by Sutton and Barto (1990; see Sutton and Barto 1998 for summary and history and Barto et al. 1983). They proposed a new model of Pavlovian conditioning called temporal difference (TD) learning. Imagine an agent transitioning from one state to the next and observing rewards received in each state. Starting in state $s(0)$, the agent's experience into the future would look like:

$$r(0) \quad r(1) \quad r(2) \quad r(3) \quad \ldots \quad r(k) \quad \ldots$$

The integer indices indicate the labels of states and do not imply a particular timescale. Temporal difference learning prescribes how the value V of each state should be updated based on this experience:

$$V_{\text{new}}(s(0)) = V_{\text{old}}(s(0)) + \lambda[r(0) + \gamma V_{\text{old}}(s(1)) - V_{\text{old}}(s(0))] \tag{13.2}$$

Here the learning rule is put in the form of updates without any specification of timescale, but one can see the relation of the Rescorla–Wagner rule in (13.1). Temporal difference learning updates values according to changes in its predictions V and immediate reward $r(0)$.

Diffuse Neuromodulatory Systems

This background sets the stage for the early 1990s where Dayan, Montague, and Sejnowski proposed that this general approach to agent-based learning problems was likely to be implemented by diffuse neuromodulatory systems in the brains of biological organisms (Quartz et al. 1993; Montague et al. 1993). More specifically, a range of single-unit recordings from putative dopamine neurons in the midbrains of monkeys gave rise to the computational hypothesis that these neurons computed reward prediction error signals in the form of temporal difference errors and encoded such computations in modulations of their spiking activity (Montague et al. 1995, 1996). This model provided an explanation for many confusing features of the electrophysiology recorded from these neurons during reward learning tasks carried out in alert monkeys. The seminal experimental results in this area were generated by Wolfram Schultz and colleagues (e.g., Romo and Schultz 1990; Ljungberg et al. 1992). This group recorded activity in dopamine neurons using a range of passive and active conditioning tasks. The neurons' behavior during these tasks and during the behavioral learning that occurs was complex and difficult to interpret. For example, in Schultz et al. (1993), the researchers concluded in their abstract:

> ... The lack of sustained activity suggests that dopamine neurons do not encode representational processes, such as working memory, expectation of external stimuli or reward, or preparation of movement.

Ironically, errors in expectation of reward are exactly what the TD model of phasic dopaminergic activity posited (Montague et al. 1996). This claim is consistent with the results shown in Fig. 13.1, which is a replotting of data from Hollerman and Schultz (1998). Here cues predict the occurrence of reward at particular times and the plot shows what happens to phasic spiking activity when these times are changed at various points during learning.

The success of this application of reinforcement learning models to real-world dopamine neuron recording provides a computationally principled starting point for work in the area because of the importance of dopamine in motivated learning, addiction, and a range of mental diseases.

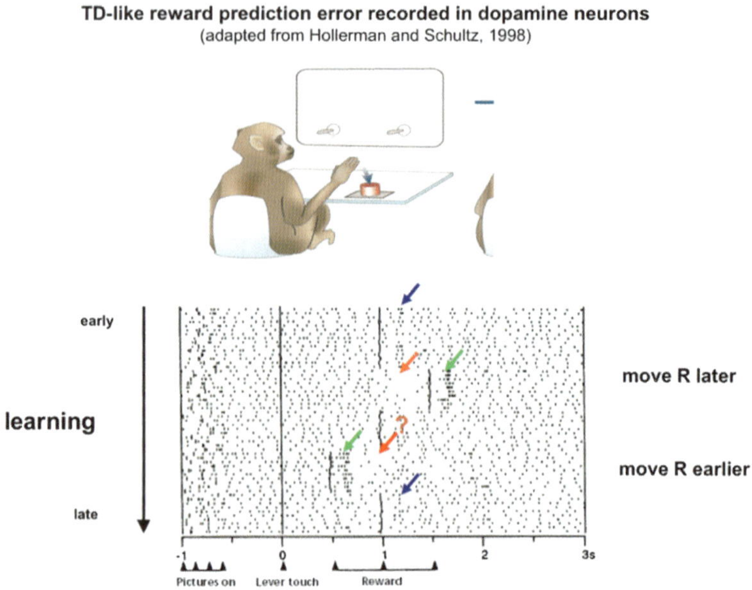

Fig. 13.1 Phasic dopamine activity encodes positive and negative reward prediction errors. Cues appear (far *left*) and the animal pushes a lever after which reward is delivered at a fixed time later. This is data from a single dopamine neuron. Time traces show spike times (*dots*) aligned on the level press. The reward times are indicated by a *vertical bar*. Moving reward times ("move R later" or "move R earlier")

Application of Reinforcement Learning Models to Addiction

In general, reinforcement learning (RL) models of addiction depict the addicted state as a *valuation disease* resulting in part from skewed reward prediction error signals encoded by dopamine transients and engendered by drug-taking episodes (Redish 2004). The temporal difference model of reinforcement learning (TDRL) associates these *experiential* prediction errors with phasic dopaminergic activity (in the 80–150 ms range) that tracks ongoing differences between expected and experienced rewards (O'Doherty et al. 2003; McClure et al. 2003; for reviews, see Schultz et al. 1997; Montague et al. 2004; Daw and Doya 2006). Within this RL framework, transient increases in dopamine induced by addictive drugs (e.g., Koob and Le Moal 1997; Pidoplichko et al. 1997) should report positive unanticipated reward prediction errors to dopaminoceptive target structures, thus exaggerating the apparent value of drug-induced states, reinforcing drug-seeking behavior, and generally causing the system to overvalue drug-related cues (Redish 2004).

This view of addiction and the role of dopamine systems has unleashed a torrent of neuroimaging studies where dopaminergic target structures are monitored using blood oxygenation level-dependent (BOLD) recordings in humans during tasks designed to test the TD model and its extensions to the addicted state. In 2003,

O'Doherty and colleagues and McClure and colleagues (O'Doherty et al. 2003; McClure et al. 2003) published analogous approaches to this issue using simple conditioning tasks in human subjects. Both groups arrived at the same conclusion—BOLD responses in striatal regions were consistent with the projecting dopamine systems emitting a reward prediction error, and in this case, having it show up encoded in a composite proxy (the BOLD measure) for neural activity in a known downstream structure.

However, it is well known that *dopamine increase during drug use is not sufficient for addiction* (Volkow et al. 2002; also see Dani and Montague 2007). So while the overvaluation model proposed by Redish is a convenient starting point for understanding the addicted state in computational terms, it is just a piece of the story. The value in this case derives from the fact that the model itself suggests new experimental directions and can deliver quantitative predictions ahead of time to help structure the design of the experiments.

Another key characteristic of addicted individuals is that they pursue and consume subjective rewards even *in the clear presence of outcomes that "might" occur* (APA 2000). Such data suggest strongly that addicts have an impaired capacity to use possible future outcomes to intervene on their habitual drug-taking. One working hypothesis is that chronic drug-taking diminishes the capacity of the nervous system to compute appropriate control signals that guide behavior. Specifically, error signals for outcomes "that might happen" may no longer be produced in the brains of addicts. An alternative, but related hypothesis is that an addicted brain continues to produce error signals related to fictive outcomes, but the influence of these signals on actual behavioral choice is significantly diminished. Just such an experiment was carried out by Chiu et al. (2008) in nicotine addicts where a temporal difference error signal and a fictive error signal (e.g., see Lohrenz et al. 2007; Hayden et al. 2009) were tracked concurrently. Using a sequential investment task in nicotine addicts, the fictive error signal was present encoded in BOLD responses in human brains, but apparently had no impact on their behavioral responses. This neural signature was equally strong in sated smokers, unsated smokers, and nonsmoking controls, but only showed a behavioral impact in the nonsmoking control group (Chiu et al. 2008). So in smokers, one consistent explanation is that their brains compute fictive errors but for some reason there is a kind of decoupling between these signals and behaviors that (should) could be guided by them.

RL Models Guide New Electrophysiology Experiments Important for Addiction

Reinforcement learning models of dopaminergic activity have clearly been helpful in explaining extant electrophysiological data (Schultz et al. 1997; Tobler et al. 2005; Bayer and Glimcher 2005); however, they also point to gaps in our understanding of the physical substrates of all the information that we know is encoded in the dopaminergic output. For example, during trials where a reward is missing or parametrically diminished, dopaminergic responses should be diminished or missing altogether. This is a prediction of the model and was seen in early experiments; however, the neural source of this information has until recently remained for the most part unknown. The habenular nuclei and their interactions with the ventral tegmental area and substantia nigra have recently been proposed as sites where negative rewards—literally the absence of expected rewards or directly aversive events—are processed. Through reciprocal inhibitory connections with ventral tegmental area and substantia nigra, the lateral habenula communicates information about the timing and likelihood of negative rewards as described just above. This work was guided by the reward prediction error model for dopamine neurons and has aided Hikosaka and colleagues in their pursuit of a deeper understanding of the underlying neural circuitry underwriting this important prediction system.

Open Questions Motivated by RL Models

RL models now form a large family of approaches to motivated learning and pathologies that disturb motivated learning (e.g., depression) or some of the primary players in RL-described neural subsystems (e.g., like the Parkinsonism that results from loss of dopamine neurons). However, many important questions remain, and the most interesting works lie in the future. For example, dopaminergic systems are low bandwidth systems with roughly 25,000–35,000 neurons on each side of the brainstem, which provide large segments of the cortical mantle and of course the striatum with dopamine. The capacity for predicting near-term reward suggests that there must be some kind of arbitration system that allows the dopaminergic modulation to be "shared" amongst competing systems that need to use or broadcast their near-term prediction error using dopamine. The biological "answer" to this need is completely unknown.

The importance of dopamine in a vast range of pathologies makes extending, breaking, and bending the model paramount, but with a model in hand prediction can be generated and this has been the value of this simple approach to dopaminergic systems.

References

American Psychiatric Association (2000) Diagnostic and statistical manual of mental disorders, Revised 4th edn. American Psychiatric Association, Washington, DC

Barto AG, Sutton RS, Anderson CW (1983) Neuronlike adaptive elements that can solve difficult learning control problems. IEEE Trans Syst Man Cybern 13:834–846

Bayer HM, Glimcher PW (2005) Midbrain dopamine neurons encode a quantitative reward prediction error signal. Neuron 47(1):129–141

Bellman RE (1957) Dynamic programming. Princeton University Press, Princeton, NJ

Chiu PH, Lohrenz TM, Montague PR (2008) Smokers' brains compute, but ignore, a fictive error signal in a sequential investment task. Nat Neurosci 11(4):514–520

Dani JA, Montague PR (2007) Disrupting addiction through the loss of drug-associated internal states. Nat Neurosci 10(4):403–404

Daw ND, Doya K (2006) The computational neurobiology of learning and reward. Curr Opin Neurobiol 16(2):199–204

Hayden BY, Pearson JM, Platt ML (2009) Fictive reward signals in the anterior cingulate cortex. Science 324(5929):948–950

Hollerman JR, Schultz W (1998) Dopamine neurons report an error in the temporal prediction of reward during learning. Nat Neurosci 1:304–309

Koob GF, Le Moal M (1997) Drug abuse: hedonic homeostatic dysregulation. Science 278(5335):52–58

Ljungberg T, Apicella P, Schultz W (1992) Responses of monkey dopaminergic neurons during learning of behavioral reactions. J Neurophysiol 67:145–163

Lohrenz T, McCabe K, Camerer CF, Montague PR (2007) Neural signature of fictive learning signals in a sequential investment task. Proc Natl Acad Sci USA 104(22):9493–9498

McClure SM, Daw ND, Montague PR (2003) A computational substrate for incentive salience. Trends Neurosci 26(8):423–428

Montague PR, Dayan P, Nowlan SJ, Pouget A, Sejnowski TJ (1993) Using aperiodic reinforcement for directed self-organization. In: Giles CL, Hanson SJ, Cowan JD (eds) Advances in neural information processing systems, vol 5. Morgan Kaufmann, San Mateo, CA, pp 969–976

Montague PR, Dayan P, Person C, Sejnowski TJ (1995) Bee foraging in uncertain environments using predictive hebbian learning. Nature 377:725–728

Montague PR, Dayan P, Sejnowski TJ (1996) A framework for mesencephalic dopamine systems based on predictive hebbian learning. J Neurosci 16(5):1936–1947

Montague PR, Hyman SE, Cohen JD (2004) Computational roles for dopamine in behavioural control. Nature 431(7010):760–767

O'Doherty JP, Dayan P, Friston K, Critchley H, Dolan RJ (2003) Temporal difference models and reward-related learning in the human brain. Neuron 38(2):329–337

Pidoplichko VI, DeBiasi M, Williams JT, Dani JA (1997) Nicotine activates and desensitizes midbrain dopamine neurons. Nature 390(6658):401–404

Quartz SR, Dayan P, Montague PR, Sejnowski TJ (1993) Expectation learning in the brain using diffuse ascending projections. Soc Neurosci Abstr 18:1210

Redish AD (2004) Addiction as a computational process gone awry. Science 306(5703): 1944–1947

Rescorla RA, Wagner AR (1972) A theory of Pavlovian conditioning: variations in the effectiveness of reinforcement and non-reinforcement. In: Black A, Prokasy W (eds) Classical conditioning. Appleton-Century-Crofts, New York, NY, pp 64–99

Romo R, Schultz W (1990) Dopamine neurons of the monkey midbrain: contingencies of responses to active touch during self-initiated arm movements. J Neurophysiol 63:592–606

Schultz W, Apicella P, Ljungberg T (1993) Responses of monkey dopamine neurons to reward and conditioned stimuli during successive steps of learning a delayed response task. J Neurosci 13:900–913

Schultz W, Dayan P, Montague PR (1997) A neural substrate of prediction and reward. Science 275:1593–1599

Sutton RS, Barto AG (1990) Time-derivative models of Pavlovian reinforcement. In: Gabriel M, Moore J (eds) Learning and computational neuroscience: foundations of adaptive networks. MIT Press, Cambridge, MA, pp 497–537

Sutton RS, Barto AG (1998) Reinforcement learning: an introduction. MIT Press, Cambridge, MA

Tobler PN, Fiorillo CD, Schultz W (2005) Adaptive coding of reward value by dopamine neurons. Science 307(5715):1642–1645

Volkow ND, Fowler JS, Wang GJ (2002) Role of dopamine in drug reinforcement and addiction in humans: results from imaging studies. Behav Pharmacol 13(5–6):355–366

Index

A

Associative memory function
- acetylcholine effects, 209–210, 212–213
- afferent input, 210, 211
- cholinergic suppression, 211–212
- encoding and retrieval, 209
- in hippocampal region CA3, 208
- input–output function, 209
- muscarinic receptors, 212
- olfactory cortex, 208–209
- piriform cortex, 210–211

C

Calcium
- long-term synaptic plasticity
  - buffers role, 147
  - conjunctive stimulation, 148–150
  - diffusion and morphology, 147–148
  - high frequency *vs.* low frequency stimulation, 144
  - LTP, frequency dependence, 144, 145
  - models characteristics, 144, 146
  - nanodomains of, 150–152
  - postsynaptic density (PSD), 144
  - postsynaptic depolarization, 144
  - within single spine, 144, 145
- neuronal firing patterns
  - channel colocalization, 154–155
  - oscillations, 152–154
  - spike frequency adaptation, 154

Calmodulin
- calcium binding protein, types of, 155
- frequency-dependent elevation, 144
- nanodomains, 150
- neurogranin binding, 152

Central pattern generators (CPGs)
- coordinated motor output, 160
- half-center oscillator, 162–163
- inherent burster, 162
- Morris–Lecar formalism, 162
- stomatogastric nervous system (STN), 162

Code sharing
- curated model repositories, 85
- methods for, 84
- personal/institutional web-site, 85
- source-code repositories, 85
- supplementary material, 84–85

Community model
- antidromic spike activation, 117–118
- behavior and function, 132
- collaborative communication, 132–133
- confidence in model structure, 130–131
- dendritic democracy and distal synaptic inputs, 124–126
- fiber activation, 120–122
- functional structure, 128–130
- granule cell pathway, 122
- modeling standards, 131–132
- Model Reuse
  - additional analysis, 115–116
  - experimental data, 114–115
- network context, 126–128
- physiological responses, 116–117
- somatic current injection, 118–120
- synaptic inputs, 122–123
- wheel reinventing, 131

Computational neuroscience (CNS)
- associative memory function
  - acetylcholine effects, 209–210, 212–213
  - afferent input, 210, 211

J.M. Bower (ed.), *20 Years of Computational Neuroscience*, Springer Series in Computational Neuroscience 9, DOI 10.1007/978-1-4614-1424-7,

Computational neuroscience (CNS) (*cont.*)
cholinergic suppression, 211–212
encoding and retrieval, 209
in hippocampal region CA3, 208
input–output function, 209
muscarinic receptors, 212
olfactory cortex, 208–209
piriform cortex, 210–211
banquet, 4–5
budget and fund raising efforts, 2
central pattern generators (CPGs)
half-center oscillator, 162–163
inherent burster, 162
Morris–Lecar formalism, 162
stomatogastric nervous system (STN), 162
CNS*92, 2
graduate programs, 2
hippocampus, 214–215
inviting senior faculty, 2–3
meeting proceedings, 3
posters of, 13, 14
back and forth creation, 28, 30
basketball motif, 27, 28
Beware theorists, 15, 16
CNS*93 poster, 15
in digital form, 28, 29
Druids all, 25, 26
funding, 16
high-resolution images, 15
internet for communication, 18
interpretations, 13, 14
intracellular trafficking, 19
misinterpretation, 17
molecular and cellular level modeling, 23–24
Ramon and K-halls, 19, 20
re-migration, 24, 25
representations, 22
in Santa Barbara, 21, 22
summer camp, 25, 26
United Cerebral Evangelical Fellowship, 20
venue, 31
wild wild west, 19, 21
reproducible research
ambiguous model numerics, 78–81
biostatistics, 75
code sharing, 84–86
computationally underspecified models, 81–82
cross-replicability, 75
declarative model descriptions (*see* Declarative model descriptions)
external replicability, 75
internal replicability, 75
reporting (*see* Research reporting)
successful reproduction definition, 82–84
theta phase precession
AMPA conductance, 216–217
entorhinal cortex, 218
grid cell firing, 217
grid cells locations, 218
hippocampal place cells, 216
intrinsic frequency, 218
oscillatory interference model, 216
persistent spiking, 220
postsynaptic depolarization, neurons, 217
sequence read-out model, 216
single cell resonance, 219
spaced firing fields, 217
vision
autonomous vision systems, 262–263, 265–266
cortical microcircuit, 256–258
edges types, 245, 246
exploratory approaches, 264–265
feedback loops, 258–260
human visual cortex, 251–252
inferential computations, 245
intermediate-level vision, 250–251
multiple-view geometry, 253, 254
natural image statistics and neural coding, 248–250
neuronal oscillations, 260–262
representation, 245
sophisticated vision, 254–256
tiling in retina, 247–248
workshops, 3–4
CPG. *See* Central pattern generators
Cross-replicability
record-keeping, 87
simulator-independent formats use, 86

D

Declarative model descriptions
ConnPlotter package, 92
extensible markup language (XML), 88
neuroConstruct, 92
NeuroML, 90–91
NineML, 91
SBML and CellML, 89–90
SED-ML, 92
systems biology markup language (SBML), 89

Discrete event system (DES), 64
Dynamic-clamp technique
high-conductance state, 175
neuronal responsiveness, 174
single-spike mode, 175
thalamic neurons, 176

G

General neural simulation system (GENESIS)
early history of, 40–42
electrical network simulator, 39
expanding simulator capabilities
ERATO project, 56
HBP, 53
ModelDB project, 54
model tuning, 49–51
MWS, 54
NeuroConstruct, 56–57
NeuroML development kit, 56
NineML, 57
NSF funding, 55
object-oriented database (OODB), 53
open source *vs.* proprietary software, 58–59
parameter search, 49–51
programming languages, 57–58
scripting and GUIs, 47–49
SenseLab project, 53
SGML, 53
XML-based representation, 55
federation and user support, 44–46
Hodgkin–Huxley model
action potentials, 36
anode break, 36
calcium-dependent potassium channels, 38
62-compartment Purkinje cell models, 38
ion-selective channels, 36
lumped parameter model, 37
postsynaptic potential (PSP) propagation, 37
transient potassium conductance, 38
MANUEL, 39
problems of, 59–61
seduction by neuroscience, 34–35
SPICE, 39

H

Hebb's rule, 188
Hodgkin–Huxley model
action potentials, 36
anode break, 36
calcium-dependent potassium channels, 38
62-compartment Purkinje cell models, 38
ion-selective channels, 36
lumped parameter model, 37
postsynaptic potential (PSP) propagation, 37
transient potassium conductance, 38

L

Long-term synaptic plasticity, calcium
buffers role, 147
conjunctive stimulation, 148–150
diffusion and morphology, 147–148
high frequency *vs.* low frequency stimulation, 144
LTP, frequency dependence, 144, 145
models characteristics, 144, 146
nanodomains of, 150–152
postsynaptic density (PSD), 144
postsynaptic depolarization, 144
within single spine, 144, 145

M

MANUEL program, 39
Modeler's workspace (MWS), 54
Morris–Lecar formalism, 162

N

Network Interchange for Neuroscience Modeling Language (NineML), 57
Neural simulation software
CSIM and Python version, 62
DES, 64
G-3 component, 63
GENESIS (*see* General neural simulation system (GENESIS))
GPU-SNN simulator, 62–63
interoperable realistic neural simulators, 61
NEST, 62
PyNN, 62
scripting language, 64–65
SPLIT, 62
XPPAUT, 62
Neuronal firing patterns
channel colocalization, 154–155
oscillations, 152–154
spike frequency adaptation, 154

O

Object-oriented database (OODB), 53
Olfactory receptor database (ORDB), 53

Olfactory system
  circuitry, 231
  concentration invariance, 233
  contrast enhancement, 233–234
  field oscillations and temporal activity patterns
    contrast enhancement by synchronization, 236
    signal-to-noise ratio, 236
    spike timing and synchronization, 234
    temporal processing, 234, 235
  spatiotemporal activity patterns and odor perception, 237, 238

P
Purkinje cell modeling
  active properties of, 111–113
  community model
    antidromic spike activation, 117–118
    behavior and function, 132
    collaborative communication, 132–133
    confidence in model structure, 130–131
    dendritic democracy and distal synaptic inputs, 124–126
    fiber activation, 120–122
    functional structure, 128–130
    granule cell pathway, 122
    modeling standards, 131–132
    Model Reuse, 114–116
    network context, 126–128
    physiological responses, 116–117
    somatic current injection, 118–120
    synaptic inputs, 122–123
    wheel reinventing, 131
  cortex, 104, 105
  dendrite, 104–106
  62 dendritic compartments, 107, 108
  Guinea Pig Purkinje cells, 109, 111
  R-DB Model lineage, 109
  requirement for, 108
  Shelton model, 109
  soma, 107
  spiny dendritic branches, 109, 110
  Zucker's model, 105–106

R
Rescorla–Wagner model
  diffuse neuromodulatory systems, 273, 274
  reinforcement learning models, 274–275
  temporal difference, 272–273
Research reporting
  ARRIVE Guidelines for Reporting Animal Research, 93
  Google Scholar, 92
  model descriptions checklist, 93–94
  Thomson Reuters Web of Knowledge, 93

S
Stomatogastric nervous system (STN), 162
Synaptic noise
  biophysical models of
    different integrative mode, 172
    enhanced responsiveness, 172
    enhanced temporal processing, 172
    location-independence, 172
    modulation of intrinsic properties, 172
  characterization of, 168–170
  dynamic-clamp technique
    high-conductance state, 175
    neuronal responsiveness, 174
    single-spike mode, 175
    thalamic neurons, 176
  optimal conductance patterns estimation, 179–181
  simplified models of, 172–173
  source of, 167–168
  VmD method
    advantage of, 178–179
    dynamic-clamp experiments, 177, 178
    Fokker–Planck equation, 176, 177
    Gaussian distribution, 177
  during wakefulness, 167, 169
Synaptic plasticity
  bistability, 196
  CaMKII switch, 190
  diffusion and stochasticity, 191
  Hebb's rule, 188
  hippocampal long-term potentiation (LTP)/LTD experiments, 188, 189
  homeostasis, 200
  input pattern selectivity, 198–200
  long-term potentiation, 192
  molecules involved in LTP, 193–194
  N-methyl-d-aspartate (NMDA) receptor, 189, 190
  non-bistable memories, 196, 198
  persistence, 189
  phosphorylation/binding states, signaling molecules, 192
  positive feedback and bistability, 195
  slow-decaying molecular capacitor, 190–191
  stochasticity and memory, 198
  systems biology, 192
Systems biology markup language (SBML), 89

T
Theta phase precession
  AMPA conductance, 216–217
  entorhinal cortex, 218
  grid cell firing, 217
  grid cells locations, 218
  hippocampal place cells, 216
  intrinsic frequency, 218
  oscillatory interference model, 216
  persistent spiking, 220
  postsynaptic depolarization, neurons, 217
  sequence read-out model, 216
  single cell resonance, 219
  spaced firing fields, 217

V
Vision
  autonomous vision systems, 262–263, 265–266
  cortical microcircuit, 256–258
  edges types, 245, 246
  exploratory approaches, 264–265
  feedback loops, 258–260
  human visual cortex, 251–252
  inferential computations, 245
  intermediate-level vision, 250–251
  multiple-view geometry, 253, 254
  natural image statistics and neural coding, 248–250
  neuronal oscillations, 260–262
  representation, 245
  sophisticated vision, 254–256
  tiling in retina, 247–248
VmD method
  advantage of, 178–179
  dynamic-clamp experiments, 177, 178
  Fokker–Planck equation, 176, 177
  Gaussian distribution, 177

Printed by Printforce, the Netherlands